THE DESIGN OF IMPEDANCE – MATCHING NETWORKS FOR RADIO – FREQUENCY AND MICROWAVE AMPLIFIERS

The Artech House Microwave Library

Introduction to Microwaves by Fred E. Gardiol
Receiving Systems Design by Stephen J. Erst
Applications of GaAs MESFETs by R.A. Soares, J. Graffeuil, and J. Obregon
GaAs Processing Techniques by R.E. Williams
GaAs FET Principles and Technology, J.V. DiLorenzo and D. Khandelwal, eds.
Modern Spectrum Analyzer Theory and Applications by Morris Engelson
Microwave Materials and Fabrication Techniques by Thomas S. Laverghetta
Handbook of Microwave Testing by Thomas S. Laverghetta
Microwave Measurements and Techniques by Thomas S. Laverghetta
Principles of Electromagnetic Compatibility by Bernhard E. Keiser
Microwave Filters, Impedance Matching Networks, and Coupling Structures by G.L. Matthaei, Leo Young, and E.M.T. Jones
Microwave Engineer's Handbook, 2 vol., Theodore Saad, ed.
Computer-Aided Design of Microwave Circuits by K.C. Gupta, R. Garg, and R. Chadha
Microstrip Lines and Slotlines by K.C. Gupta, R. Garg, and I.J. Bahl
Microstrip Antennas by I.J. Bahl and P. Bhartia
Microwave Circuit Design Using Programmable Calculators by J. Lamar Allen and Max Medley, Jr.
Stripline Circuit Design by Harlan Howe, Jr.
Microwave Transmission Line Filters by J.A.G. Malherbe
Electrical Characteristics of Transmission Lines by W. Hilberg
Multiconductor Transmission Line Analysis by Sidney Frankel
Microwave Diode Control Devices by Robert V. Garver
A Practical Introduction to Impedance Matching by Robert Thomas
Active Filter Design by A.B. Williams
Adaptive Electronics by Wolfgang Gaertner
Laser Applications by W.V. Smith
Electronic Information Processing by W.V. Smith
Logarithmic Video Amplifiers by Richard S. Hughes
Avalanche Transit-Time Devices, George Haddad, ed.
Gallium Arsenide Bulk and Transit-Time Devices, Lester Eastman, ed.
Ferrite Control Components, 2 vol., Lawrence Whicker, ed.

THE DESIGN OF IMPEDANCE – MATCHING NETWORKS FOR RADIO – FREQUENCY AND MICROWAVE AMPLIFIERS

Pieter L.D. Abrie

Copyright © 1985

ARTECH HOUSE, INC.
610 Washington Street
Dedham, MA 02026

All rights reserved. Printed and bound in the United States of America. No part of this book may be reproduced or utilized in any form or by any means, electronic or mechanical, including photocopying, recording, or by any information storage and retrieval system, without permission in writing from the publisher.

International Standard Book Number: 0-89006-172-6
Library of Congress Catalog Card Number: 85-047746

Dedication

This book is dedicated to everyone who shared in the pleasure of its creation and publication.

CONTENTS

Chapter 1 Network Characterization and Analysis with Y-, Z-, T- and S-parameters　　1

1.1　Introduction　　1
1.2　Y-Parameters　　1
1.3　The Indefinite Admittance Matrix　　5
1.4　Z-Parameters　　6
1.5　T-Parameters　　8
1.6　Scattering Parameters　　9
　1.6.1　S-Parameter Definitions　　9
　1.6.2　The Physical Meanings of the Normalized Incident and Reflected Components of an N-Port　　15
　1.6.3　The Physical Interpretations of the Scattering Parameters　　17
　1.6.4　Constraints Imposed on the Normalized Components by the Terminations of an N-Port　　19
　1.6.5　Derivation of Expressions for the Gain Ratios and Reflection Parameters of a Two-Port　　21
　1.6.6　Conversion of S-Parameters to Others Parameters　　24
　1.6.7　The Indefinite S-Matrix　　24
　1.6.8　Extension of the Single-Frequency S-Parameter Definitions to the Complex Frequency Plane　　26
　1.6.9　Constraints on the Scattering Matrix of a Lossless N-Port　　29

Questions and Problems　　33

References and Additional Reading　　36

Chapter 2 Radio-Frequency Components　　37

2.1　Introduction　　37
2.2　Capacitors　　37
2.3　Inductors　　40
　2.3.1　The Influence of Parasitic Capacitance on an Inductor　　41
　2.3.2　Low-Frequency Losses in Inductors　　43
　2.3.3　The Skin Effect　　44
　2.3.4　The Proximity Effect　　46
　2.3.5　Magnetic Materials　　47
　2.3.6　The Design of Air-Cored Single-Layer Solenoidal Coils　　51
　2.3.7　The Design of Inductors with Magnetic Cores　　56

2.4 Transmission Lines   59
   2.4.1   Coaxial Cables   60
   2.4.2   Microstrip Lines   61
   2.4.3   Twisted-Pair Transmission Lines   65

Questions and Problems   65

References and Additional Reading   67

Chapter 3 Narrowband Impedance-Matching with LC Networks   69

3.1 Introduction   69
3.2 Parallel Resonance   70
3.3 Series Resonance   74
3.4 L-Sections   76
3.5 Π- and T-sections   81
   3.5.1   The Π-Section   82
   3.5.2   The T-Section   86
3.6   The Design of Π- and T-Sections when the Terminations
          are Complex   87
3.7   Four-Element Matching Networks   89
3.8   Calculation of the Insertion Loss of LC Matching
          Networks   90
3.9   Calculation of the Bandwidth of Cascaded LC Matching
          Networks   92

Questions and Problems   93

Chapter 4   Coupled Coils and Transformers   95

4.1 Introduction   95
4.2 The Ideal Transformer   95
4.3 Equivalent Circuits for the Practical Transformer   97
4.4 Wideband Impedance Matching with Transformers   100
4.5 The Single-Tuned Transformer   102
4.6 The Tapped Coil   103
4.7 The Parallel Double-Tuned Transformer   109
4.8 The Series Double-Tuned Transformer   115
4.9 Measurement of the Coupling Factor   118
   4.9.1   Measurement of the Coupling Factor by Short-Circuiting
          the Secondary Winding of the Transformer   118
   4.9.2   Measurement of the Coupling Factor by Measuring the
          Open-Circuited Voltage Gain of the Transformer   119
   4.9.3   Measurement of the Coupling Factor by Measuring the
          S-Parameters of the Transformer   119

Questions and Problems   120

References   123

Chapter 5   Transmission-Line Transformers   **125**

5.1   Introduction   125

5.2   Transmission-Line Transformer Configurations   127

5.3   The Analysis of Transmission-Line Transformers   135

5.4   The Design of Transmission-Line Transformers   142

   5.4.1   Determining the Optimum Characteristic Impedance and Diameter of the Transmission Line to be Used   143

   5.4.2   Determining the Minimum Value of The Magnetizing Inductance of the Transformer at the Lowest Frequency in the Pass Band   144

   5.4.3   Determining the Type and Size of the Magnetic Core to the Used   147

   5.4.4   Compensation of Transmission-Line Transformers for Non-optimum Characteristic Impedances   149

   5.4.5   The Design of Low-Pass LC Networks to Extend the Bandwidth of a Transmission-Line Transformer   154

Questions and Problems   158

References and Additional Reading   160

Chapter 6   Wideband LC and RLC Impedance-Matching Networks   **161**

6.1   Introduction   161

6.2   Determining an Impedance Function for a Set of Impedance *versus* Frequency Coordinates   162

6.3   The Analytical Approach to Impedance Matching   170

   6.3.1   Darlington Synthesis of Impedance-Matching Networks   172

   6.3.2   LC Transformers   177

   6.3.3   The Gain-Bandwidth Constraints Imposed by a Parallel RC and Series Load   180

   6.3.4   The Direct Synthesis of Impedance Matching-Networks when the Load (or Source) Is Reactive   182

   6.3.5   Synthesis of Networks for Matching a Reactive Load to a Purely Resistive or Reactive Source by Using the Principle of Parasitic Absorbtion   186

   6.3.6   The Analytic Approach to Designing Commensurate Distributed Impedance-Matching Networks   189

     6.3.6.1   Richards' Transformation   190

     6.3.6.2   Kuroda and Norton's Identities   193

6.4 The Iterative Design of Impedance-Matching Networks   195

    6.4.1   The Line-Segment Approach to Matching a Reactive Load to a Purely Resistive Source   198

    6.4.2   The Reflection Coefficient Approach to Solving Double-Matching Problems   205

    6.4.3   The Transformation Q Approach to the Design of Impedance-Matching Networks   215

        6.4.3.1   Constraints on the Input Impedance of a Lossless Matching Network if the Gain is to Remain Constant at a Specified Frequency   216

        6.4.3.2   Extention of the Transformation Q Impedance-Matching Technique   218

        6.4.3.3   Optimization of the Transformation Q Factors of a Matching Network   220

        6.4.3.4   An Algorithm for the Design of Impedance-Matching Networks by Using the Transformation Factors of the Network   230

6.5 The Design of RLC Impedance-Matching Networks   231

Questions and Problems   235

References and Additional Reading   239

Chapter 7   Microwave Lumped Elements, Distributed Equivalents and the Parasitics Associated with Microstrip Transmission Lines   241

7.1 Introduction   241

7.2 Lumped Microwave Resistors   242

7.3 Evaluation of the Limitations of a Series Transmission Line Used as a Lumped Element   242

7.4 Lumped Microwave Inductors   245

7.5 Lumped Microwave Capacitors   251

7.6 Distributed Equivalents for Shunt Inductors and Capacitors   252

7.7 A Transmission Line Equivalent for a Symmetric Low-Pass T-or π-Section   257

7.8 Parasitic Effects of Microstrip Discontinuities at the Lower Microwave Frequencies   263

7.9 A Compensation Technique for Microstrip Discontinuities   268

Questions and Problems   271

References   273

Chapter 8 The Design of Radio-Frequency and Microwave Amplifiers   **275**

8.1   Introduction   275
8.2   Amplifier Stability   275
8.3   The Optimal Stabilization of an Amplifier by Resistive Loading   280
8.4   Constant Gain Circles   284
   8.4.1   Circles of Constant Mismatch   284
   8.4.2   Constant Operating Power Gain Circles   285
   8.4.3   Constant Available Power Gain Circles   288
8.5   Tunability   289
8.6   Unilateralness   290
8.7   A Technique for Designing Amplifiers with Non-Unilateral Inherently Stable Transistors   291
8.8   The Dynamic Range of an Amplifier   293
   8.8.1   Evaluation and Optimization of the Noise Performance of an Amplifier — A Procedure for Determining the Optimum Combination of Available Power Gain and Noise Figure for a Multistage Amplifier   293
   8.8.2   Evaluation of the Linearity of an Amplifier   298
8.9   The Design of Multistage Amplifiers   300
   8.9.1   A Procedure for Designing Small-Signal Amplifiers Based on the Operating Power Gain   300
   8.9.2   A Procedure for Designing Small-Signal Amplifiers Based on the Available Power Gain   302
   8.9.3   A Procedure for Designing a Wideband Amplifier for a Specified Noise-Figure and Transducer Power Gain   304
8.10   Reflection Amplifiers   311
8.11   Balanced Amplifiers   314
8.12   Considerations Applying to Power Amplifiers   315

Questions and Problems   319

References and Additional Reading   323

Appendices

Appendix A   PLNM Fortran   325

Appendix B   ZVR Fortran   331

Appendix C   LSM Fortran   338

Appendix D   RCDM Fortran   346

Appendix E   *S*-Parameter Expressions Relevant to the Design of
             RF and Microwave Amplfiers   359

Appendix F   SYZ Basic   364

Index   369

PREFACE

In this book the problem of designing impedance-matching networks for radio-frequency and microwave amplifiers is addressed. Today, a personal computer is affordable to most amplifier designers and the results that can be obtained by using it are greatly superior to those obtainable by graphic means. Therefore, a computer-aided design approach is followed throughout this book. To enhance understanding, the approach is conceptual and the text, to a large degree, is self-contained. The material is intended for students and practicing engineers alike. The only background knowledge assumed is a basic knowledge of circuit and sometimes network theory, complex variable algebra, and matrix manipulations.

In Chapter 1 the characterization of active devices, and the analysis of radio-frequency and microwave circuits with single-frequency parameters are addressed. Z-, T-, Y-, and S-parameters are considered in detail. The latter two sets of parameters are applied to two-port networks in order to find expressions for the power gains and input and output impedances in terms of the Y- and S-parameters. The characteristics of the scattering matrix of passive, lossless networks are also investigated.

If impedance-matching networks are designed without sufficient knowledge of the characteristics of practical components, the designed networks frequently turn out to be unrealizable. Thus, characteristics of passive radio-frequency components are investigated in Chapter 2. Because it is a major problem in the design of realizable circuits, considerable attention is given to the design of air-cored solenoidal inductors with specified Q-factors and resonant frequencies. The design of inductors with magentic cores are also considered.

The design of narrowband, lumped, lossless impedance-matching networks is the subject of Chapter 3. A diagrammatic approach to designing these networks is followed, and a simple procedure for calculating the insertion lossand bandwidth of any cascaded matching network is outlined. L-, T-, and Π-sections are considered.

In Chapter 4 the characteristics of the ideal magnetically-coupled transformer are compared to those of practical transformers, and the design of these transformers is discussed. It is shown that the coupling factor of a transformer is the dominant factor determining its relative bandwidth when the parasitic capacitance can be ignored. The design of wideband, as well as narrowband, single-tuned and double-tuned transformers is considered.

The analysis and design of transmission-line transformers for wideband radio-frequency applications is considered in Chapter 5. At radio-frequencies, the design of power amplifiers is mostly a matter of designing the transmission-line transformers necessary for transforming the load and source resistances to lower values, and performing the necessary combining and splitting functions. Compensation techniques for extending the bandwidth of these transformers are also considered.

The problem of designing wideband impedance-matching networks between arbitrary load and source impedances to approximate any specified transducer power-gain *versus* frequency response as well as possible is addressed in Chapter 6. Analytical and iterative techniques for designing lumped and distributed networks are considered. Because of the relative simplicity of iterative techniques and their superior results, analytical gain-bandwidth theory is not considered in this text. Computer programs implementing Carlin's line segment approach to single-matching problems, and the reflection coefficient approach of Carlin and Yarman to the double-matching problem, form part of the text. A new iterative technique which yields excellent results with very little effort is also introduced. The chapter concludes with a section on the design of RLC impedance-matching networks.

At microwave frequencies, it is often not possible to realize a lumped network design. The main reasons for this are the excessive phase shift across high-inductance inductors and the finite incremental characteristic impedances of lumped components. Similarly, the range of lumped series inductors which can be transformed over wide bandwidths to series transmission lines is limited. The different types of microwave capacitors and inductors are considered in Chapter 7 and limits to the inductance of realizable lumped inductors are derived. At the same time, the range of shunt inductors and shunt capacitors which can be transformed to equivalent shunt transmission lines with negligible error is also investigated. It is shown that significantly better results are obtained through partially transforming low-pass Π- and *T*-sections to distributed equivalents, instead of using only single inductors or capacitors. The chapter concludes with a section on the parasitic influence of discontinuities in a distributed circuit at lower microwave frequencies (below X-band) and compensation techniques are considered. It is shown that a tapering technique introduced a few years ago by Malherbe for stripline circuits can be applied successfully to reduce the parasitic effects of most of these discontinuities, including those resulting from *T*-junctions, at lower microwave frequencies.

The design of radio-frequency and microwave amplifiers consists basically of determining the terminations which will yield the desired performance and the design of impedance-matching networks for that purpose. In the last chapter, theory relevant to setting up the impedance specifications is derived and the effectiveness of the techniques outlined is illustrated with several examples. A significant advantage of the techniques described compared to other techniques is that the desired performance can be obtained without costly optimization even when the transistors used are non-unilateral. The design of single-ended single-stage and multistage amplifiers, reflection amplifiers, power amplifiers, and balanced amplifiers are considered.

Numerous questions and problems are included in the text. These questions and problems should not only be used to develop the skills required of an impedance-matching network and amplifier designer, but should be examined briefly before each chapter is studied in order to orientate the reader.

I hope that the reader will enjoy studying this book as much as I enjoyed writing it. When the material outlined is applied, the pleasure in getting results is guaranteed.

Were it not for the excellent work of many of the workers in the field, this text would not exist. I therefore take pleasure in acknowledging their contributions. Where the techniques outlined here are novel, the necessary research was done at, and sponsored by, the University of Pretoria, South Africa.

Pieter L.D. Abrie
Pretoria
January 1985

CHAPTER 1

NETWORK CHARACTERIZATION AND ANALYSIS WITH Y-, Z-, T-, AND S-PARAMETERS

1.1 INTRODUCTION

Because of the complexity added by parasitic elements and resulting from the interconnection of different networks, broadband equivalent circuits for active components are not frequently used to design or analyze RF and microwave circuits. The better alternative is to use one of the many sets of single-frequency parameters.

The parameters most frequently used are the Y-, Z-, T-, and S-parameters. The first three sets of parameters relate the terminal voltages and currents in different ways, while the S-parameters are closely related to the power incident to and reflected from a device, component, or network.

Because of the relative ease with which it can be done and the useful information directly obtained from it, devices are usually characterized by measuring their S-parameters and circuits are analyzed by calculating their S-parameters. The other parameters are often used to simplify the computations necessary for circuit analysis.

Each of these sets of parameters will be considered in detail in the following sections.

1.2 Y-PARAMETERS

The Y-parameters of an N-port network are defined by the expression

$$\overline{I} = \overline{Y}\ \overline{V} \tag{1.1}$$

where

$$\overline{I} = \begin{bmatrix} I_1 \\ I_2 \\ \cdot \\ \cdot \\ I_N \end{bmatrix} \tag{1.2}$$

$$\bar{V} = \begin{bmatrix} V_1 \\ V_2 \\ \cdot \\ \cdot \\ V_N \end{bmatrix}$$

(1.3)

$$\bar{Y} = \begin{bmatrix} y_{11} \ y_{12} \ldots y_{1N} \\ y_{21} \ y_{22} \ldots \\ \cdot \ \cdot \ \cdot \\ y_{N1} \ y_{N2} \ldots y_{NN} \end{bmatrix}$$

(1.4)

and I_i is the current flowing into the ith terminal, and V_i is the voltage across the ith port of the network.

Each element of the Y-parameter matrix can be calculated or measured by using the relationship

$$y_{ij} = \left. \frac{I_i}{V_j} \right|_{V_h = 0} \quad , \quad h \in [1,2,3, \ldots , N], \ h \neq j$$

(1.5)

that is, y_{ij} is given by the ratio of the current flowing into the ith terminal (output signal) and the voltage across the jth port (input signal), with all the other voltages set equal to zero.

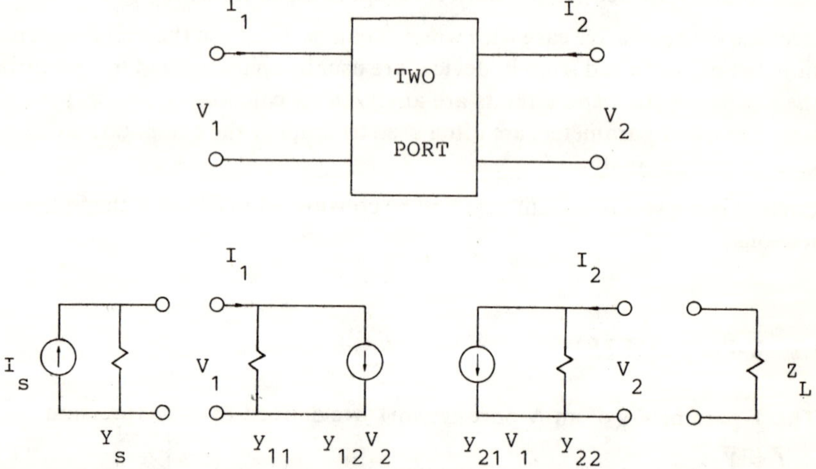

Figure 1.1   An Equivalent Circuit for a Two-Port (in terms of its Y-parameters)

By using (1.1), the terminal currents corresponding to any given set of terminal voltages can be determined. From a linear response viewpoint, the network is therefore completely characterized when the N^2 elements of the Y-parameter matrix are known.

As with any other set of parameters, the *Y*-parameters of an *N*-port can be used to calculate the impedances and gain ratios corresponding to any set of terminations.

By using the equivalent circuit in Fig. 1.1b, it can be shown easily that the following expressions apply to a two-port network terminated as shown:

$$Y_{IN} = \frac{I_1}{V_1} = y_{11} - \frac{y_{12}\,y_{21}}{y_{22} + Y_L} \tag{1.6}$$

$$Y_{OUT} = \frac{I_2}{V_2} = y_{22} - \frac{y_{12}\,y_{21}}{y_{11} + Y_s} \tag{1.7}$$

$$A_V = \frac{V_2}{V_1} = -\frac{y_{21}}{y_{22} + Y_L} \tag{1.8}$$

$$A_I = \frac{I_o}{I_1} = -\frac{I_2}{I_1} = A_V\,Y_L/Y_{IN} \tag{1.9}$$

$$G_\omega = \frac{P_L}{P_{IN}} = \left|\frac{y_{21}}{y_{22} + Y_L}\right|^2 \frac{G_L}{Re\,(Y_{IN})} \tag{1.10}$$

$$G_T = \frac{P_L}{P_{AV\text{-}E}} = \left|\frac{y_{21}\,Y_s}{[y_{22} + Y_L][Y_s + Y_{IN}]}\right|^2 4\,G_L\,R_s \tag{1.11}$$

$$G_A = \frac{P_{AV\text{-}O}}{P_{AV\text{-}E}} = \left|\frac{y_{21}}{y_{11} + Y_s}\right|^2 \frac{G_s}{Re\,(Y_{OUT})} \tag{1.12}$$

In these equations, $Y_L = G_L + jB_L$ is the load admittance, Y_s the source admittance, P_L the power dissipated in the load, P_{IN} the power entering the input port of the network, $P_{AV\text{-}E}$ the power available from the source, $P_{AV\text{-}O}$ the available output power at the output terminals of the two-port, Y_{IN} the input admittance, and Y_{OUT} the output admittance of the two-port.

The available power of a source is defined as the power dissipated in a load which conjugately matches the source, and is given by the expression

$$P_{AV\text{-}E} = \frac{|E|^2}{4\,R_s} = \frac{|I_s|^2}{4\,G_s} \tag{1.13}$$

where E is the source voltage, I_s the equivalent (Norton) source current, and R_s and G_s are defined by

$$Y_s = G_s + jB_s \tag{1.14}$$

$$Z_s = R_s + jX_s \tag{1.15}$$

where Z_s is the source impedance and Y_s its inverse.

When a circuit is analyzed, the *Y*-parameters are frequently used to find a single set of parameters characterizing two networks connected in parallel (same voltages, currents adding). This is illustrated in Fig. 1.2.

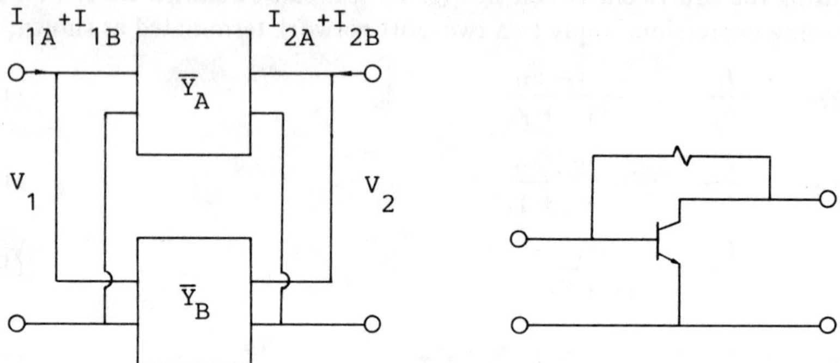

Figure 1.2   Two Networks Connected in Parallel

The *Y*-parameters of two networks connected in parallel simply equal the sum of the *Y*-parameters of each individual network:

$$\overline{Y}_T = \overline{Y}_A + \overline{Y}_B \tag{1.16}$$

Example 1.1

The equation for the input admittance of a two-port network will be derived as an example.

The input admittance is defined by

$$Y_{IN} = I_1 / V_1 \tag{1.6}$$

To find the input admittance it is therefore necessary to find an expression for V_1 in terms of I_1: Ohm's law and Kirchhoff's current law applied to the input port yield

$$V_1 = [I_1 - y_{12} V_2]/y_{11} \tag{1.17}$$

The output voltage is given by

$$V_2 = -I_2 / Y_L = -[y_{21} V_1 + y_{22} V_2]/Y_L \tag{1.18}$$

that is,

$$V_2 = -\frac{y_{21}}{y_{22} + Y_L} V_1 \tag{1.19}$$

After some manipulation, substitution of (1.19) into (1.17) yields

$$Y_{IN} = y_{11} - \frac{y_{12}\, y_{21}}{y_{22} + Y_L} \tag{1.6}$$

1.3   THE INDEFINITE ADMITTANCE MATRIX

The indefinite admittance matrix is a useful tool by which the Y-parameters of network can be determined, if they are known for the same network connected differently. For example, if the common-emitter parameters of a bipolar transistor are known, this matrix can be used to determine the common-base or common-collector parameters.

An admittance matrix is indefinite when none of the network terminals have yet been connected to ground, and the total current flowing into it is therefore equal to the sum of the currents flowing into each terminal.

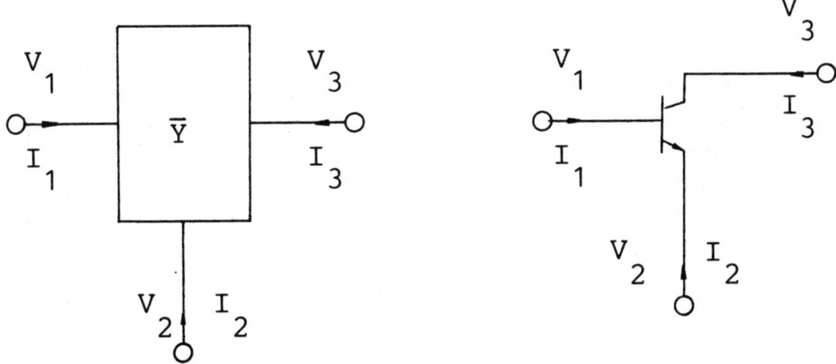

Figure 1.3   An Indefinite Three-Port

It can easily be shown that the sum of the elements in each row or each column of an indefinite admittance matrix is equal to zero. Considering a three-port network, this implies that if four of the nine parameters are known, all the parameters are known.

The proof that the sum of the elements of each row must equal zero follows easily be choosing the terminal voltages to be equal. Each of the currents will then be zero, too, and extraction of each individual equation from (1.1) yields the desired result.

That the sum of the elements in each column should also equal zero follows by setting two of the voltages equal to zero and adding the three currents, the sum of which must be equal to zero.

Example 1.2

As an example of using the indefinite admittance matrix, the common-base parameters of a transistor will be determined in terms of its common-emitter parameters.

The indefinite admittance parameters, which correspond to the common-emitter parameters, can be identified by setting V_2 in Fig. 1.3b and the following equation equal to zero:

$$\begin{bmatrix} I_1 \\ I_2 \\ I_3 \end{bmatrix} = \begin{bmatrix} y_{11} & y_{12} & y_{13} \\ y_{21} & y_{22} & y_{23} \\ y_{31} & y_{32} & y_{33} \end{bmatrix} \begin{bmatrix} V_1 \\ V_2 \\ V_3 \end{bmatrix} \tag{1.20}$$

Since the current in the emitter (I_2) is not of interest when the common-emitter configuration is considered, (1.20) then reduces to

$$\begin{bmatrix} I_1 \\ I_3 \end{bmatrix} = \begin{bmatrix} y_{11} & y_{13} \\ y_{31} & y_{33} \end{bmatrix} \begin{bmatrix} V_1 \\ V_3 \end{bmatrix} \begin{bmatrix} y_{11e} & y_{12e} \\ y_{21e} & y_{22e} \end{bmatrix} \begin{bmatrix} V_1 \\ V_3 \end{bmatrix}$$

With the common-emitter parameters known, y_{11}, y_{13}, y_{31}, and y_{33} are also known, and the rule for the zero column and row can now be applied to determine the other parameters. The only remaining step is to identify the common-base parameters in (1.20). Similarly to the common-emitter parameters, this is done by setting V_1 in (1.20) equal to zero and eliminating the equation giving the base current (I_2) as a function of the voltages. It follows that

$$\begin{bmatrix} y_{11b} & y_{12b} \\ y_{21b} & y_{22b} \end{bmatrix} = \begin{bmatrix} y_{22} & y_{23} \\ y_{32} & y_{33} \end{bmatrix} \tag{1.22}$$

The common collector parameters are given by

$$\begin{bmatrix} y_{11c} & y_{12c} \\ y_{21c} & y_{22c} \end{bmatrix} = \begin{bmatrix} y_{11} & y_{12} \\ y_{21} & y_{22} \end{bmatrix} \tag{1.23}$$

1.4   Z-PARAMETERS

The Z-parameters of an N-port network are defined by the expression

$$\bar{V} = \bar{Z}\,\bar{I} \tag{1.24}$$

where

$$\bar{Z} = \begin{bmatrix} z_{11} & z_{12} & \ldots & z_{1N} \\ z_{21} & z_{22} & \ldots & z_{2N} \\ z_{N1} & z_{N2} & \ldots & z_{NN} \end{bmatrix} \tag{1.25}$$

and $\bar{V}$ and $\bar{I}$ are defined by (1.3) and (1.2), respectively.

Each element in (1.25) can be computed or measured by using the relationship

$$z_{ij} = \left. \frac{V_i}{I_j} \right|_{I_h = 0} \quad , \quad h \in [1,2,3, \ldots, N],\ h \neq j \tag{1.26}$$

that is, z_{ij} is the ratio of the voltages across the jth port (output signal) and the current at the ith port (input signal) with all the other ports idle (or open-circuited).

Equation (1.24) can be used to find the terminal voltages corresponding to any given set of terminal currents.

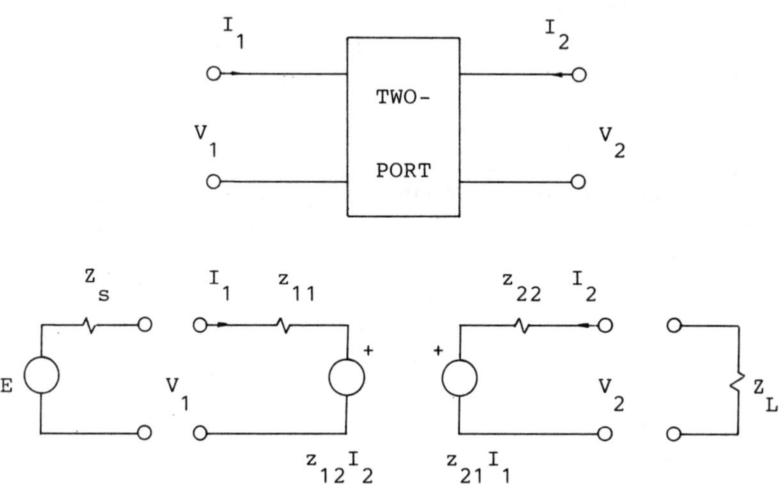

Figure 1.4   An Equivalent Circuit for a Two-Port Network (in terms of its Z-Parameters)

Comparison of (1.24) and (1.1) reveals that the Z-parameters of a network are related to its Y-parameters in the following way:

$$\overline{Z} = \overline{Y}^{-1}$$

(1.27)

Z-parameters are frequently used to find an equivalent set of parameters for two networks connected in series (same currents, voltages adding) as illustrated in Fig. 1.5.

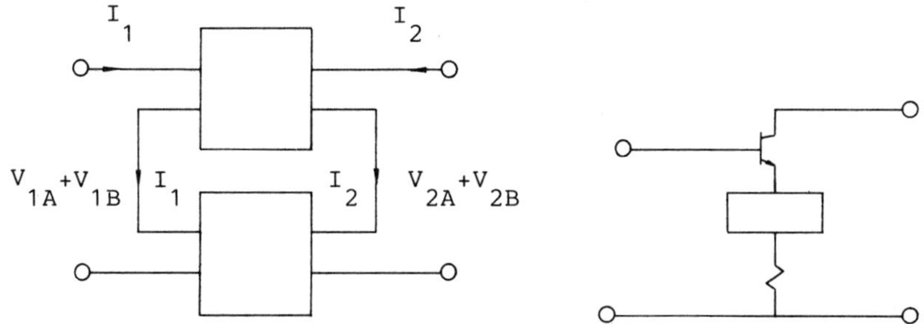

Figure 1.5   Two Networks Connected in Series

The Z-parameters of two networks connected in series are given in terms of the individual Z-parameters by

$$\overline{Z}_T = \overline{Z}_A + \overline{Z}_B \tag{1.28}$$

1.5 T-PARAMETERS

The transmission parameters (A, B, C, D parameters) of a two-port are defined by the equation

$$\begin{bmatrix} V_1 \\ I_1 \end{bmatrix} = \begin{bmatrix} A & B \\ C & D \end{bmatrix} \begin{bmatrix} V_2 \\ I_2 \end{bmatrix} \tag{1.29}$$

with the voltage and current as defined in Fig. 1.6. Note that I_2 is the output current and not the current entering the output terminal as in the case of the Y- and Z-parameters.

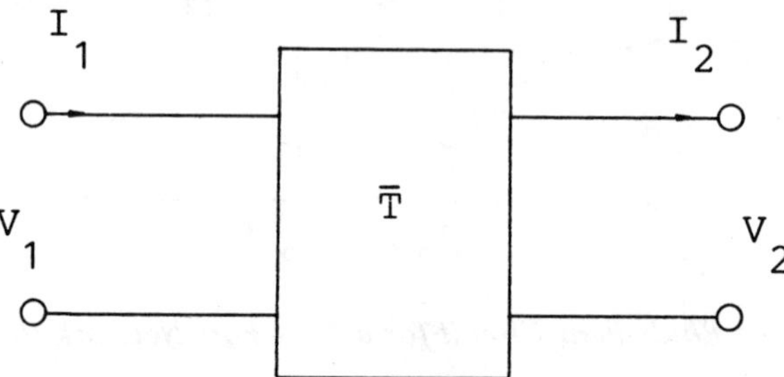

Figure 1.6   *The Voltage and Current Relevant to the Definition of the Transmission-Parameters*

The expressions for the individual elements of the transmission matrix can be obtained by setting either V_2 or I_2 in (1.28) equal to zero after extracting the individual equations from the matrix equation.

T-parameters can be converted to Y-parameters by using the following set of equations:

$$y_{11} = D/B \tag{1.30}$$

$$y_{12} = C - AD/B \tag{1.31}$$

$$y_{21} = -1/B \tag{1.32}$$

$$y_{22} = A/B \tag{1.33}$$

The inverse expressions are

$$A = -y_{22}/y_{21} \tag{1.34}$$

$$B = -1/y_{21} \tag{1.35}$$

$$C = y_{12} - y_{11}y_{22}/y_{21} \tag{1.36}$$

$$D = -y_{11}/y_{21} \tag{1.37}$$

Transmission parameters are frequently used to find an equivalent set of parameters for two cascaded networks. The transmission matrix for the equivalent network is given in terms of the matrices for the individual networks by

$$\overline{T}_T = \overline{T}_A \times \overline{T}_B \tag{1.38}$$

This is illustrated in Fig. 1.7.

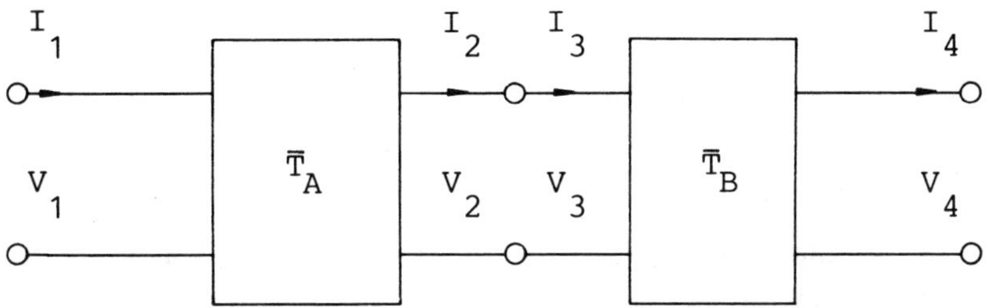

Figure 1.7   Two Cascaded Two-Port Networks

1.6   SCATTERING PARAMETERS

Because of the ease with which it can be measured and the physical meanings attached to it, S-parameters are used extensively to characterize components and devices, and to analyze circuits.

The definitions relevant to these parameters, their physical meanings, and their application in analyzing circuits will be considered in the following sections. Both single-frequency S-parameters and S-parameters in the complex frequency plane will be considered. Because lossless networks are of considerable interest in this text, the limits on the S-matrix of a lossless network will also be examined.

1.6.1   S-Parameter Definitions

In a similar way to the reflection coefficients in transmission-line theory, S-parameters are defined in terms of incident and reflected components. However, in S-parameter theory an incident component is defined as that component which would exist if the port under consideration were conjugately matched to the normalizing impedance at that port. The normalizing impedances are the equivalents of the short-circuit and open-circuit terminations used to characterize a network in terms of its Y-, Z-, or T-parameters. They can be defined to have any arbitrary value (as long as the resistive part is a not equal to zero), but 50Ω impedances are used in most cases.

In terms of the current and voltage at each terminal, the incident and reflected components are defined by the following set of equations:

$$\overline{E}_0 = \overline{V} + \overline{Z}_0 \overline{I} \tag{1.39}$$

$$\overline{I}_i = [\overline{Z}_0 + \overline{Z}_0^*]^{-1} \overline{E}_0 \tag{1.40}$$

$$\overline{I} = \overline{I}_i - \overline{I}_r \tag{1.41}$$

$$\overline{V}_i = \overline{Z}_0^* \overline{I}_i \tag{1.42}$$

$$\overline{V} = \overline{V}_i + \overline{V}_r \tag{1.43}$$

$$\overline{a} = \frac{1}{\sqrt{2}} [\overline{Z}_0 + \overline{Z}_0^*]^{1/2} \overline{I}_i \tag{1.44}$$

$$\overline{b} = \frac{1}{\sqrt{2}} [\overline{Z}_0 + \overline{Z}_0^*]^{1/2} \overline{I}_r . \tag{1.45}$$

with Z_{0j} the normalizing impedance at port j, $\overline{Z}_0^*$ the matrix with conjugate elements of those of $\overline{Z}_0$,

$$\overline{Z}_0 = \begin{bmatrix} Z_{01} & 0 & 0 & \cdots & 0 \\ 0 & Z_{02} & 0 & \cdots & 0 \\ 0 & 0 & Z_{03} & & \\ \cdots & & & & \\ 0 & 0 & 0 & \cdots & Z_{0N} \end{bmatrix} \tag{1.46}$$

with Z_{0j} the normalizing impedance at port j, $\overline{Z}_0^*$ the matrix with conjugate elements of those of $\overline{Z}_0$,

$$\frac{1}{\sqrt{2}} [\overline{Z}_0 + \overline{Z}_0^*]^{1/2} = \begin{bmatrix} \sqrt{R_{01}} & 0 & 0 & \cdots & 0 \\ 0 & \sqrt{R_{02}} & 0 & \cdots & 0 \\ 0 & 0 & \sqrt{R_{03}} & \cdots & 0 \\ \cdots & & & & \\ 0 & 0 & 0 & \cdots & \sqrt{R_{0N}} \end{bmatrix} \tag{1.47}$$

and I_{ji} and V_{ji} the incident current and voltage at port j; I_{jr} and V_{jr} the reflected currect and voltage, a_j the normalized incident component, and b_j the normalized reflected component at port j.

The voltage and current relationships are illustrated in Fig. 1.8 for a two-port network.

Note that the incident voltage is equal to the product of the conjugate of the normalizing impedance and the incident current. The equivalent relationship in transmission-line theory is

$$V_i = Z_0 I_i$$

It can be shown easily that, similar to transmission-line theory, the relationship between the reflected currents and voltages is

$$\overline{V}_r = \overline{Z}_0 \overline{I}_r \tag{1.48}$$

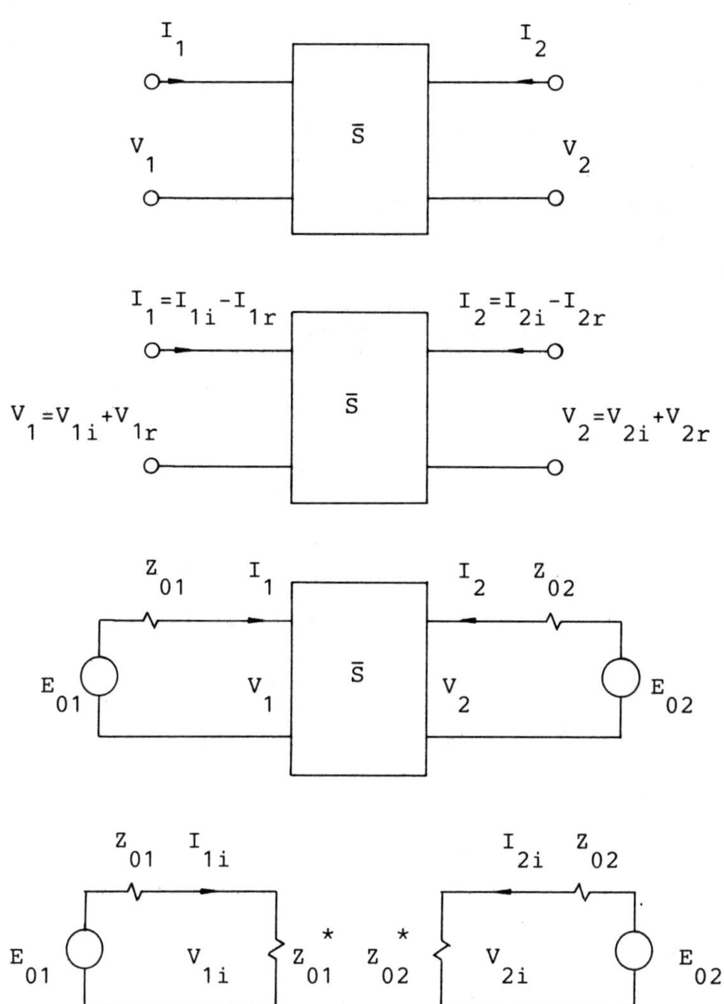

Figure 1.8 [a], [b]:   The Voltage and Current Relevant to the S-Parameter Definitions; [c] The Two-Port of [a] and [b] Augmented by the Normalizing Impedances; [d] An Equivalent Circuit for Calculating the Incident Current and Voltage

There are three different types of S-parameters, which are defined in the following way:

$$\overline{I}_r = \overline{S}^I \, \overline{I}_i \tag{1.49}$$

$$\overline{V}_r = \overline{S}^V \, \overline{V}_i \tag{1.50}$$

$$\overline{b} = \overline{S} \, \overline{a} \tag{1.51}$$

These parameter sets are respectively the current, voltage, and normalized S-parameters.

For a two-port network (1.51) reduces to

$$\begin{bmatrix} b_1 \\ b_2 \end{bmatrix} = \begin{bmatrix} s_{11} & s_{12} \\ s_{21} & s_{22} \end{bmatrix} \begin{bmatrix} a_1 \\ a_2 \end{bmatrix}$$

The definitions given above are summarized together with other useful relationships in Fig. 1.9.

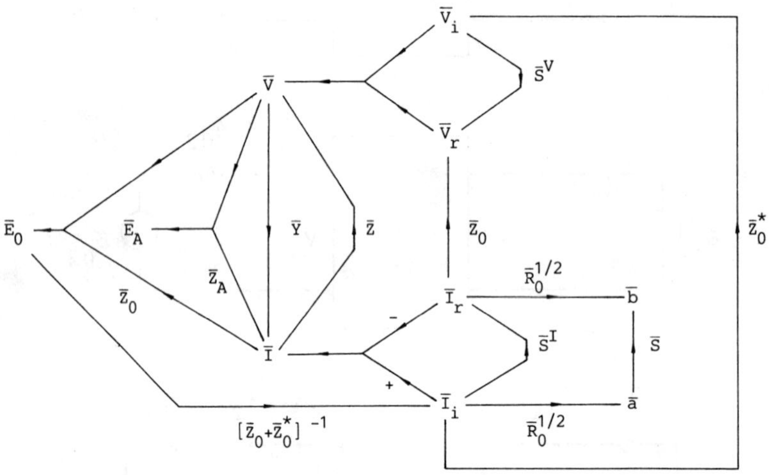

Figure 1.9   Diagram of S-Parameter Relationships

The impedance matrix $\overline{Z}_A$ in Fig. 1.9 is defined by

$$\overline{Z}_A = \begin{bmatrix} Z_{A1} & 0 & 0 & \cdots & 0 \\ 0 & Z_{A2} & 0 & \cdots & 0 \\ \cdots & & & & \\ 0 & 0 & 0 & \cdots & Z_{AN} \end{bmatrix} \tag{1.53}$$

and the matrix $\bar{E}_A$ by

$$\bar{E}_A = \begin{bmatrix} E_{A1} \\ E_{A2} \\ \cdot \\ \cdot \\ E_{AN} \end{bmatrix} \tag{1.54}$$

where E_{Aj} refers to the source voltage at the jth port of the N-port augmented by the actual source and load impedances of interest $(\bar{Z}_A)$. These definitions are illustrated in Fig. 1.10 for a two-port network.

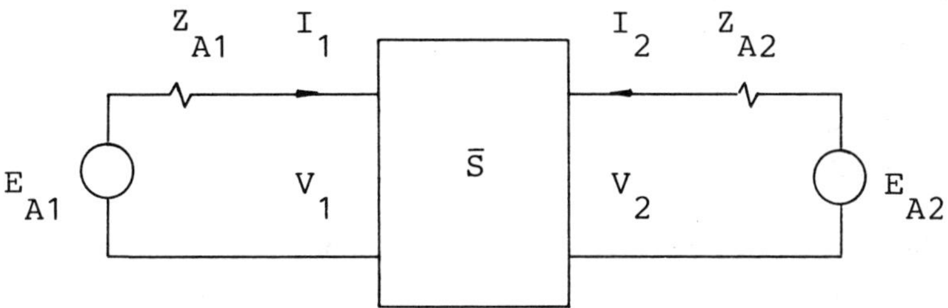

Figure 1.10   The Two-Port Augmented by the Actual Load and Source Ter-
minations [E_{A2} is usually equal to zero]

It can be shown that $\bar{E}_0$ (the source voltage of the N-port augmented by its normalizing impedances as illustrated in Fig. 1.8) is given in terms of $\bar{E}_A$ (the source voltage of the N-port augmented by the actual impedances and source voltage of interest) by the expression

$$\bar{E}_0 = [\bar{I}_N - (\bar{Z}_0 - \bar{Z}_A)(\bar{I}_N - \bar{S}')(\bar{Z}_0 + \bar{Z}_0^*)^{-1}]^{-1} \bar{E}_A \tag{1.55}$$

where $\bar{I}_N$ is the unit diagonal matrix:

$$\bar{I}_N = \begin{bmatrix} 1 & 0 & 0 & \ldots & 0 \\ 0 & 1 & 0 & \ldots & 0 \\ 0 & 0 & 1 & \ldots & 0 \\ \ldots & & & & \\ 0 & 0 & 0 & \ldots & 1 \end{bmatrix} \tag{1.56}$$

Example 1.3

As an example of using the diagram in Fig. 1.9, consider the derivation of the equality

$$\bar{V}_r = \bar{Z}_0 \bar{I}_r \tag{1.48}$$

It follows by inspection of the diagram that in order to find a relationship between $\bar{V}_r$ and $\bar{I}_r$ it is necessary to relate $\bar{V}$ to $\bar{I}$. Of all the possible ways, the easiest would be to use the expression

$$\bar{E}_0 = \bar{V} + \bar{Z}_0 \bar{I} \tag{1.39}$$

$\bar{E}_0$ can then be replaced in terms of $\bar{I}_i$, $\bar{V}$ in terms of $\bar{V}_i$ and $\bar{V}_r$, $\bar{V}_i$ in terms of $\bar{Z}_0^*$ and $\bar{I}_i$, and $\bar{I}$ in terms of $\bar{I}_r$ and $\bar{I}_i$.

After a few manipulations on the equation thus obtained, (1.48) follows.

Example 1.4

In order to bring more reality to the definitions given above, consider finding the incident and reflected components when the terminal voltage and current of a two-port are given by

$$V_1 = 1.0 \ V$$
$$V_2 = 0.5 \ V$$
$$I_1 = 0.1 \ A$$
$$I_2 = -0.2 \ A$$

and the normalizing impedances are chosen to be

$$Z_{01} = 5\Omega$$
$$Z_{02} = 10\Omega$$

The first step is to find the source voltage in the equivalent circuit shown in Figure 1.8d in order to find the incident current and voltage. Inspection of the diagram yields that

$$\bar{E}_0 = \bar{V} + \bar{Z}_0 \bar{I} \tag{1.39}$$

$$= \begin{bmatrix} 1.0 \\ 0.5 \end{bmatrix} + \begin{bmatrix} 5 & 0 \\ 0 & 10 \end{bmatrix} \begin{bmatrix} 0.1 \\ -0.2 \end{bmatrix}$$

$$= \begin{bmatrix} 1.5 \\ -1.5 \end{bmatrix}$$

The incident components can now be obtained by using the equivalent circuit in Fig. 1.8d:

$$\begin{bmatrix} I_{1i} \\ I_{2i} \end{bmatrix} = \begin{bmatrix} E_{01}/[2R_{01}] & 0 \\ 0 & E_{02}/[2R_{02}] \end{bmatrix} = \begin{bmatrix} 0.150 \\ -0.075 \end{bmatrix}$$

$$\begin{bmatrix} V_{1i} \\ V_{2i} \end{bmatrix} = \begin{bmatrix} Z_{01}^* & I_{1i} \\ Z_{02}^* & I_{2i} \end{bmatrix} = \begin{bmatrix} 0.75 \\ -0.75 \end{bmatrix}$$

The normalized incident components follow by application of (1.44):

$$\begin{bmatrix} a_1 \\ a_2 \end{bmatrix} = \begin{bmatrix} \sqrt{R_{01}} \ I_{1i} \\ \sqrt{R_{02}} \ I_{2i} \end{bmatrix} = \begin{bmatrix} 0.3354 \\ -0.2372 \end{bmatrix}$$

The reflected components can be obtained by applying (1.41), (1.48), and (1.45):

$$\begin{bmatrix} I_{1r} \\ I_{2r} \end{bmatrix} = \begin{bmatrix} I_{1i} - I_1 \\ I_{2i} - I_2 \end{bmatrix} = \begin{bmatrix} 0.050 \\ 0.125 \end{bmatrix}$$

$$\begin{bmatrix} V_{1r} \\ V_{2r} \end{bmatrix} = \begin{bmatrix} Z_{01} I_{1r} \\ Z_{02} I_{2r} \end{bmatrix} = \begin{bmatrix} 0.25 \\ 1.25 \end{bmatrix}$$

$$\begin{bmatrix} b_1 \\ b_2 \end{bmatrix} = \begin{bmatrix} \sqrt{R_{01}}\, I_{1r} \\ \sqrt{R_{02}}\, I_{2r} \end{bmatrix} = \begin{bmatrix} 0.1118 \\ 0.3953 \end{bmatrix}$$

1.6.2 The Physical Meanings of the Normalized Incident and Reflected Components of an N-Port

The normalized incident and reflected components are defined in (1.44) and (1.45) in terms of the incident and reflected components of the terminal current. It is useful to have expressions for these components in terms of the terminal voltage and current. The inverse relationships are also of interest.

The required expression for a_j can be obtained easily by using the relationship between the incident current and E_0:

$$a_j = R_{0j}^{1/2}\, I_{ji}$$

$$= R_{0j}^{1/2}\, E_{0j} / [R_{0j} + R_{0j}]$$

$$= \frac{V_j + Z_{0j}\, I_j}{2\, \sqrt{R_{0j}}} \tag{1.57}$$

The expression for the normalized reflected component can be derived by using this result in the following way:

$$b_j = R_{0j}^{1/2}\, I_{jr}$$

$$= R_{0j}^{1/2}\, [I_{ji} - I_j]$$

$$= R_{0j}^{1/2}\, I_{ji} - R_{0j}^{1/2}\, I_j$$

$$= \frac{V_j + Z_{0j}\, I_j}{2\, \sqrt{R_{0j}}} - R_{0j}^{1/2}\, I_j$$

$$= \frac{V_j - Z_{0j}^{*}\, I_j}{2\, \sqrt{R_{0j}}} \tag{1.58}$$

The inverse relationships follow easily by manipulating (1.57) and (1.58):

$$I_j = \frac{a_j - b_j}{\sqrt{R_{0j}}} \tag{1.59}$$

$$V_j = \frac{Z_{0j}^{*}\, a_j + Z_{0j}\, b_j}{\sqrt{R_{0j}}} \tag{1.60}$$

When

$$Z_{0j} = Z_{0j}^* = R_{0j} \tag{1.61}$$

(1.60) simplifies to

$$V_j = \sqrt{R_{0j}}\, [a_j + b_j] \tag{1.62}$$

An expression for the power entering any port can be derived in terms of the normalized components by using (1.59) and (1.60) in conjunction with the expression for the input power:

$$
\begin{aligned}
P_{IN,j} &= 0.5\left[V_j I_j^* + V_j^* I_j \right] \tag{1.63} \\[2mm]
&= 0.5\left[\frac{Z_{0j}\, a_j^* + Z_{0j}^*\, b_j^*}{2\sqrt{R_{0j}}}\, \frac{a_j - b_j}{\sqrt{R_{0j}}} + \right. \\[2mm]
&\qquad\quad \left. \frac{Z_{0j}^*\, a_j + Z_{0j}\, b_j}{2\sqrt{R_{0j}}}\, \frac{a_j^* - b_j^*}{\sqrt{R_{0j}}} \right] \\[2mm]
&= |a_j|^2 - |b_j|^2 \tag{1.64}
\end{aligned}
$$

The power entering any port is, therefore, simply equal to the difference between the square of the normalized incident and reflected components at that port.

The last statement can be taken a step further. It can be shown easily that $|a_j|^2$ is the available power at the jth port of the N-port augmented by its normalizing impedances (refer to Fig. 1.8c and 1.8d). From this fact and (1.63), it follows that $|b_j|^2$ is the reflected power at the jth port of the augmented N-port and, consequently, the power entering any port of a network is equal to the difference between the available and reflected power at the jth port of the N-port augmented by its reference impedances.

It is important to realize that the available power in the N-port augmented by the normalizing impedances is not equal to the available power in the N-port augmented by the actual source and load impedances, unless the two sets of impedances are identical.

The simple expressions for the voltage, current, and power in terms of the normalized incident and reflected components are summarized in Table 1.1.

Table 1.1

The Expressions for the Terminal Current, Terminal Voltage, and the Power Entering a Port Summarized in Terms of the Normalized Incident and Reflected Components

$$I_j = [a_j - b_j]/\sqrt{R_{0j}} \tag{1.59}$$

$$V_j = [Z_{0j}^* a_j + Z_{0j} b_j]/\sqrt{R_{0j}} \tag{1.60}$$

$$\quad = \sqrt{R_{0j}}\,[a_j + b_j] \text{ if } Z_{0j} = Z_{0j}^* \tag{1.62}$$

$$P_j = |a_j|^2 - |b_j|^2 \tag{1.64}$$

1.6.3 The Physical Interpretations of the Scattering Parameters

Consider the definitions of the elements of a two-port scattering matrix. The input reflection parameter s_{11} is defined by

$$s_{11} = \left.\frac{b_1}{a_1}\right|_{a_2=0} \tag{1.65}$$

and the forward transmission parameter s_{21} by

$$s_{21} = \left.\frac{b_2}{a_1}\right|_{a_2=0} \tag{1.66}$$

The constraints on the current and voltage at the output terminals, when $a_2 = 0$, can be determined by using (1.57):

$$0 = a_2 = \frac{V_2 + Z_{02} I_2}{2\sqrt{R_{02}}} \tag{1.57}$$

leading to

$$V_2 = Z_{02}\,[-I_2] \tag{1.67}$$

So that a_2 equals zero, the load impedance across the output port must, therefore, be equal to the normalizing impedance at that port and the electromotive force (EMF) must be equal to zero. This is illustrated in Fig. 1.11a.

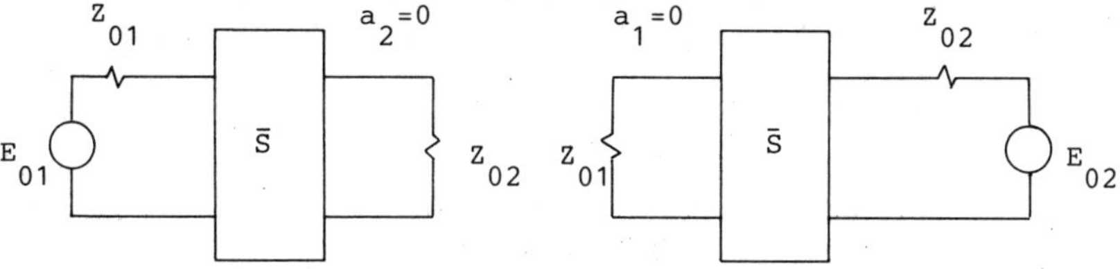

Figure 1.11   The conditions under which [a] $a_2 = 0$ and [b] $a_1 = 0$

At this stage, (1.57) and (1.58) can be substituted into (1.65) and (1.66) to find an expression for the parameters in terms of the terminal current and voltage:

$$s_{11} = \left.\frac{V_1 - Z_{01}^* I_1}{V_1 + Z_{01} I_1}\right|_{a_2=0}$$

$$\quad = \left.\frac{Z_{IN} - Z_{01}^*}{Z_{IN} + Z_{01}}\right|_{a_2=0} \tag{1.68}$$

where Z_{IN} is the input impedance of the two-port terminated as shown in Fig. 1.11a and

$$s_{21} = \sqrt{\frac{R_{01}}{R_{02}}} \, \frac{V_2 - Z_{02}^* I}{V_1 + Z_{01} I_1} \bigg|_{a_2 = 0}$$

$$= \sqrt{\frac{R_{01}}{R_{02}}} \, \frac{Z_{02}[-I_2] - Z_{02}^* I_2}{E_{01}} \bigg|_{a_2 = 0}$$

$$= -\sqrt{R_{01} R_{02}} \, \frac{I_2}{E_{01}} \bigg|_{a_2 = 0} \tag{1.69}$$

The equivalence between (1.68) and the equivalent expression

$$T_{IN} = \frac{Z_{IN} - Z_{01}}{Z_{IN} + Z_{01}} \tag{1.70}$$

for a reflection coefficient in transmission-line theory is obvious. When

$$Z_{01} = R_{01}$$

as is often the case, the two expressions will be identical.

When the normalizing resistances are equal, the forward transmission parameter s_{21} is simply the voltage gain V_L / E_{01} of the two-port augmented with its normalizing impedances with E_{02} set equal to zero.

Because the S-parameters are defined in terms of the normalized incident and reflected components, and the square of these components were shown to be the incident and reflected power at the relevant port of the two-port augmented with its normalizing impedances, respectively, it follows that

$$\left| s_{11} \right|^2 = \left| \frac{b_2}{a_1} \right|_{a_2 = 0}$$

$$= \frac{P_{1r}}{P_{AV-E_{01}}} \quad a_2 = 0 \tag{1.71}$$

and

$$\left| s_{21} \right|^2 = \left| \frac{b_2}{a_1} \right|_{a_2 = 0}$$

$$= \frac{|b_2|^2 - |a_2|^2}{|a_1|^2} \quad a_2 = 0$$

$$= \frac{P_L}{P_{AV-E_{01}}} \quad a_2 = 0$$

$$= G_T \quad a_2 = 0 \tag{1.72}$$

where $P_{AV\text{-}E_{01}}$ is the power available from the source when the two-port is augmented by the normalizing impedances, and P_{1r} is the power reflected from the input port when it is augmented by the normalizing impedances and E_{02} is set equal to zero.

The meanings of $|s_{11}|^2$ and $|s_{21}|^2$ are illustrated in Fig. 1.12.

Similar expressions apply to the output reflection parameter s_{22} and the reverse transmission parameter s_{12}.

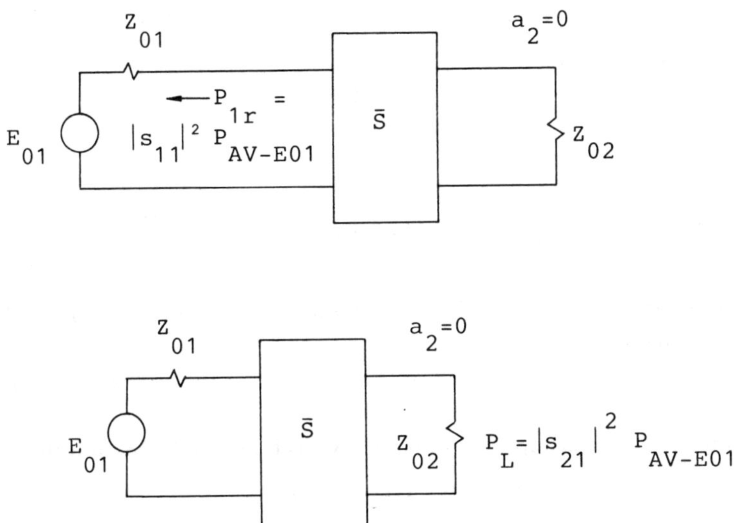

Figure 1.12   The Physical Meanings of the Scattering Parameters Illustrated

When the normalizing impedances are also the impedances in the actual network of interest, the transducer power gain and the ratio of the reflected power at the input to the available power from the source are given directly by s_{21} and s_{11}, respectively.

With the normalizing impedances are purely resistive, and s_{11} and s_{22} displayed on a Smith chart, the input and output impedances of the network augmented by its normalizing impedances can be read directly.

1.6.4   Constraints Imposed on the Normalized Components by the Terminations of an N-Port

In order to derive expressions for the gains and impedances of an N-port with arbitrary terminations, it is necessary to derive expressions for the constraints imposed by the terminations on the normalized incident and reflected components.

Consider port n of the N-port terminated in an impedance Z_{An} in series with a voltage source E_{An} as shown in Fig. 1.13.

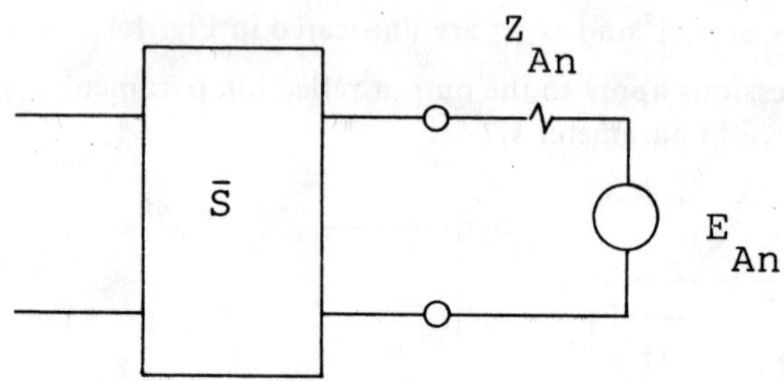

Figure 1.13   The N-Port under Consideration

The termination forces the following relationship between the terminal voltage and current:

$$E_{An} = V_n + Z_{An} I_n \tag{1.73}$$

By using this relationship in conjunction with (1.57) and (1.58), it follows that

$$2 \sqrt{R_{0n}}\, a_n = V_n + Z_{0n} I_n = E_{An} - (Z_{An} - Z_{0n}) I_n$$

leading to

$$2 \sqrt{R_{0n}}\, a_n - E_{An} = - (Z_{An} - Z_{0n}) I_n \tag{1.74}$$

and

$$2 \sqrt{R_{0n}}\, b_n = V_n - Z_{0n}^* I_n = E_{An} - (Z_{An} - Z_{0n}^*) I_n$$

leading to

$$2 \sqrt{R_{0n}}\, b_n - E_{An} = - (Z_{An} + Z_{0n}^*) I_n \tag{1.75}$$

Dividing (1.74) by (1.75) yields

$$\frac{2 \sqrt{R_{0n}}\, a_n - E_{An}}{2 \sqrt{R_{0n}}\, b_n - E_{An}} = \frac{-[Z_{An} - Z_{0n}] I_n}{-[Z_{An} + Z_{0n}^*] I_n}$$

leading to

$$a_n = \frac{Z_{An} - (Z_{0n}^{**})}{Z_{An} + (Z_{0n}^*)}\, b_n + \frac{\sqrt{R_{0n}}}{Z_{An} + (Z_{0n}^*)}\, E_{An} \tag{1.76}$$

With

$$E_{An} = 0,$$

the second term in (1.76) is equal to zero, and the following relationship applies:

$$S_n = a_n / b_n = \frac{Z_{An} - (Z_{0n}^{**})}{Z_{An} + (Z_{0n}^{*})} \tag{1.77}$$

This expression clearly has the form of a reflection parameter with normalizing impedance Z_{0n}^{*}. The termination can therefore be considered as the interconnection of a one-port network with a port of the N-port. Furthermore, the component incident on the N-port (a_n) is reflected from the one-port and the component reflected from the N-port (b_n) is incident on the one-port.

The normalizing impedance for the single-port is the conjugate of that for the N-port.

1.6.5 Derivation of Expressions for the Gain Ratios and Reflection Parameters of a Two-Port

Consider the two-port with terminations as shown in Fig. 1.14 and the associated S-parameter expression:

$$\begin{bmatrix} b_1 \\ b_2 \end{bmatrix} = \begin{bmatrix} s_{11} & s_{12} \\ s_{21} & s_{22} \end{bmatrix} \begin{bmatrix} a_1 \\ a_2 \end{bmatrix} \tag{1.78}$$

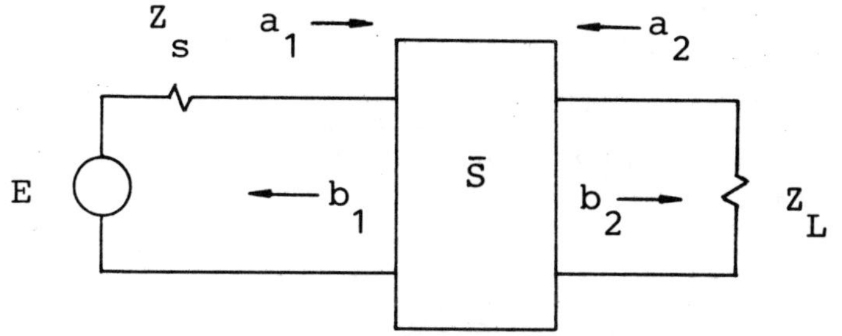

Figure 1.14   The Two-Port under Consideration

In (1.78) a_1 is an independent variable, the magnitude and phase of which are determined by the source voltage E and the fixed normalizing impedance Z_{01}.

According to (1.77), b_2 is constrained to

$$b_2 = a_2 / \frac{Z_L - (Z_{02}^{**})}{Z_L + (Z_{02}^{*})}$$

$$= a_2 / S_L \tag{1.79}$$

With a_1 the independent variable and b_2 known in terms of a_2, (1.78) amounts to two equations with two unknowns and values for a_2, b_1 and b_2 can be determined in terms of the scattering parameters and a_1. The results are

$$a_1 = 1 \tag{1.80}$$

$$b_1 = S_{11} + \frac{s_{12}s_{21}S_L}{1 - s_{22}S_L} \tag{1.81}$$

$$a_2 = \frac{s_{21}S_L}{1 - s_{22}S_L} \tag{1.82}$$

$$b_2 = a_2 / S_L \tag{1.83}$$

At this stage, the reflection parameters and the gain ratios of interest can be determined. The expressions most frequently used are repeated below.

$$s_{11}' = \frac{Z_{IN} - Z_{01}^*}{Z_{IN} + Z_{01}} = \frac{b_1}{a_1} = s_{11} + \frac{s_{12}s_{21}S_L}{1 - s_{22}S_L} \tag{1.84}$$

$$s_{22}' = \frac{Z_{OUT} - Z_{02}^*}{Z_{OUT} + Z_{02}} = \frac{b_2}{a_2} = s_{22} + \frac{s_{12}s_{21}S_s}{1 - s_{11}S_s} \tag{1.85}$$

$$S_s = \frac{Z_s - (Z_{01}^{**})}{Z_s + (Z_{01}^*)} \tag{1.86}$$

$$G_w = \frac{|b_2|^2 - |a_2|^2}{|a_1|^2 - |b_1|^2}$$

$$= \frac{|s_{21}|^2 [1 - |S_L|^2]}{|1 - s_{22}S_L|^2 - |s_{11}(1 - s_{22}S_L) + s_{12}s_{21}S_L|^2} \tag{1.87}$$

$$G_T = \frac{|b_2|^2 - |a_2|^2}{P_{AV\text{-}E}}$$

$$= \frac{|s_{21}|^2 [1 - |S_L|^2][1 - |S_s|^2]}{|[1 - s_{11}S_s][1 - s_{22}S_L] - s_{12}s_{21}S_s S_L|^2} \tag{1.88}$$

$$G_{T, s_{12} = 0} = G_{T, u}$$

$$= \frac{1 - |S_s|^2}{|1 - s_{11}S_s|^2} |s_{21}|^2 \frac{1 - |S_L|^2}{|1 - s_{22}S_L|^2} \tag{1.89}$$

where $G_{T\,u}$ is the unilateral transducer power gain,

$$G_A = \frac{P_{AV\text{-}O}}{P_{AV\text{-}E}} \tag{1.90}$$

$$= \frac{|s_{21}|^2 [1 - |S_s|^2]}{|1 - |s_{22}|^2 + |S_s|^2 [|s_{11}|^2 - |\Delta|^2] - 2\,Re(C_1 S_s)|} \tag{1.91}$$

where $P_{AV\text{-}O}$ is the maximum available power at the output terminals of the transistor,

$$\Delta = s_{11} s_{22} - s_{12} s_{21} \tag{1.92}$$

and

$$C_1 = s_{11} - \Delta \, s_{22}^* \tag{1.93}$$

$$A_V = \sqrt{\frac{R_{02}}{R_{01}}} \; \frac{a_2 + b_2}{a_1 + b_1}$$

$$= \sqrt{\frac{R_{02}}{R_{01}}} \; \frac{s_{21}\,[1 + S_L]}{1 + s_{11} - s_{22}\,S_L - s_{11}\,s_{22}\,S_L + s_{12}\,s_{21}\,S_L} \tag{1.94}$$

In order for (1.94) to apply, the normalizing impedances must be purely resistive.

Example 1.5

As an example of the application of (1.80) to (1.83), consider the derivation of (1.88).

An expression for

$$P_L = |b_2|^2 - |a_2|^2$$

follows directly from (1.82) and (1.83):

$$P_L = \left| \frac{s_{21}}{1 - s_{22}\,S_L} \right|^2 [1 - |S_L|^2] |a_1|^2 \tag{1.95}$$

In order to derive an expression for $P_{AV\text{-}E}$ it is necessary to use (1.76):

$$a_1 = \frac{Z_s - (Z_{01}^{**})}{Z_s + (Z_{01}^*)} \, b_1 + \frac{\sqrt{R_{01}}}{Z_s + Z_{01}^*} \, E \tag{1.96}$$

$$P_{AV\text{-}E} = E^2 / [4 R_s]$$

$$= \frac{|Z_s + Z_{01}^*|^2}{4 \, R_{01} \, R_s} \, |a_1 - S_s \, b_1|^2$$

$$= \frac{|a_1 - S_s \, b_1|^2}{1 - |S_s|^2} \tag{1.97}$$

Substitution of b_1 in terms of a_1 in this equation yields

$$P_{AV\text{-}E} = \frac{|[1 - s_{11}\,S_s][1 - s_{22}\,S_L] - s_{21}\,s_{12}\,S_s\,S_L|^2}{[1 - |S_L|^2]|1 - s_{22}\,S_L|^2} \tag{1.98}$$

Combination of (1.98) and (1.95) yields the desired expression.

1.6.6   Conversion of S-Parameters to Other Parameters

The schematic representation of the S-parameter and related relationships in Fig. 1.8 can be used to derive expressions for the conversion of normalized S-parameters to the other S-parameters as well as Z- or Y-parameters. The results are

$$\overline{S} = \overline{R}_0^{1/2} \, \overline{S}^I \, \overline{R}_0^{-1/2} \tag{1.99}$$

$$\overline{S}^I = \overline{Z}_0^{-1} \, \overline{S}^V \, \overline{Z}_0^* \tag{1.100}$$

$$\overline{S}^I = [\overline{Z} + \overline{Z}_0]^{-1} [\overline{Z} - \overline{Z}_0^*] \tag{1.101}$$

$$\overline{S}^V = -[\overline{Y} + \overline{Y}_0]^{-1} [\overline{Y} - \overline{Y}_0^*] \tag{1.102}$$

$$\overline{Y} = \overline{Y}_0 [\overline{I}_n - \overline{S}] [\overline{I}_n + \overline{S}]^{-1} \tag{1.103}$$

with

$$\overline{Y}_0 = \overline{Z}_0^{-1} \tag{1.104}$$

When the normalizing impedances are all purely resistive and equal

$$\overline{S} = \overline{S}^I = \overline{S}^V \tag{1.105}$$

Example 1.6

Consider the derivation of (1.102) as an example.

It follows by inspection of the diagram in Fig. 1.9 that

$$\overline{I} = \overline{Y} \, \overline{V}$$

is equivalent to

$$\overline{I}_i - \overline{I}_r = \overline{Y} [\overline{V}_i + \overline{V}_r]$$

which is equivalent to

$$\overline{Z}_0^{*-1} \, \overline{V}_i - \overline{Z}_0^{-1} \, \overline{V}_r = \overline{Y} \, \overline{V}_i + \overline{Y} \, \overline{V}_r$$

This equation can be maniplated to

$$\overline{V}_r = -[\overline{Y} + \overline{Y}_0]^{-1} [\overline{Y} - \overline{Y}_0^*] \, \overline{V}_i$$

which yields the required expression.

1.6.7   The Indefinite S-Matrix

Similar to the indefinite admittance matrix, the sum of the elements in each row or column of the indefinite S-matrix is equal to a constant. In this case the constant is unity.

In order to prove that the sum of the elements in each row must equal one, consider the three-port shown in Fig. 1.15a.

Under the conditions shown, all the incident components are equal, and

$$b_j = s_{j1} \, a_1 + s_{j2} \, a_2 + s_{j3} \, a_3$$

which simplifies to

$$b_j = [s_{j1} + s_{j2} + s_{j3}] \, a_1$$

Substitution of b_j and a_1 in terms of the reflected and incident currents yields

$$I_{jr} = [s_{j1} + s_{j2} + s_{j3}] \, I_{1i}$$

and since the terminal currents must equal zero when all the source voltages are equal, I_{jr} must equal I_{1i}; it follows that

$$s_{j1} + s_{j2} + s_{j3} = 1 \tag{1.106}$$

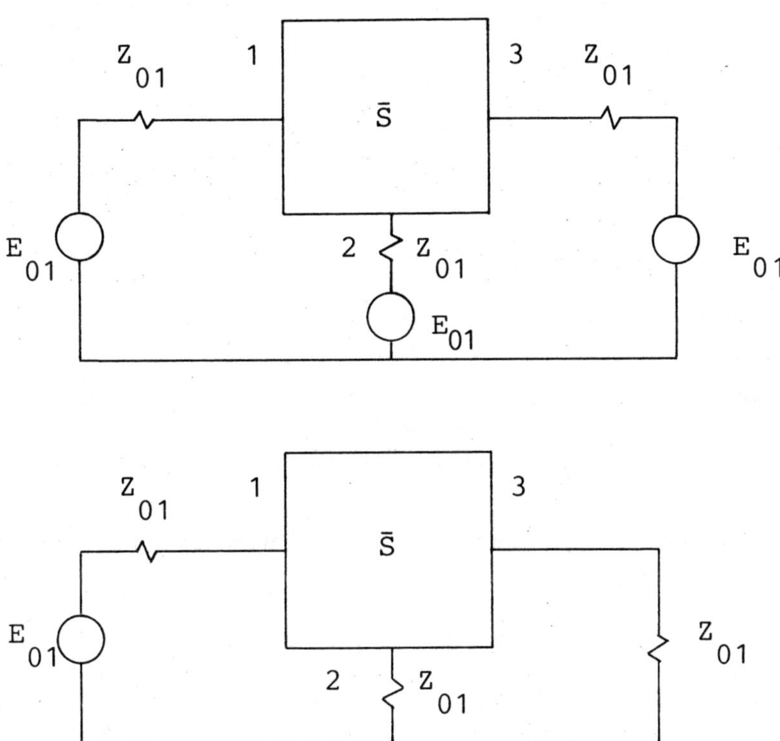

Figure 1.15   *Circuits Used to Prove that the Sum of the Elements in (a) any Row, or (b) any Column of an Indefinite S-matrix is Equal to One*

The circuit in Fig. 1.15b can be used to prove that the sum of the elements of the first column of the indefinite matrix is equal to one. Because the incident components at terminals two and three are equal to zero, the necessary condition

$$I_1 + I_2 + I_3 = 0$$

simplifies to

$$I_{1i} = I_{1r} + I_{2r} + I_{3r} \tag{1.107}$$

With $a_2 = 0 = a_3$,

$$b_1 = s_{11}\, a_1 + s_{12}\, a_2 + s_{13}\, a_3$$
$$b_2 = s_{21}\, a_1 + s_{22}\, a_2 + s_{23}\, a_3$$
$$b_3 = s_{31}\, a_1 + s_{32}\, a_2 + s_{33}\, a_3$$

simplifies to

$$b_1 = s_{11}\, a_1$$
$$b_2 = s_{21}\, a_1$$
$$b_3 = s_{31}\, a_1$$

and, therefore,

$$I_{1r} = s_{11}\, I_{1i}$$
$$I_{2r} = s_{21}\, I_{1i}$$
$$I_{3r} = s_{31}\, I_{1i} \tag{1.108}$$

Equation (1.107) combined with (1.108) yields

$$s_{11} + s_{21} + s_{31} = 1 \tag{1.109}$$

By moving the voltage source in Fig. 1.15b to the other two ports and following the same procedure, it can also be shown that the sum of the elements in each of the other two columns of the indefinite S-matrix is equal to one.

1.6.8 Extension of the Single-Frequency S-Parameter Definitions to the Complex Frequency Plane

A necessary condition for a matrix to be the S-parameter matrix of a linear, lumped, passive network normalized to N minimum reactance functions (that is, impedance functions with no poles on the real-frequency axis) is that none of its elements may have any poles in the closed right-hand side (RHS) of the complex frequency plane. The definitions given for a and b in sec. 1.6.1 are adequate for any single frequency application, and also in the complex plane when the normalizing impedance functions ($Z_{0j}(s)$) do not have any finite poles (that is, purely resistive normalizing impedances, or impedances of the form $R_{0j} + sL_{0j}$). However, when these impedance functions are more complex it is necessary to extend the definitions of the normalized incident and reflected components. The following definitions are relevant to the more general case:

$$Z_0(s) = \begin{bmatrix} Z_{01}(s) & 0 & \ldots & 0 \\ 0 & Z_{02}(s) & \ldots & 0 \\ \ldots & & \ldots & 0 \\ 0 & 0 & \ldots & Z_{0N}(s) \end{bmatrix} \tag{1.110}$$

where $Z_{0j}(s)$ is the normalizing impedance at port j,

$$\overline{r}(s) = \begin{bmatrix} r_1(s) & 0 & \ldots & 0 \\ 0 & r_2(s) & \ldots & 0 \\ \ldots & & & \\ 0 & 0 & \ldots & r_N(s) \end{bmatrix} \tag{1.111}$$

$$= \overline{h}(s)\,\overline{h}(-s) \tag{1.112}$$

where

$$\overline{r}_j(s) = 0.5\,[\overline{Z}_{0j}(s) + \overline{Z}_{0j}(-s)] \tag{1.113}$$

and

$$\overline{h}(s) = \begin{bmatrix} m_1(s)/n_1(s) & 0 & \ldots & 0 \\ 0 & m_2(s)/n_2(s) & \ldots & 0 \\ \ldots & & & \\ 0 & 0 & \ldots & m_N(s)/n_N(s) \end{bmatrix} \tag{1.114}$$

where $m_j(s)$ and $n_j(s)$ are polynomials and the zeros of $n_j(s)$ (poles of $h_j(s)$) are constrained to the open LHP and the zeros of $m_j(s)$ (zeros of $h_j(s)$) to the closed RHP.

$$\overline{a}(s) = \overline{h}(-s)\,\overline{I}_i(s) \tag{1.115}$$

$$\overline{b}(s) = \overline{h}(s)\,\overline{I}_r(s) \tag{1.116}$$

where $\overline{a}(s)$ is the matrix of normalized incident components, $\overline{b}(s)$ the normalized reflected components, with $\overline{I}_i$ and $\overline{I}_r$ as defined in sect. 1.6.1.

Note that the elements of $\overline{r}_j(s)$ are even functions and are the effective series resistance parts of the corresponding normalizing impedances.

With these definitions for the normalized and reflected components, it follows that

$$a_j(s) = \frac{V_j(s) + Z_{0j}(s)\,I(s)}{2\,h_j(s)} \tag{1.117}$$

$$b_j(s) = \frac{V_j(s) - Z_{0j}(-s)\,I(s)}{2\,h_j(-s)} \tag{1.118}$$

$$s_{jj}(s) = \frac{h_j(s)}{h_j(-s)}\,\frac{Z_{INj}(s) - Z_{0j}(-s)}{Z_{INj}(s) + Z_{0j}(s)} \tag{1.119}$$

$$s_{jk}(s) = -2\,h_j(s)\,h_k(s)\,\frac{I_j(s)}{E_{0k}} \tag{1.120}$$

These relationships are identical to those derived previously for single-frequency applications as long as

$$h_j (s) = R_{0j}^{1/2} = h_j (-s)$$

This relationship will apply in all cases where the normalizing impedances are purely resistive or of the form $R_{0j} + sL_{0j}$.

Independent of the complexity of the normalizing impedances, the incident and reflected power are still given by $|a_j|^2$ and $|b_j|^2$, respectively.

Example 1.6

As an example, $h(s)$ will be determined for the normalizing impedances shown in Fig. 1.16.

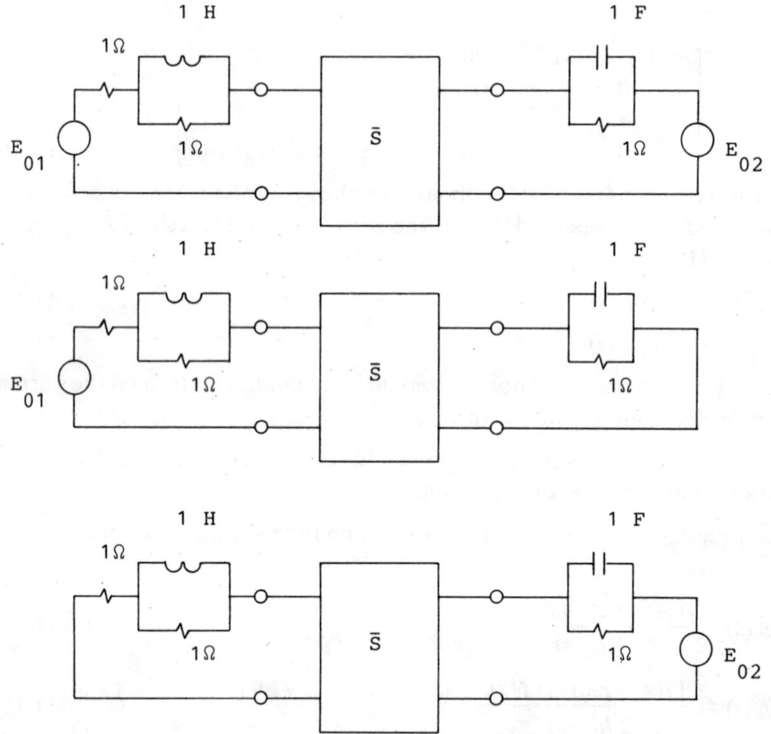

Figure 1.16 [a] *The Normalizing Impedances under Consideration;* [b] *The Equivalent Circuit Used to Determine* $s_{11}(s)$ *and* $s_{21}(s)$; *and* [c] *The Equivalent Circuit Used to Determine* $s_{22}(s)$ *and* $s_{12}(s)$

Since

$$Z_{01}(s) = 1 + s/[1+s],$$

it follows that

$$r_1(s) = 0.5\left[Z_{01}(s) + Z_{01}(-s)\right]$$

$$= \frac{1 - 2s^2}{1 - s^2}$$

$$= \frac{1 - \sqrt{2}s}{1+s}\,\frac{1 + \sqrt{2}s}{1-s}$$

and, therefore,

$$h_1(s) = [1 - \sqrt{2}s]/[1+s]$$

Similarly,

$$h_2(s) = s/[1+s]$$

1.6.9 Constraints on the Scattering Matrix of a Lossless N-Port

The average power entering a passive lossless device must be equal to zero. This imposes the following constraints on the scattering matrix:

$$\begin{aligned}
0 = P_{AV} &= 0.5\left[\overline{V}^{*\prime}(j\omega)\,\overline{I}(j\omega) + \overline{I}^{*\prime}(j\omega)\,\overline{V}(j\omega)\right] \\
&= \left[\overline{a}^{*\prime}(j\omega)\,\overline{a}(j\omega)\,\overline{b}^{*\prime}(j\omega)\,\overline{b}(j\omega)\right] \\
&= \overline{a}^{*\prime}(j\omega)\left[\overline{I}_n - \overline{S}^{*\prime}(j\omega)\,\overline{S}(j\omega)\right]\overline{a}(j\omega)
\end{aligned} \qquad (1.121)$$

leading to

$$\overline{S}^{*\prime}(j\omega)\,\overline{S}(j\omega) = \overline{I}_n \qquad (1.122)$$

In these equations the superscript $^{*\prime}$ indicates the transposed conjugate of the relevant matrix.

It is clear from (1.122) that the inverse of the scattering matrix of a lossless network is constrained to be equal to its transposed conjugate; that is,

$$\overline{S}^{-1}(j\omega) = \overline{S}^{*\prime}(j\omega) \qquad (1.123)$$

A matrix whose inverse is equal to its transposed conjugate is called a unitary matrix. A necessary and sufficient condition for a matrix to be unitary is that its columns (or rows) should be mutually orthogonal unit vectors [1]. In terms of the elements of the scattering matrix, this implies that the following equations must be satisfied:

$$\sum_{j=1}^{N} s_{ij}(j\omega)\,s_{kj}^{*}(j\omega) = \delta_{ik} \qquad (1.124)$$

$$\sum_{j=1}^{N} s_{ji}(j\omega)\,s_{jk}^{*}(j\omega) = \delta_{ik} \qquad (1.125)$$

where δ_{ik} is the Kronecker delta ($\delta_{ik} = 0$, V $i \neq k$; $\delta_{ik} = 1$, V $i = k$).

The unitary constraint on the magnitude of each row or column vector of the scattering matrix forces the following two relationships on the elements of each row and column, respectively,

$$\sum_{j=1}^{N} s_{ij}(j\omega)\, s_{ij}^{*}(j\omega) = \sum_{j=1}^{N} |s_{ij}(j\omega)|^2 = 1 \tag{1.126}$$

and

$$\sum_{j=1}^{N} s_{ji}(j\omega)\, s_{ji}^{*}(j\omega) = \sum_{j=1}^{N} |s_{ji}(j\omega)|^2 = 1 \tag{1.127}$$

The magnitude of each element of the S-matrix of a lossless (and also passive) network is therefore bounded by unity, i.e.,

$$|s_{ij}(j\omega)| \leq 1 \tag{1.128}$$

By applying (1.119) and (1.120) to a two-port network, it follows that

$$|s_{11}(j\omega)|^2 = 1 - |s_{12}(j\omega)|^2 \tag{1.129a}$$

$$|s_{11}(j\omega)|^2 = 1 - |s_{21}(j\omega)|^2 \tag{1.129b}$$

$$|s_{22}(j\omega)|^2 = 1 - |s_{21}(j\omega)|^2 \tag{1.129c}$$

$$|s_{22}(j\omega)|^2 = 1 - |s_{12}(j\omega)|^2 \tag{1.129d}$$

$$s_{11}(j\omega)\, s_{12}^{*}(j\omega) = -\, s_{21}(j\omega)\, s_{22}^{*}(j\omega) \tag{1.130}$$

$$s_{11}(j\omega)\, s_{21}^{*}(j\omega) = -\, s_{12}(j\omega)\, s_{22}^{*}(j\omega) \tag{1.131}$$

Combining (1.129b) and (1.129c) yields

$$|s_{11}(j\omega)| = |s_{22}(j\omega)| \tag{1.132}$$

while (1.129a) and (1.129b) can be combined to show that

$$|s_{12}(j\omega)| = |s_{21}(j\omega)| \tag{1.133}$$

(1.133) can be extended to

$$s_{12}(j\omega) = s_{21}(j\omega) \tag{1.134}$$

wherever the network considered is passive and reciprocal.

Equation (1.134) can be proved easily by using the reciprocity theorem.

By combining (1.131) and (1.134), it can be shown that

$$s_{22}(j\omega) = -\, \frac{s_{21}(j\omega)}{s_{21}^{*}(j\omega)}\, s_{11}^{*}(j\omega) \tag{1.135}$$

These relationships will prove useful in later chapters.

Example 1.7

As an example, the S-parameters of the lossless two-port in Fig. 1.17 will be derived, and some of the relationships given above will be illustrated.

Since the normalizing impedances are purely resistive and, therefore,

$$h_j\,(s) = R_{0j}^{1/2} = h_j\,(-s)$$

the input reflection and forward transmission parameters can be determined by using (1.68) and (1.69), respectively.

The equivalent circuit correspondingly to $a_2=0$ and the chosen normalizing impedances are shown in Fig. 1.17b.

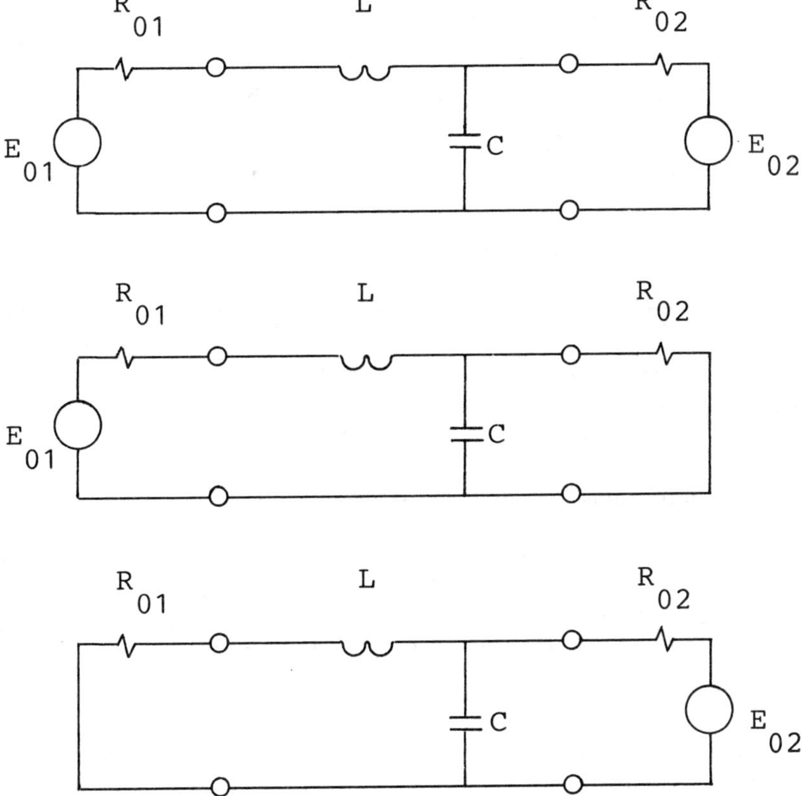

Figure 1.17   [a] The Lossless Two-Port under Consideration; [b] The Equivalent Circuit Used to Determine $s_{11}\,(j\omega)$ and $s_{21}\,(j\omega)$; and [c] The Equivalent Circuit Used to Determine $s_{22}\,(j\omega)$ and $s_{12}\,(j\omega)$.

The input impedance necessary for determining $s_{11}(j\omega)$ is given by

$$Z_{IN}(s) = \frac{1}{G_{02} + sC} + sL$$

and $s_{11}(j\omega)$ is therefore given by

$$s_{11}(j\omega) = \frac{Z_{IN}(j\omega) - Z_{01}^{*}}{Z_{IN}(j\omega) + Z_{01}}\bigg|_{a_2=0} \tag{1.68}$$

$$= \frac{\dfrac{1}{G_{02} + j\omega C} + j\omega L - R_{01}}{\dfrac{1}{G_{02} + j\omega C} + j\omega L + R_{01}}$$

$$= \frac{R_{02} - R_{01} - \omega^2 L C R_{02} + j\omega[L - R_{01} R_{02} C]}{R_{02} + R_{01} - \omega^2 L C R_{02} + j\omega[L + R_{01} R_{02} C]} \tag{1.136}$$

Similarly, the output reflection parameter is given by

$$s_{22}(j\omega) = \frac{\dfrac{R_{01} + j\omega L}{1 + (R_{01} + j\omega L) j\omega C} - R_{02}}{\dfrac{R_{01} + j\omega L}{1 + (R_{01} + j\omega L) j\omega C} + R_{02}}$$

$$= \frac{[R_{02} - R_{01} - \omega^2 L C R_{02}] + j\omega[L - R_{01} R_{02} C]}{R_{02} + R_{01} - \omega^2 L C R_{02} + j\omega[L + R_{01} R_{02} C]} \tag{1.137}$$

Comparison of (1.136) and (1.137) yields that

$$|s_{11}(j\omega)| = |s_{22}(j\omega)|$$

as expected.

The forward transmission parameter is given by

$$s_{21}(j\omega) = \sqrt{R_{01} R_{02}} \; \frac{-I_{02}}{E_{01}}\bigg|_{a_2=0} \tag{1.69}$$

$$= \frac{\sqrt{R_{01} R_{02}}}{R_{01} + R_{02}} \; \frac{R_{02}}{R_{02} - \omega^2 L C R_{02} + j\omega[L + C R_{01} R_{02}]} \tag{1.138}$$

Similary, the reverse transmission parameter is given by

$$s_{12}(j\omega) = \sqrt{R_{01} R_{02}} \; \frac{-I_{01}}{E_{02}}\bigg|_{a_1=0}$$

$$= \frac{\sqrt{R_{01} R_{02}}}{R_{01} + R_{02}} \; \frac{R_{02}}{R_{02} - \omega^2 L C R_{02} + j\omega[L + C R_{01} R_{02}]} \tag{1.139}$$

Equations (1.138) and (1.139) are clearly identified and, therefore,

$$s_{12}(j\omega) = s_{21}(j\omega)$$

as expected. The same result can be obtained directly by application of the reciprocity theorem.

Although the algebra involved is tedious, it is a simple matter to show that

$$|s_{21}|^2 = 1 - |s_{11}|^2$$

and that

$$|s_{12}|^2 = 1 - |s_{22}|^2$$

Questions and Problems

1. Derive (1.6) through (1.12) by using the equivalent circuit in Fig. 1.1b.

2. Determine the common-base parameters of a bipolar transistor with the following common-emitter parameters:

   $y_{11e} = (4.0+j2.0)\,\text{mS}$
   $y_{12e} = (0.0-j0.05)\,\text{mS}$
   $y_{21e} = (40.0+j10)\,\text{mS}$
   $y_{22e} = (1.0+j1.0)\,\text{mS}$

3. Derive (1.30) through (1.37).

4. What procedure would you follow to find the Y-parameters of the circuit shown in (a) Fig. 1.18a and (b) Fig. 1.18b?

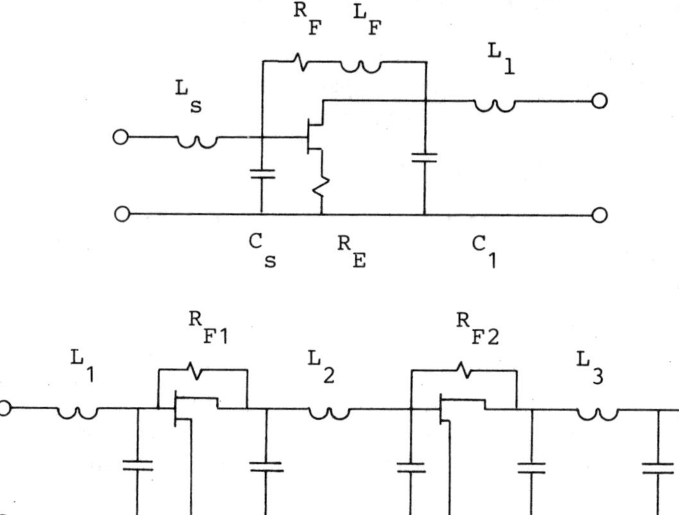

Figure 1.18   *The Two Circuits Relevant to Problem 1.4*

5. Use the diagram in Fig. 1.9 to prove (1.55).

6. What are the physical meanings of $|a_i|^2$, $|b_i|^2$, $|s_{11}|^2$, and $|s_{21}|^2$?

7. Prove that $|a_1|^2$ is the available power in an two-port augmented by its normalizing impedances.

8. Prove that $|b_1|^2$ is the power reflected from the input port of a two-port network augmented by its normalizing impedances.

9. Under what condition will the available power in a two-port network augmented by the actual load and source impedances be equal to that of the two-port augmented by the normalizing impedances?

10. The terminal voltages and currents of a two-port are

    $V_1 = 0.2V$
    $V_2 = 1.0V$
    $I_1 = 0.05A$
    $I_2 = -0.2A$

    If the normalizing impedances are

    $Z_{01} = 20\Omega$ and $Z_{02} = 10\Omega$

    determine the incident and reflected current and voltage, and the normalized components of the two-port.

11. Use the normalized components to calculate the power entering the network as well as that dissipated in the load. Use the normalized components to calculate the terminal current and voltage

12. Derive (1.84) through (1.94).

13. The S-parameters of a transistor at 2 GHz are ($20 \log |s_{ij}|$; normalizing impedances of 50Ω):

    $s_{11} = -0.355$ dB, $-23°$
    $s_{12} = -30.5$ dB, $83°$
    $s_{21} = 7.20$ dB, $159°$
    $s_{22} = -2.715$ dB, $-6°$

    Calculate the input and output impedances and the transducer power gain, if a 100Ω load is connected to it and the internal impedance of the source is equal to 50Ω.

14. Convert the S-parameters of the transistor in *Problem 1.13* to Y-parameters and find the S-parameters for the transistor if the normalizing impedances are changed to $Z_{01} = 50\Omega$ and $Z_{02} = 100\Omega$. Compare the results with those obtained in the previous problem.

15. The program SYZ BASIC in Appendix F can be used to find the inverse of a two-dimensional matrix with complex elements ($\overline{Z} = \overline{Y}^{-1}$; $\overline{Y} = \overline{Z}^{-1}$), to convert Y-parameters to T-parameters, T-parameters to Y-parameters, S-parameters to Y-parameters, and Y-parameters to S-parameters. Use it to find the Y- and S-parameters of the circuit in Fig. 1.19. *Note:* The S-parameters of the transistor are the same as those given in *Problem 1.13.*

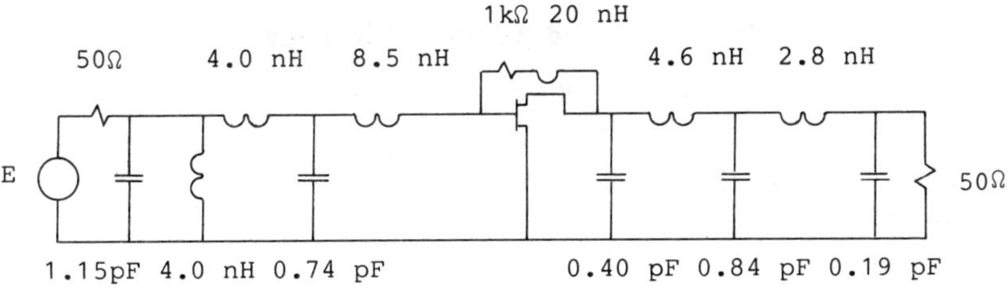

Figure 1.19   The Circuit Relevant to Problem 1.15

16. Use (1.97) to prove that the transducer power gain of a one-port network is given by

$$G_T = \frac{[1 - |S_L|^2][1 - |S_s|^2]}{|1 - S_s\, T_L|^2}$$

where S_s is the reflection parameter of the source and S_L that of the load.

17. Prove (1.99) through (1.105).

18. If the S-parameters given above are the parameters of a common-source GaAs transistor, determine the S-parameters corresponding to the common-gate configuration.

19. Is the following matrix unitary?

$$\frac{1}{2-\omega^2 + j2\omega} \begin{bmatrix} -\omega^2 & 0.5 \\ 0.5 & \omega^2 \end{bmatrix}$$

20. Use the reciprocity theorem to prove that the following equalities apply to any passive reciprocal network:

$$s_{21}(s) = s_{12}(s)$$

$$y_{21}(s) = y_{12}(s)$$

$$z_{21}(s) = z_{12}(s)$$

21. Prove that $|s_{21}|^2 = 1 - |s_{11}|^2$ for any lossless passive network.

22. Instead of using the T-parameters to find an equivalent set of parameters for two cascaded networks, the transmission matrix defined by the equation:

$$\begin{bmatrix} a_1 \\ b_1 \end{bmatrix} = \overline{P} \begin{bmatrix} a_2 \\ b_2 \end{bmatrix}$$

and its inverse can be used. Derive the equations relating the elements of $\overline{P}$ and the corresponding S-matrix.

References and Additional Reading

1. Chen, W.K., *Theory and Design of Broadband Matching Networks,* Oxford: Pergamon Press, 1976.

2. Carson, R.S., *High Frequency Amplifiers,* New York: John Wiley and Sons, 1975.

CHAPTER 2
RADIO-FREQUENCY COMPONENTS

2.1 INTRODUCTION

In order to design realizable impedance-matching networks, some knowledge of the limitations of practical components is essential. The characteristics of practical capacitors, inductors, magnetic materials, and transmission lines will be discussed in this chapter.

The capacitors used in an impedance-matching network can usually be obtained from one of the many manufacturers of these components. Unfortunately, this does not always apply to inductors. The design of inductors (air-cored and those with magnetic cores) will, therefore, also be discussed in this chapter.

2.2 CAPACITORS

Capacitors differ in capacitance, resonant frequency, losses, temperature stability, tolerances, packaging, and size. Most of these characteristics are determined by the dielectric material used. The parasitic inductance is, however, also a function of the packaging and the lead lengths of the capacitor.

The equivalent circuit for a practical capacitor is shown in Fig. 2.1.

The parasitic inductance causes the impedance of the capacitor to be lower than expected. The impedance at the series resonant frequency is equal to the series resistance of the capacitor. Above this frequency the impedance becomes inductive.

The effective capacitance below the resonant frequency is given by

$$C_{eff} = C_0 / [1 - (f/f_r)^2] \tag{2.1}$$

where

C_0 = the capacitance at low frequencies,

and

$$f_r = 1 / [2\pi(LC_0)^{1/2}],$$

where

f_r = the resonant frequency of the capacitor.

The resonant frequencies for some capacitors (with very short lead lengths) are given in Table 2.1 [2, 3].

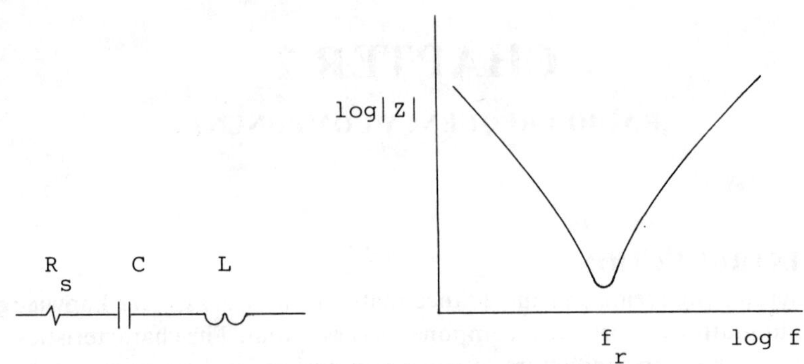

Figure 2.1   [a] Equivalent Circuit for a Capacitor; [b] Effect of the Parasitic Inductance and Resistance on the Impedance of a Capacitor

Table 2.1

Resonant Frequencies for Some Capacitors

Capacitance	100 pF	1 nF	10 nF
Mica; Disk Ceramic	170 MHz	60 MHz	20 MHz
ATC 100® Chip capacitors	1 GHz	230 MHz	—

As can be seen from Table 2.1, even chip capacitors have parasitic inductance. There are two reasons for this: the first is the finite dimensions (and therefore inductance) of the capacitor plates, and the second the finite distance across the plates.

That there must be some inductance associated with the finite separation of the capacitor plates is obvious if Maxwell's law

$$\overline{V} \times \overline{H} = \overline{i} + \partial \overline{D} / \partial t$$

is inspected. According to this equation, even a displacement current has magnetic flux and, therefore, inductance associated with it.

The losses in a capacitor are usually specified by the quality factor (Q), where

$$Q = X_s / R_s \tag{2.2}$$

R_s is the series resistance of the capacitor and X_s the effective reactance of the capacitor.

The quality factor, (or Q-factor) is frequency and temperature dependent. It is, therefore, important to specify the measuring frequency and the power level at which the measurement was made.

While the losses of the components are specified in terms of the Q-factor, the losses of dielectric materials are specified in terms of the dissipation factor (DF) or the loss tangent (tan δ).

The dissipation factor specifies the ratio of the power dissipated to the power stored in the material:

$$DF = P_{diss} / P_{stored} \qquad (2.3)$$

The relative power dissipation of dielectric materials is directly proportional to the dissipation factor. High losses are associated with high dielectric constants.

The dissipation factors for three commonly used materials are given in Table 2.2. [3]. Note the decrease in losses as the relative dielectric constant drops, as well as the increase in dissipation at higher frequencies.

Table 2.2

Dielectric Constants (ϵ_r) and Dissipation Factors for some Commonly Used Materials

Material	ϵ_r	DF (Low frequencies)	DF (100 MHz)
BaT$_1$O$_3$	1200	0.01	0.03
NPO	30	0.0001	0.002
Porcelain	15	—	0.00007

It can be shown easily that if the parasitic inductance of a capacitor can be ignored, the dissipation factor and the Q-factor are related in the following way:

$$DF = 1 / Q \qquad (2.4)$$

The losses of the dielectric materials and capacitors are sometimes specified in terms of the loss tangent (tan δ). The definition of the loss tangent is the same as that of the dissipation factor.

Dissipation factors are not only frequency dependent, but increase with temperature and, therefore, with power level. The power dissipation inside a typical chip capacitor need only be on the order of 40mW to increase the temperature to that of commonly used soldering irons [3]. At high temperatures the dissipation factor can be an order of magnitude higher than at room temperature. As the temperature inside a capacitor increases, the dissipation factor increases, which causes a further increase in temperature with more losses. This thermal runaway phenomenon is particularly important at low impedance and high power level points in a circuit.

The series resistance and Q-factors of two high quality capacitors at room temperature are given at two different frequencies in Table 2.3 [3]. Even for good capacitors, the Q-factor is surprisingly low at high frequencies.

Table 2.3

The Quality Factor and Resistance of Two Capacitors at High Frequencies

Frequency	100 MHz	500 MHz
10pF	2200 (0.055Ω)	180 (0.169Ω)
100pF	700 (0.018Ω)	60 (0.055Ω)

Not only the dissipation factor, but also the capacitance of a capacitor are effected by a change in temperature. The change in capacitance can be very small (NPO) and linear (class 1 ceramics), or large and nonlinear (class 2 ceramics). The temperature dependence of some materials are shown in Table 2.1. Class 1 ceramics with positive (up to 150 ppm/$^\circ$C) and negative (up to -5500 ppm/$^\circ$C) temperature coefficients are available [1].

As a final remark on capacitors, it should be noted that the capacitance of capacitors with high dieletric constants is usually also voltage sensitive. The capacitance of class 2 ceramics can change by more than 20% if the voltage is varied from 0 to 150% of the rated value [1].

Summary

The following points are important when choosing a capacitor for a particular purpose:

1. The parasitic inductance;
2. The tolerance of the capacitor;
3. The Q-factor at the desired frequency and power level;
4. The influence of voltage on the capacitor (capacitance changes, as well as breakdown voltages);
5. The influence of temperature on the capacitor (ambient, as well as increases due to the power dissipation in the capacitor);
6. The size and packaging of the capacitor.

2.3 Inductors

The performance of practical inductors are degraded by parasitic capacitance and resistive losses.

The parasitic capacitance (Fig. 2.2a) causes the resistance of the inductor to be higher than expected. This effect is very pronounced near the resonant frequency (f_r).

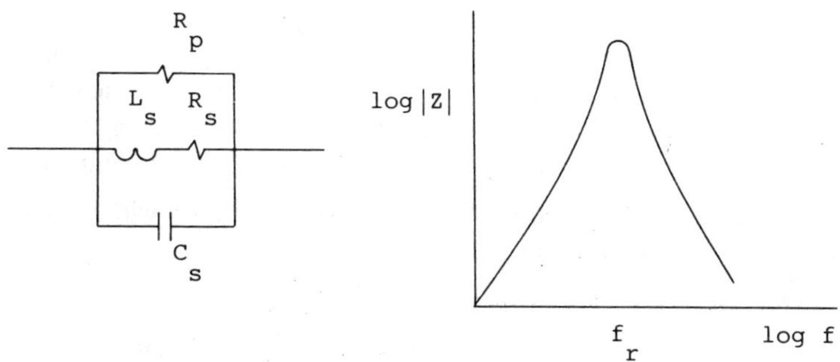

Figure 2.2 [a] *The Equivalent Circuit of a Practical Inductor; and* [b] *The Effect of Parasitic Capacitance and Losses on its Impedance.*

Inductor losses are copper losses (R_s) and, if magnetic material is used, hysteresis and eddy current losses (R_p). Both of these types of losses are frequency dependent. The copper losses increase above its dc value because of the skin and proximity effects.

By using magnetic material, the size of the inductor can be reduced drastically and the parasitic capacitance will, therefore, also be considerably less. Unfortunately, there will also be some losses in the material. These losses are mainly hysteresis losses in the case of ferrite materials.

The effect of parasitic capacitance on the Q-factor and inductance of inductors, the skin and proximity effects, the design of air-cored solenoidal coils, the properties of magnetic materials, and the design of inductors with ferrite cores will be discussed in the following sections.

2.3.1   The Influence of Parasitic Capacitance on an Inductor

By using the equivalent circuit shown in Fig. 2.2, it can be shown easily that the effective inductance (L_{eff}) of an inductor is given by

$$L_{eff} = L_s / [1-(f/f_r)^2] \qquad (2.5)$$

This equation applies only if the approximation

$$1+1/Q_s^2 \simeq 1$$

where

$$Q_s = wL_s / R_s \qquad (2.6),$$

can be made.

As can be seen from (2.5), the inductance increases rapidly as the resonant frequency (f_r) is approached.

Under the same conditions, the effective resistance (R_p ignored) is given by

$$R_{eff} = R_s / [1-(f/f_r)^2] \tag{2.7}$$

Because the effective resistance has increased due to the parasitic capacitance present, the losses in the coil are higher if the input current to the inductor is considered to be the same. This happens because the current in the parasitic capacitor is out of phase with that in the inductive part of the inductor.

The effective Q-factor of the coil will therefore be lower than without parasitic capacitance. The effective Q-factor is given by:

$$Q_{eff} = Q_s [1-(f/f_r)^2] \tag{2.8}$$

When $f = 0.707 f_r$, the effective Q-factor will be half that of the inductive part of the inductor.

These effects can be minimized by keeping the parasitic capacitance as low as possible.

The capacitance of an air-cored solenoidal coil is given in Fig. 2.3 as a function of the length-to-diameter ratio and the mean radius of the coil [8].

The capacitance of the coil is not a function of the number of turns as might be suspected. It is a strong function of the coil size (radius) and a weak function of the coil shape (length-to-diameter ratio, l/D). The capacitance can therefore be minimized by making the coil as small as possible. An initial value of two can be used for the length-to-diameter ratio.

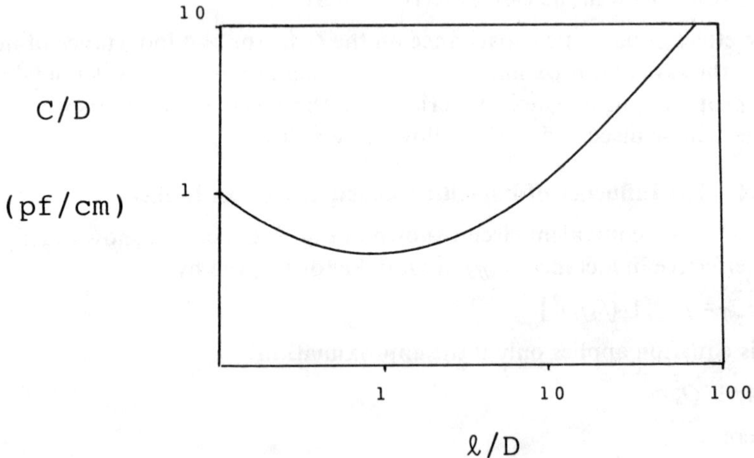

Figure 2.3   The Self-Capacitance of a Single-Layer Solenoidal Coil

Source: Medhurst, R.G., "High-Frequency Resistance and Capacity of Single-Layer Solenoids," *Wireless Engineer,* March 1947.

For high inductance, the turns of a coil should be spaced as closely as possibe. It will be shown later that this distance is determined by the desired Q-factor of the coil.

When the coil capacitance is known, the resonant frequency can be found by using the equation

$$f = \frac{1}{2\pi (LC)^{1/2}} \tag{2.9}$$

Typical resonant frequencies for some inductance values are given here as a guide to what can be achieved easily [2]:

    100 nH: 400-800 MHz
    1 μH: 100-200 MHz
    10 μH: 25-60 MHz

2.3.2   Low Frequency Losses in Inductors

The resistive losses in a conductor are approximately constant at low frequencies. The resistance is a function of the material used and the wire diameter.

The diameters and the resulting resistances of copper wire with wire gauges from 12 to 32 are given in Table 2.4. Note that the wire diameter doubles whenever the wire gauge increases by a factor of 6.

Table 2.4

**The Wire Diameter and Resistance for Wire Gauges 12-32
(20°C; copper material)**

Gauge	Bare Diameter	Double enamel-coated diameter	Resistance
	AWG (SWG) mm	AWG (SWG) mm	AWG (SWG) Ω/km
12	2.052 (2.64)	2.13 (2.73)	5.488 (3.091)
14	1.628 (2.03)	1.71 (2.12)	8.576 (5.223)
16	1.291 (1.63)	1.37 (1.71)	15.24 (8.162)
18	1.024 (1.22)	1.10 (1.29)	21.96 (14.51)
20	0.812 (0.914)	0.879 (0.984)	34.30 (25.80)
22	0.644 (0.711)	0.701 (0.774)	60.97 (42.64)
24	0.511 (0.559)	0.564 (0.617)	87.82 (69.07)
26	0.405 (0.457)	0.452 (0.512)	133.9 (103.2)
28	0.321 (0.376)	0.366 (0.424)	212.9 (152.6)
30	0.255 (0.315)	0.295 (0.361)	338.5 (217.4)
32	0.202 (0.274)	0.241 (0.316)	538.5 (286.6)

It can be seen from the table that the diameter of AWG No. 12 wire is approximately 2mm and that of AWG No. 22, 0.2mm. The resistance of No. 12 wire is $5.5\Omega/\text{km}$ and that of No. 32 wire, $538\Omega/\text{km}$. The increase of approximately 100 in resistance correlates well with the decrease in the diameter by a factor of 10 ($R \propto 1/A$, where A is the cross-section area of the wire).

2.3.3   The Skin Effect

A conductor can be viewed as a guide for the electrical and magnetic fields around it, as is shown in Fig. 2.4. The current flowing in the conductor is caused by the changing magnetic flux that penetrates into the conductor. This current opposes the magnetic field that causes it. The result is that the magnetic field decreases in strength (exponentially) as it penetrates the conductor.

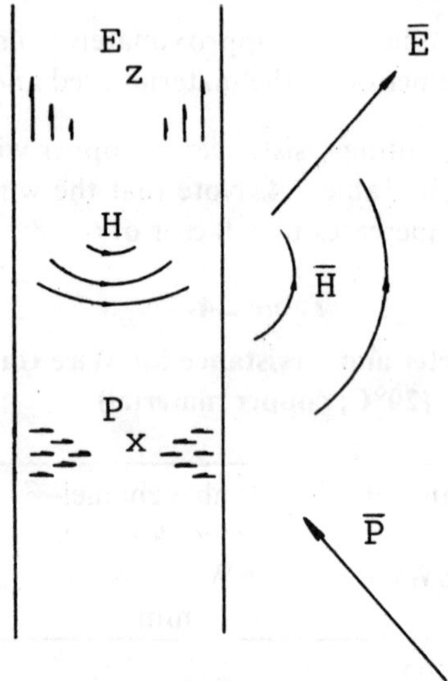

Figure 2.4:   The Electric, Magnetic, and Poynting Fields Around and Inside a Circular Conductor [after Skilling]

The induced electrical field within the conductor is given as a function of the penetration depth x by the equation:

$$E_z = E_{z0}\, e^{Tx} \tag{2.10}$$

E_{z0} is the electric field strength at the surface of the conductor (in the direction of the conductor).

The propagation constant of the electrical field in the wire, is

$$T = (jw\mu\gamma)^{1/2}$$
$$= (\pi f\mu\gamma)^{1/2} (1+j)$$
$$= \alpha+j\beta \tag{2.11}$$

where γ is the resistivity of the conductor.

The inverse of the attenuation constant α, is defined as the skin depth δ:

$$\delta = 1/\alpha = 1/(\pi f\mu\gamma)^{1/2} \tag{2.12}$$

Therefore, the amplitude of the electrical field at a distance x inside the conductor is

$$E(x) = E(0)\, e^{-x/\delta} \tag{2.13}$$

Because of the decrease in the field strength, the current density will be higher closer to the surface of the conductor. When the conductor is at least six skin-depths (or depths of penetration) in diameter, all the current can be considered to flow uniformly in a layer one skin-depth deep along the surface of the conductor.

The resistance of the conductor can then be calculated within 10% by using the following equations:

$$R_{ac} = \{\pi R^2/[\pi R^2-\pi (R-\delta)^2]\}\, R_{dc} \tag{2.14}$$
$$= \{\pi R^2/[\pi R^2-\pi (R^2-2\delta R+\delta^2)]\}\, R_{dc}$$
$$= \{\pi R^2/[2\pi\delta R-\pi\delta^2]\}\, R_{dc} \tag{2.15}$$

where $2R$ is the outside diameter of the conductor.

At high frequencies, where $\delta \ll 2R$, this equation simplifies to

$$R_{ac} = [R/(2\delta)]\, R_{dc} \tag{2.16}$$

Because the skin depth is inversely proportional to the frequency, the resistance R will increase in proportion to the root of the frequency if $\delta \ll d$ (where d is the diameter of the conductor).

The skin depths for some materials are given in Table 2.5 as a function of the frequency.

As an illustration of the change in skin depth with frequency, consider the skin depth for copper at various frequencies:

0.66mm at 10 kHz
66μm at 1 MHz
6.6 μm at 100 MHz

Because the skin depth is very small at high frequencies, it is important to ensure that conductor surfaces are smooth if the lowest possible resistance with a specific material is required.

Table 2.5

The Skin Depth of Some Materials as a Function of Frequency

Material	Skin Depth (cm)
Brass	$12.7/(f)^{1/2}$
Aluminum	$8.3/(f)^{1/2}$
Gold	$7.7/(f)^{1/2}$
Copper	$6.6/(f)^{1/2}$
Silver	$6.2/(f)^{1/2}$
Mu-metal	$0.4/(f)^{1/2}$

Where materials with low conductivities are used (usually to ensure temperature stability), it becomes worthwhile to plate conductors with silver above 100 MHz.

To get an idea of the increase in resistance with frequency caused by the skin effect, consider the resistance of one meter of AWG No. 22 wire as a function of frequency:

0.06Ω at dc
0.60Ω at 1 MHz
5.95Ω at 100 MHz

Note that the resistance at 100 MHz is approximately $(100)^{1/2}$ times that at 1 MHz.

It is obvious from these figures that the increase in resistance caused by the skin effect, cannot be ignored at high frequencies.

2.3.4   The Proximity Effect

A conductor carrying alternating current has a changing magnetic field around it. If another conductor is brought close to it, the changing magnetic field through or around it (when $d>5\delta$, the penetration depth of the field is small compared to the diameter) will cause eddy current losses in it. These losses are reflected in the first conductor in the form of increased resistance.

Similar to the skin effect, the increase in resistance is proportional to the root of the frequency at high frequencies ($d>5\delta$).

When only two conductors are in close proximity the influence of the proximity effect is relatively small compared to that of the skin effect, but when more conductors are used it should be taken into account.

Because a solenoidal coil consists of many conductors close to one another,

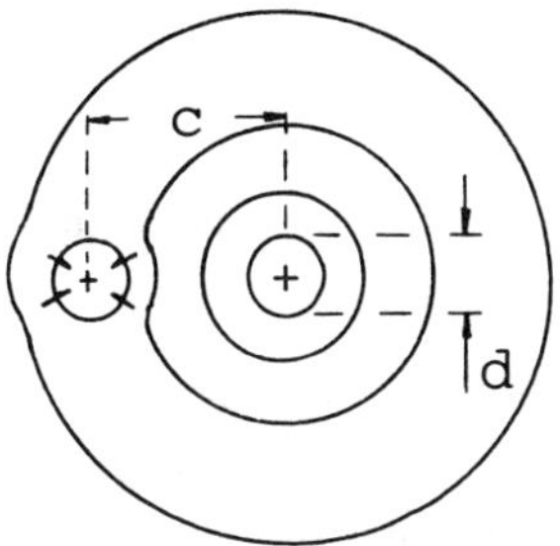

Figure 2.5   The Proximity Effect

the proximity effect can significantly affect its resistance at high frequencies. As an example of this, the resistance of a single-layer solenoidal coil with turns

touching and length-to-diameter ratio of 0.7 is almost six times that of the same wire when straightened out (that is, if more than ten turns are used).

When the turns of a coil are spaced will apart, the proximity effect can be ignored.

2.3.5   Magnetic Materials

The inductance of an air-cored coil can be increased significantly by using a magnetic material as the core. The reason for this is that the magnetic flux density increases substantially when the relative permeability of the material is high.

Typical values for the relative permeability (μ_r) of ferrite materials at radio frequencies are 10-150. The higher value applies to materials with cut-off frequencies on the order of 20 MHz and the lower value applies when the cut-off frequency is approximately 1 GHz. Above the cut-off frequency, the relative permeability decreases sharply.

Apart from the relative permeability and its frequency dependence, losses in magnetic materials must also be taken into account, especially at high voltage points.

When ferrite materials are used, these losses are mainly hysteresis losses. When materials with higher conductivities are used, the eddy-current losses in the material also become significant.

Losses in a ferrite core are proportional to the energy stored in it. The energy stored is proportional to the energy density and the volume of the core. The volume is approximately equal to the product of the cross-sectional area and the mean path length. Therefore, losses in a ferrite core are given by an equation of the form

$$P_{loss} = k\,(\mu_r, f,)\,B_{max}^2\,Al \tag{2.17}$$

where

A = The average cross-sectional area of the core;

l = The mean path length of the core;

B_{max} = The maximum rms flux density in the core;

k = A constant dependent on the frequency, relative permeability, flux density, and material used.

The power losses in a ferrite core are best specified in terms of the ratio $\mu_r\,R_p/L$ and not by (2.17). R_p is the loss resistance in parallel with the inductance (L) of the magnetic-cored inductor.

This ratio is independent of the core dimensions and is only a function of the material used and the its maximum flux density. That the ratio $\mu_r R_p/L$ should be independent of the core size can be established as follows.

Because R_p represents the losses in the core, the power loss in the core is given by

$$P_{loss} = V_p^2 / R_p \tag{2.18}$$

where V_p is the rms voltage across the inductor.

This voltage is related to the maximum flux density B_{max} as follows:

$$V_p = j\omega\,(N\Phi) = j\omega NA B_{max} \tag{2.19}$$

where N is the number of turns.

By using these two equations the resistance R_p is found to be

$$\begin{aligned}
R_p &= V_p^2/P_{loss} \\
&= \frac{\omega^2 N^2 A^2 B_{max}^2}{P_{loss}} \\
&= \frac{\omega^2 N^2 A^2 B_{max}^2}{k\,Al\,B_{max}^2} \\
&= [\omega^2/k]\,N^2 A/l
\end{aligned} \tag{2.20}$$

The resistance R_p is therefore proportional to the square of the number of turns and the cross-sectional area of the coil. It is inversely proportional to the mean path length.

This is also true for the inductance which is given by the equation:

$$L = \mu_0 \mu_r N^2 A/l \tag{2.21}$$

The ratio $\mu_r\,R_p/L$ is, therefore, independent of the core dimensions.

By using equations (2.20) and (2.21), it follows that

$$\mu_r R_p / L = \omega^2 / (k \mu_0) \tag{2.22}$$

Because k is a function of the flux density and the frequency, the ratio $\mu_r R_p / L$ is also a function of the flux density and the frequency.

Curves for this ratio as a function of frequency are shown in Fig. 2.6 [9]. These curves apply to small-signal conditions; that is, for small values of B_{max}.

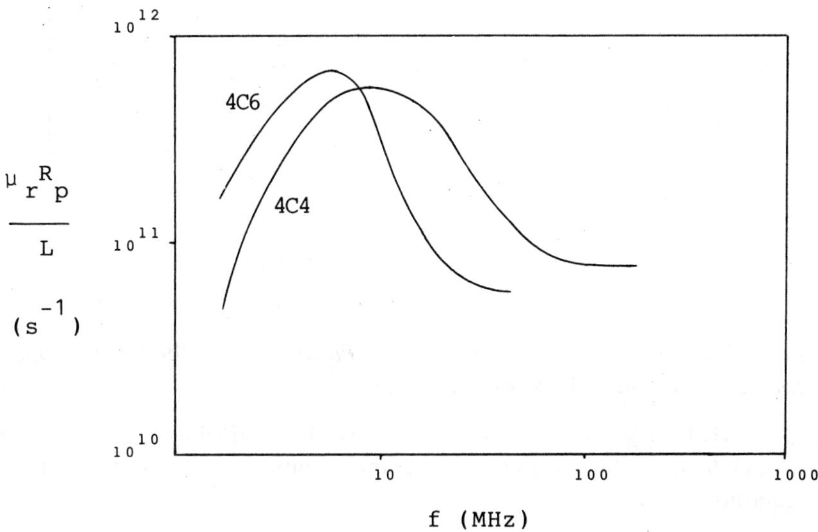

Figure 2.6   Curves of the Ratio $\mu_r R_p / L$ ($\omega\mu_r/\tan\delta$) Plotted against Frequency for Two Ferrite Materials ($B_{max} \rightarrow 0$)

Source: Hilbers, A.H., "On the Design of H.F. Wideband Power Transformers (ECO 6907)," Philips Electronic Components and Materials Division, Eindhoven, 1969.

By using these curves and a value of 120 for the relative permeability, it can be shown easily that the highest unloaded Q ($Q_u = R_p / (\omega L)$) that can be expected at 6 MHz by using 4C6 material is approximately 125.

When the flux density increases, the losses in the core increase as well. Curves for the ratio $\mu_r R_p / L$ as a function of the product $B_{max} f$ are shown for 4C4 material at different frequencies in Fig. 2.7.

The product $B_{max} f$ is used because it is independent of the frequency if the maximum voltage across the inductor (V_p) is assumed to be constant.

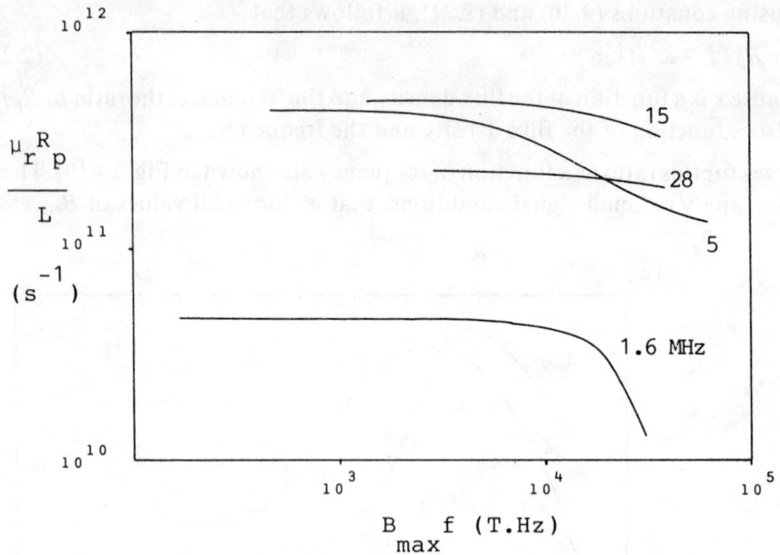

Figure 2.7:   Curves of $\mu_r R_p / L$ ($\omega\mu_r / \tan \delta$) Plotted against the Product $B_{max} f$ for 4C4 Material at Various Frequencies

Source: Hilbers, A.H., "On the Design of H.F. Wideband Transformers (ECO 6907)," Philips Electronic Components and Materials Division, Eindhoven, 1969.

By using the curve for 1.6 MHz, it follows that the losses double from their small-signal value when the flux density is approximately 14mT (140 Gauss).

As a final remark on magnetic materials, it should be noted that the relative permeability of magnetic materials is temperature dependent. Materials with higher permeabilities are influenced more by temperature changes.

Because the temperature of the material changes when heat is dissipated in it, the relative permeability will also change when more power is dissipated in it.

Summary

The following points should be taken into account when a magnetic material is selected for a particular purpose:

1.   The highest frequency of operation.
2.   The maximum allowable amount of losses.
3.   The size of the inductor and, therefore, the relative permeability.
4.   The temperature dependence of the magnetic material.

2.3.6 The Design of Single-Layer Solenoidal Coils

Single-layer solenoidal coils are often used at radio frequencies. Their use is limited by the inductance values and unloaded Q-factors obtainable, as well as the associated parasitic capacitance.

The inductance of a single-layer solenoidal coil is given approximately by the equation:

$$L = N^2 r / [22.9\, l/r + 25.4]\, (\mu\text{H}) \tag{2.23}$$

where

$r =$ The mean radius of the coil (in cm);
$l =$ The length of the coil (in cm);
$N =$ The number of turns.

The parasitic capacitance of these coils is given in Fig. 2.3 as a function of the length-to-diameter ratio (l/D) and the radius of the coil. The capacitance is small when the coil radius is small.

The unloaded Q of air-cored coils is a function of the frequency, inductance, dc resistance, skin effect, proximity effect, and the self-capacitance of the coil.

At frequencies where the self-capacitance can be neglected, the unloaded Q is given by [8]

$$Q_u = kr\,(f)^{1/2} \tag{2.24}$$

The factor k depends on the length-to-diameter ratio of the coil and the relative spacing of the turns. Its value is plotted in Fig. 2.8 for various coil shapes and wire spacing ratios (d/c). Thus, c is the distance between the centers of two adjacent turns and d is the diameter of the wire used.

The following facts can be deduced from the curves in Fig. 2.8 and Eq. (2.24):

1. Higher unloaded Q-factors can be obtained by using coils with larger diameters and length-to-diameter ratios (l/D).

2. The turns of an air-cored solenoidal coil should be spaced close enough to ensure that the d/c ratio is larger than 0.4 d, and in shorter coils ($l/D \simeq 1$) they should be spaced far enough apart to ensure that the d/c ratio is smaller than 0.8 d.

When larger coils are used the turns can tough without any significant reduction in the unloaded Q (less than 25%).

By using the curves in Fig. 2.8 and the equations given, solenoidal coils can be designed to have a specified inductance and unloaded Q. The parasitic capacitance can be determined by using the curve in Fig. 2.3 and the design can be done as summarized below.

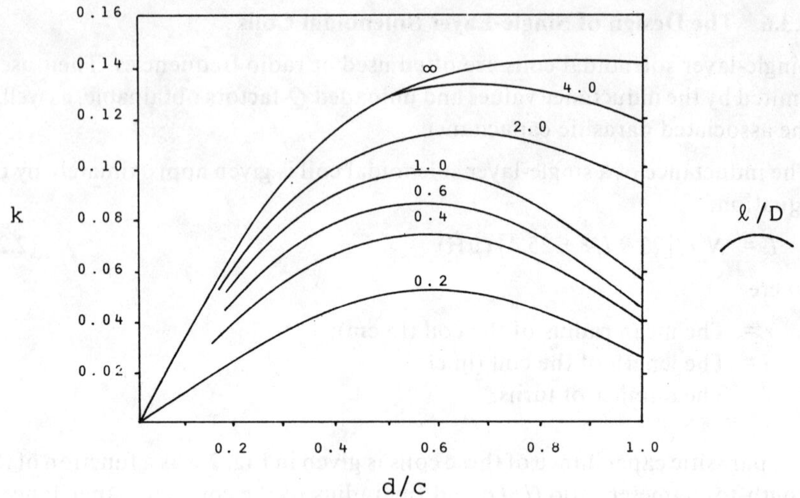

Figure 2.8   Curves for Calculating the Unloaded Q of Single-Layer Solenoidal Coils at High Frequencies

Source: Medhurst, R.G., "High-Frequency Resistance and Capacity of Single-Layer Solenoids," *Wireless Engineer,* March 1947.

Design Procedure for a Specified Inductance Value and Quality Factor

1) Choose the length-to-diameter ratio (l/D) equal to one.

2) Calculate the radius (r) of the coil (in cm) by using the equation:

$$r = Q_u/(k(f))^{1/2} \tag{2.25}$$

where Q_u is the unloaded Q required, and

$k = 0.1$ for $l/D = 1.0$ (see Fig. 2.8)

3) Find the parasitic capacitance of the coil by using Fig. 2.3. Calculate the resonant frequency by using the equation:

$$f_{r1} = [1/(LC)]^{1/2}/(2\pi) \tag{2.26}$$

where

$C/D = 0.45\text{pF}/\text{cm}$ for $l/D = 1$

4) If the resonant frequency is too low, the specifications cannot be reached and it will have to be changed.

5) Calculate the required number of turns by using the equation

$$N = [L(22.9l/r + 25.4)/r]^{1/2} \tag{2.27}$$

6)   Calculate the required wire thickness by using the d/c ratio used in step 2:

$$d = d/c \cdot l/(N\text{-}1) = l/D \cdot d/c \cdot 2r \: / \: (N\text{-}1) \tag{2.28}$$

where

d = The wire diameter to be used,

and

$d/c = 0.55$ for $l/D = 1$ (see Fig. 2.8).

7)   If the required wire thickness is small, a coil former will be needed. If the coil is to be self-supporting, it can be redesigned.

In order to increase the wire diameter, it will be necessary to increase the size of the coil. When the resonant frequency is a potential problem, the l/D ratio can be increased. The resonant frequency will decrease if the radius is increased.

Where the resonant frequency is not a problem, the radius of the coil can be increased in order to increase the wire diameter. The maximum value of the radius is

$$r_{max} = C_m/(2C) \tag{2.29}$$

where

C_m = The maximum self-capacitance allowable,

and

C = The capacitance per centimeter as given by Fig. 2.3.

With $l/D = 1$, $C = 0.45$ pF/cm.

Example 2.1

As an example of the application of the procedure outlined, a $1\,\mu$H coil was designed to have a minimum unloaded Q of 300 at 50 MHz and resonant frequency above 250 MHz. The results of the different steps were:

1. $l/D = 1$
2. $r = 0.42$cm
3. $f_r = 256$ MHz
4. —
5. $N = 13$
6. $d = 0.36$mm
7. Because the wire diameter is small, it will be necessary to use a coil former.

It is not possible to increase the wire diameter by increasing the coil radius, in this case ($f_r = 250$ MHz). It is however, possible to increase it by increasing the l/D ratio of the coil.

Unfortunately, it is not possible to increase the wire thickness sufficiently to make the coil self-supporting.

The results for different l/D ratios are compared in Table 2.6. Note that the wire diameter can be doubled if the length-to-diameter ratio is chosen to be equal to four.

Although the wire thickness is a strong function of the length-to-diameter ratio, the resonant frequency of coils with length-to-diameter ratios from 0.6 to 4 does not vary significantly if they are designed to have the same unloaded Q-factor.

The volumes of the coils in Table 2.6 increase with increasing l/D ratio. When a small coil is required, the length-to-diameter ratio can, therefore, be chosen to be equal to 0.6.

Table 2.6

The Dimensions, Unloaded Q, and Resonant Frequency for a 1 μH Coil as a Function of the l/D Ratio

l/D	r	N	d	d/c	f_r	Q_u
	(cm)		(mm)		MHz	
0.6	0.48	10	0.31	0.55	252	300
1.0	0.42	13	0.36	0.55	256	300
2.0	0.37	18	0.52	0.63	255	300
4.0	0.32	26	0.63	0.63	242	300

The capacitance, k-factor, and optimum d/c ratio for coils with the l/D ratios used in Table 2.6, are tabulated in Table 2.7 for convenience.

Table 2.7

The Self-Capacitance, d/c Ratio, Optimum Value of k, and the Ratio of the k and the Self-Capacitance per cm for Coils with Different l/D Ratios

l/D	C	d/c	k_{opt}	k/C
	pf/cm		Hz/cm	pF/Hz
0.6	0.44	0.55	0.088	0.200
1.0	0.45	0.55	0.100	0.222
2.0	0.53	0.63	0.115	0.216
4.0	0.68	0.63	0.133	0.196

When resonant circuits with high Q-factors are designed, the unloaded Q-factors of the coils and capacitors used must be as high as possible. In order to determine the maximum realizable unloaded Q possible for a single-layer air-cored coil, it is necessary to determine the optimum l/D ratio. Because

$$Q = k\,r\,(f)^{1/2} \tag{2.24}$$

the length-to-diameter ratio influences the unloaded Q directly through the associated value of the constant k and indirectly (through r) because of the limit that exists on the self-capacitance of the coil.

The maximum radius corresponding to a particular l/D ratio can be determined by using (2.29).

By substituting the value for r as given by (2.29) into (2.24), the maximum Q corresponding to a particular l/D ratio is found to be

$$Q_{max} = [k_m/C]\,(f)^{1/2}\,C_{max} \tag{2.30}$$

where

k_m = The maximum value of k corresponding to the particular l/D ratio;

C = The capacitance per cm as given by the curve in Fig. 2.3;

C_{max} = The maximum value of the self-capacitance as determined from the specified resonant frequency.

The influence of the l/D ratio on the unloaded Q is clearly limited to the first term in (2.30). The k/C ratios for different l/D ratios are compared in the last column of Table 2.8. It follows from this comparison that the highest Q will be obtained when the length-to-diameter ratio of the coil is equal to one.

At this stage, the highest Q realizable with a single-layer solenoidal air-cored coil can be determined for any particular inductance value if the operating frequency and the self-resonant frequency are specified. The following procedure can be followed in order to determined the Q.

Design Procedure for Maximum Q and Specified Inductance

1) Choose $l/D = 1$

2) Determine the maximum value of the self-capacitance (C_{max}).

If the coil is to be used in a parallel resonant circuit, the self-resonant frequency can be chosen close to the resonant frequency of the circuit.

Calculate the maximum allowable radius of the coil by using the equation:

$$r_{max} = C_{max}/(2C) = C_{max}/0.9 \text{ (cm)} \tag{2.31}$$

with C_{max} specified in picofarads (pF).

If the value of the radius is unrealistically high, reduce it to an acceptable value.

3)   Determine the maximum realizable unloaded Q by using (2.24):

$$Q_{max} = 0.10 \, r_{max} \, (f)^{1/2} \tag{2.32}$$

4)   Calculate the required thickness of the wire:

$$N^2 = 71.2 \, L/r_{max} \tag{2.33}$$

where the inductance (L) is specified in μH and r_{max} in cm.

$$c = l/(N\text{-}1) = 2r_{max}/(N\text{-}1) \tag{2.34}$$

$$d = 0.55 \, c \text{ (cm)} \tag{2.35}$$

If the wire thickness turns out to be unrealistic, change the radius.

Example 2.2

The highest possible Q will be determined for a coil of 10 μH at 5 MHz with self-resonant frequency at 10 MHz by following the procedure outlined above.

1     $l/D = 1$

2     $C_{max} = 1/[(2\pi 10 \cdot 10^6)^2 \, 10 \cdot 10^{-6}]$

        $= 25.3 \text{pF}$

      $r_{max} = 25.3/0.9$

        $= 28.1 \text{cm}$

3     $Q_{max} = 0.10 \, r_{max} \, (f)^{1/2}$

        $= 6286!$ $\hspace{4cm}$ (2.32)

4     $N^2 = 71.2 \cdot 10/28.1$

        $= 25.3$

      $N = 5.0$

      $c = 2 \cdot 28.1/(5\text{-}1)$ $\hspace{4cm}$ (2.34)

        $= 14.1 \text{cm}$

      $d = 0.55 \, c$ $\hspace{4cm}$ (2.35)

        $= 7.73 \text{ cm}!$

If the coil size is limited to 3cm $\times$ 3cm $\times$ 3cm, the maximum realizable Q will be 335.

2.3.7   The Design of Inductors with Magnetic Cores

Smaller inductors with less parasitic capacitance can be designed by using magnetic materials.

The core can be a rod, a toroid, a balun, or stacked toroids.

Rods are often used if the inductor is to be tuned, while toroids and baluns are used for fixed-value inductors. Stacked cores can be used as an alternative to a balun.

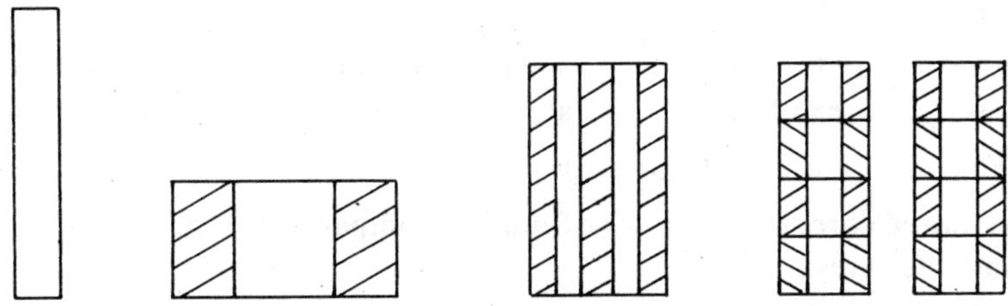

Figure 2.9   Different Types of Magnetic Cores: [a] *Rod Core;* [b] *Toroid;* [c] *Balun Core;* [d] *Stacked Toroids*

The type of material used is a function of the frequency range over which the inductor is to be used, the desired unloaded Q, the available space, and the temperature range.

The unloaded Q is determined by the flux density in the core and, therefore, by the maximum voltage across the inductor, the number of turns, and the frequency.

Materials with a high relative permeability are usually very sensitive to changes in temperature.

With the material and type of core selected, the size of the core must be determined. The core must be large enough for the flux density to be sufficiently low to ensure that the desired unloaded Q is realized and that the required number of turns can be accommodated.

The selection of the minimum core size for toroidal (single and stacked) inductors will be discussed in the next two sections. If a balun is to be used, the results for stacked cores can be applied to get an idea of the size required.

The Design of an Inductor with a Toroidal Core

The inductance of a toroidal-core inductor is given by

$$L = \partial\phi / \partial i = \mu_0 \mu_r N^2 A / l \tag{2.36}$$

The value of the product $\mu_r R_p / L$ can be determined from the unloaded Q (Q_u) by using the following equation:

$$\mu_r R_p / L = \mu_r \omega R_p / (\omega L) = 2\pi \mu_r f Q_u \tag{2.37}$$

The flux density corresponding to this ratio (B_{maxA}) can be determined, where given, from the manufacturer's specifications (see Fig. 2.7 for an example).

The flux density in the core is given by

$$B_{max} = V_{max}/(\omega A N) \tag{2.38}$$

The flux density in the core must be less than or equal to the maximum allowable value B_{maxA}.

If the number of turns in (2.36) is replaced by using (2.38), the product of the cross-section area of the core (A) and the mean path length (l) is found to be

$$Al = [\mu_r\mu_0/(\omega B_{maxA}^2)]V_{max}^2/(\omega L) \tag{2.39}$$

The required core size can now be found by comparing this product with that of available cores.

If a core with the required Al-product is not available, a core with a larger Al-product can be chosen. The number of turns required must then be calculated by using (2.36). The alternative is to use more than one core (smaller) to obtain the required Al-product.

With the core dimensions known, the number of the turns required can be found by using (2.38).

The following procedure can be followed to design an inductor with a toriodal core.

Design Procedure for an Inductor with A Toroidal Core

1)   Select a suitable material. Take the frequency range, temperature range, required unloaded Q, and inductor size into account.

2)   Calculate the $\mu_r R_p/L$ ratio at the lowest frequency by using the equation:

$$\mu_r R_p/L = 2\pi\mu_r f Q_u \tag{2.37}$$

where Q_u is the desired value of the unloaded Q.

3)   Find the flux density corresponding to the calculated $\mu_r R_p/L$ ratio from the manufacturer's specifications.

4)   Calculate the required Al-product:

$$Al = \mu_r\mu_0/(\omega B_{max}^2 A)V_{max}^2/(\omega L) \tag{2.39}$$

5)   Compare this product to that of available cores. Select a core with Al-product equal or close to it. If the difference in Al-product is significant, choose the core with an Al-product greater than that required.

Alternatively, smaller cores can be combined to obtain the required Al-product (see the next section).

6)   Calculate the required number of turns by using the equation:

$$L = \mu_r\mu_0 N^2 A/l \tag{2.36}$$

7)   Check if there is enough space to accommodate the required number of

turns of the conductor with the required thickness. If the core is too small, a larger toroid must be used.

Example 2.3

As an example of the application of the procedure outlined here, the core size for a magnetic-cored inductor with 31.4Ω reactance at 2 MHz, and loss resistance equal to 392Ω, will be determined. The maximum rms voltage across the inductor will be 20V and 4C4 materil is available. Note that $\mu_r = 120$.

The $\mu_r R_p / L$ ratio for the inductor is

$$\frac{\mu_r R_p}{L} = \frac{120 \cdot 392}{31.4/(2\pi \cdot 2 \cdot 10^6)} = 1.88 \cdot 10^{10} \, s^{-1}$$

By using the 1.6 MHz curve given for 4C4 material in Fig. 2.7, the $B_{max} f$ product corresponding to a $\mu_r R_p / L$ ratio of $1.8 \cdot 10^{10} \, s^{-1}$ is found to be $2 \cdot 10^4$ T $\cdot$ Hz. The maximum allowable flux density in the core is, therefore, 0.01 T.

The Al-product of the required core can be found by using (2.39). The required Al-product is $1.53 \; 10^{-6} \, m^3$.

By comparing this value to that in the list of some Al-products given in *Problem 2.18*, it can be seen that the core with Al-product equal to $1.8\mu m^3$ ($A=31.5\mu m^3$; $l=57mm$; $23\times14\times7 \, mm^3$) can be used.

The number of turns required is

$$N = [Ll/(\mu_0 \, \mu_r A)]^{1/2} = 5.5 \text{ turns}$$

The selected core can accommodate the required number of turns with ease.

The Design of an Inductor with a Stacked-Toroidal Core

The design of an inductor with a stacked toroidal cores is similar to that of an inductor with a single core, except for the fact that the cross-sectional area (A) used in the previous section must now be taken as $N_c A$, where N_c is the number of toroids used (an even number) and A is the cross-sectional area of a single toroid. The mean path length is that of a single toroid.

The inductance of a stacked core inductor is given by the equation:

$$L = \mu_r \, \mu_0 \, N^2 \, (N_c \, A / l) \tag{2.40}$$

The maximum flux density is

$$B_{max} = V_{max} / (\omega \, (N_c A) N) \tag{2.41}$$

and the required Al-product is obtained by

$$N_c Al = [\mu_r \mu_0 / (\omega B_{max}^2 A)] V_{max}^2 / (\omega L) \tag{2.42}$$

2.4 TRANSMISSION LINES

The transmission lines used at radio frequencies are usually coaxial cables (flexible (F) or semi-rigid (SR)), microstrip lines, or twisted pairs.

The important characteristics of these lines are the characteristic impedance, the insertion loss, and the power-handling capability.

The characteristics of coaxial cables, microstrip lines, and twisted pairs will be discussed briefly.

2.4.1 Coaxial Cables

The characteristic impedance of a coaxial cable is given by the equation:

$$Z_0 = 138/(\epsilon_r)^{1/2} \cdot \log_{10}(b/a) \tag{2.43}$$

where

    a = The outer diameter of the inner conductor (cm),

and

    b = The inner diameter of the outer conductor (cm).

The attenuation of the cable is given by [11]:

$$A = [3.615/Z_0](K_1/a+K_2/b)\ Tf^{1/2} + 9.121\ \epsilon_r^{1/2} \tan(\Delta) \cdot f \tag{2.44}$$

where

    A = The attenuation in dB/100m;

    K = The square root of the ratio of the resistivity of copper to that of the particular conductor;

    f = Operating frequency in MHz;

    $T = [1+0.0039(t-20)]^{1/2}$, where t is the operating temperature in degrees Celsius;

    a = The inner conductor outer diameter in cm;

    b = The outer conductor inner diameter in cm;

  $\tan(\delta)$ = The loss tangent of the inner conductor insulation.

The attenuation is increasing with frequency because of the skin effect and the losses in the dielectric material.

The power-handling capability of a coaxial cable is limited by the maximum allowable temperature. This is a function of the insulation used (200°C for polytetrafluoroethylene), the diameter of the cable, and the environmental temperature.

The power-handling capability and attenuation along some coaxial cables are given in Table 2.8 [11]. Note the decrease in power-handling capability with increasing frequency.

Table 2.8

**The Attenuation and Power-Handling Capabilities of Some Coaxial Cables
at Different Frequencies**

Frequency	A (dB/m) ($P(W)$)			
	1 MHz	10 MHz	100 MHz	500 MHz
50Ω; 1.7mm; F	0.04 (1k)	0.14 (300)	0.44 (90)	—
50Ω; 2.8mm; F	0.03 (1k)	0.08 (800)	0.27 (250)	—
50Ω; 1.1mm; SR	—	—	0.35 (68)	0.75 (32)
50Ω; 2.2mm; SR	—	—	0.18 (330)	0.43 (140)
50Ω; 6.4mm; SR	—	—	0.11 (1.17k)	0.25 (515)

Semi-rigid coaxial cable is often used for transmission-line transformers in the
VHF and UHF ranges.

Coaxial lines with characteristic impedances of 50Ω and 25Ω are freely available. Lower impedances can be obtained by connecting cables in parallel,
while higher impedances can be obtained by connecting lines with lower
impedances in series (using semi-rigid cable). By doing this the effective capacitance is decreased. The series connection is shown in Fig. 2.10.

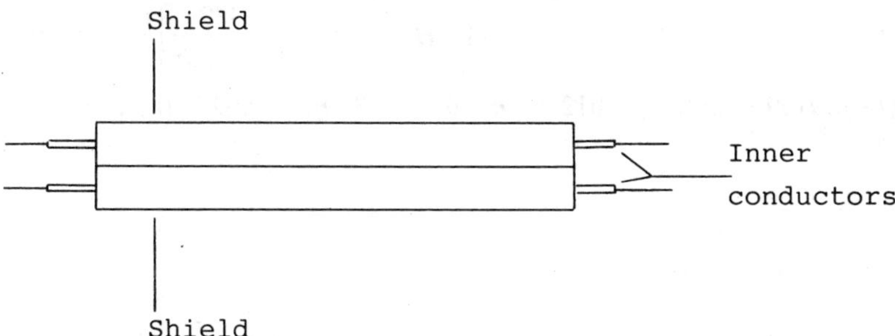

*Figure 2.10   Increasing the Characteristic Impedance of a Coaxial Cable by
Connecting Two Cables in Series*

2.4.2 Microstrip Transmission Lines

The characteristic impedance ($Z_0 (f)$) and the effective relative dielectric constant ($\epsilon_{r,\ eff} (f)$) of a microstrip line is a function of the width-to-height ratio
(W/h), the conductor thickness (t), cover height (H_2), and the frequency (f).
The characteristic impedance is also a function of the effective dielectric
constant.

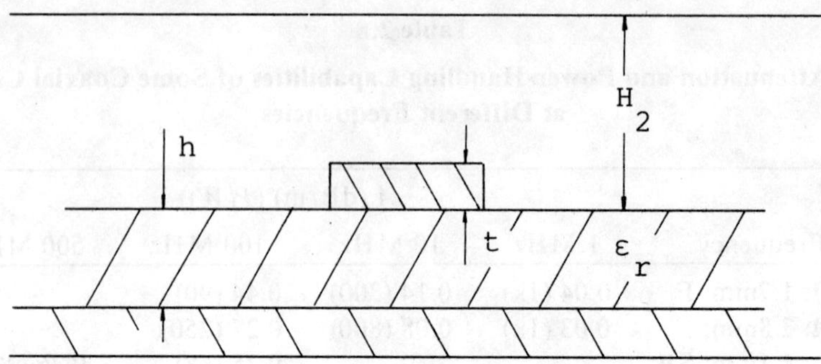

Figure 2.11   The Geometry of a Microstrip Line

The characteristic impedance (Z_0 (f)) and effective relative dielectric constant (ϵ_{eff} (f)) can be computed by using the following set of equations [13, 14]:

$$W_{eff} = W + \frac{t}{\pi} \left\{ 1 + \ln 4 - 0.5 \ln \left[\left(\frac{t}{h} \right)^2 + \left(\frac{t}{\pi W} \right)^2 \right] \right\} \tag{2.45}$$

$$f(W/h) = 6 + (2\pi - 6) \, \text{EXP} \left[- \left(\frac{30.666}{W/h} \right)^{0.7528} \right] \tag{2.46}$$

$$Z_{0a\infty} = 60 \ln \frac{f(W/h)}{W/h} \left[+ \sqrt{1 + \left(\frac{2h}{W} \right)^2} \right]^{1/2} \tag{2.47}$$

$$P = 270 \left\{ 1 - \tanh \left[1.192 + 0.706 (1 + H_2/h)^{1/2} - \frac{1.389}{1 + H_2/h} \right] \right\} \tag{2.48}$$

$$Q = 1.0109 - \tanh^{-1} \left\{ [0.012 \, W/h + 0.177 \, (W/h)^2 - 0.027 \, (W/h)^3] / [1 + H_2/h]^2 \right\} \tag{2.49}$$

$$Z_{0a} = Z_{0a\infty} - P \, Q \tag{2.50}$$

$$b = -0.564 \left(\frac{\epsilon_r - 0.9}{\epsilon_r + 3.0} \right)^{0.053} \tag{2.51}$$

$$a \doteq 1 + 1/49 \cdot \ln \{ (W/h)^2 [(W/h)^2 + 1/52^2] / [(W/h)^4 + 0.432] \} + 1/18.7 \cdot \ln \{ 1 + [W/(18.1h)]^3 \} \tag{2.52}$$

$$j = a \, b \tag{2.53}$$

$$q = \{ [1 + 10h/W]^j - 2 \ln 2 / \pi \cdot t/h / (W/h)^{1/2} \} \cdot \tanh [1.043 + 0.121 \, H_2/h - 1.164 (/ H_2/h)] \tag{2.54}$$

$$\epsilon_{eff} = \frac{\epsilon_r + 1}{2} + q \, \frac{\epsilon_r - 1}{2} \tag{2.55}$$

$$Z_0 = Z_{0a} / \sqrt{\epsilon_{eff}} \tag{2.56}$$

$$v_p = c / \sqrt{\epsilon_{eff}} \tag{2.57}$$

where v_p is the phase velocity in the microstrip,

$$f_p = Z_0/[2\mu_0 h] \tag{2.58}$$

$$G = \pi^2/12 \cdot [\epsilon_r - 1]/\epsilon_{eff} \cdot [Z_0/60]^{1/2} \tag{2.59}$$

$$\epsilon_{r-eff}(f) = \epsilon_r - \frac{\epsilon_r - \epsilon_{eff}}{1 + G\,[f/f_p]^2} \tag{2.60}$$

$$s = \frac{c^2}{4f^2\,[\epsilon_{r-eff}(f) - 1]} \tag{2.61}$$

$$y = s/3 - [W/3]^2 \tag{2.62}$$

$$W_{eff}(0) = 120\pi h/[Z_0\,\sqrt{\epsilon_{eff}}] \tag{2.63}$$

$$p = [W/3]^3 + s/2 \cdot [W_{eff}(0)\ W/3] \tag{2.64} \qquad r = (p^2 + y^3)^{1/2} \tag{2.65}$$

$$W_{eff}(f) = W/3 + [r + p]^{1/3} - [r - p]^{1/3} \tag{2.66}$$

$$Z_0(f) = \frac{120\,\pi h}{W_{eff}(f)\,[\epsilon_{r-eff}(f)]^{1/2}} \tag{2.67}$$

The frequency dependence (dispersion) of the characteristic impedance and the effective dielectric constant of a microstrip line result from the non-TEM nature (inhomogeneity) of the mode of propagation along the microstrip.

As an example of the application of (2.45) to (2.67), the width-to-height ratios and effective dielectric constants of a 50Ω line on an alumina ($\epsilon_r = 10.2$) and a Teflon ($\epsilon_r = 2.5$) substrate at 2 GHz with $H_2/h = 20.0$ were calculated. The results are

$$W/h = 0.85$$

with

$$\epsilon_{r,\,eff} = 6.6945$$

and

$$W/h = 2.75$$

with

$$\epsilon_{r,eff} = 2.0775$$

respectively,

At microwave frequencies it also becomes necessary to take into account the losses in microstrip lines. The main source of these losses is usually conductor loss. The conductor loss attenuation constant α_c is given by the following set of equations [15, 16]:

$$\alpha_c = \frac{8.68\,R_s\,M}{2\pi Z_0\,h}\left[1 + \frac{h}{W_{eff}} + \frac{h}{\pi W_{eff}}\left(\ln\frac{4\pi W}{t} + \frac{t}{W}\right)\right], \quad W/h < 1/(2\pi)$$

$$\alpha_c = \frac{8.68\, R_s\, M\, N}{2\pi Z_0\, h}\, , \quad 1/(2\pi) < W/h < 2$$

$$\alpha_c = \frac{8.68\, R_s\, N}{Z_0\, h}\, \left\{ \frac{W_{eff}}{h} + \frac{2}{\pi}\, \ln\left[2\pi\, \text{EXP}\, \left(\frac{W_{eff}}{2\,h} + 0.94 \right) \right] \right\}^{-2} \times$$

$$\left[\frac{W_{eff}}{h} + \frac{W_{eff}/(\pi h)}{W_{eff}/(2h) + 0.94} \right] , \quad W/h > 2 \tag{2.68}$$

with α_c in dB/cm.

$$M = 1 - \left[\frac{W_{eff}}{4h} \right]^2 \tag{2.69}$$

$$N = 1 + h/W_{eff} + \frac{h}{\pi\, W_{eff}} \left[\ln\, \frac{2h}{t} - t/h \right] \tag{2.70}$$

$$R_s = [\pi f \mu_0 / \sigma]^{1/2} \tag{2.71}$$

where σ is the conductivity of the strip conductor.

At high frequencies, the copper losses are higher than those predicted by the equations above. This is due to the coarse interface between the dielectric material and the conductor. These losses are included in the dielectric losses by some manufacturers.

With the loss tangent ($\tan\delta$) known, the attenuation constant corresponding to the dielectric losses in a microstrip line can be calculated by using the following equation [15, 17]:

$$\alpha_d = 27.3\, \frac{\epsilon_r [\epsilon_{r-eff} - 1] \tan\delta}{\lambda_0 \sqrt{\epsilon_{r-eff}} [\epsilon_r - 1]}\quad (\text{dB}/\text{cm}) \tag{2.72}$$

Materials with dissipation factors of 0.00085 at 1 MHz and 0.0018 at 10 GHz are available. With such low values for the dissipative factor, the dielectric loss is usually small compared to the conductor loss. Silicon is an example of a material where the dielectric loss cannot be neglected.

As an example of the dissipative loss in a microstrip line, the insertion loss of an 8-in 50Ω line on an Epsilam-10® substrate is specified by the manufac- to be approximately 0.1 dB at 100 MHz and 0.21 dB at 500 MHz (0.19 dB/wavelength).

The power-handling capability of a microstrip line is a function of the insertion loss, the breakdown voltage of the dielectric material, and the maximum allowable temperature of the line. If the thermal resistance of the substrate is known as a function of the line width, the maximum power-handling capability can be computed easily.

2.4.3 Twisted Pairs

Transmission lines with a wide range of characteristic impedances can be realized by twisting lengths of wire together.

The characteristic impedances of these twisted-pair lines decrease when thicker wire is used. For example, the characteristic impedance is 35Ω when No. 20 (AWG) enamel-insulated wire with 3 twists per centimeter is used and 120Ω when No. 30 vinyl-coated wire (0.05cm outside diameter) with 3.6 twists per centimeter is used [12].

Increasing the number of twists per centimeter also decreases the characteristic impedance of these transmission lines. For example, the characteristic impedance obtained by twisting two No. 20 enamel-insulated wires together decreases from approximately 42Ω to 30Ω when the number of twists is doubled from two to four [12].

A line with 50Ω characteristic impedance can be obtained by twisting two No. 22 enamel-insulated wires together to have 2.5 twists per centimeter.

Characteristic impedances lower than 10Ω are often required in the HF range. These impedances can be realized by twisting together many wires (using two-wire lines) with smaller diameters.

It is difficult to calculate the losses in these transmission lines because the dielectric losses, skin effect, proximity effect, and the fact that the current is flowing in both directions along the line must be taken into account. It is, therefore, easier to determine the attenuation of these lines practically.

The losses in twisted-wire transmission lines are usually not a problem below 100 MHz.

Questions and Problems

1. At high frequencies the reactance of a capacitor is lower than expected. What is the reason for this?

2. At what frequency would you expect a 10nF siver-mica capacitor (with very short lead lengths) to resonate?

3. The resonator frequency of a 1nF disk ceramic capacitor is 60 MHz. What is its effective capacitance at 40 MHz?

4. Is there any parasitic inductance associated with a chip capacitor?

5. (a) What is the definition of the Q-factor of a component?
   (b) What is the definition of the dissipation factor?
   (c) How are these two factors related?

6. Is the phenomenon of thermal runaway possible in a capacitor?

7.   What would you expect the Q-factor of a good capacitor to be at (a) 1 MHz, (b) 100 MHz, and (c) 500 MHz?

8.   Is it true that the parasitic capacitance of an inductor actually increases its losses? Find the ratio of the operating and resonant frequencies for which the Q-factor of an inductor will be 10% lower than expected.

9.   Two single-layer solenoidal coils have exactly the same dimensions, but one has more turns than the other. Which coil has more capacitance? Which coil has the lowest resonant frequency?

10.   At what frequency would you expect a 1μH inductor to resonate?

11.   What is the skin depth for copper at 100 MHz, at 10 kHz?

12.   Why is the resistance of a solenoidal coil higher at high frequencies than at low frequencies (three reasons).

13.   The losses in magnetic materials can be specified by plotting the ratio $\mu_r R_p / L$ as a function of the product $B_{max} f$ as a function of frequency. Show that the ratio $\mu_r R_p / L$ is independent of the core dimensions and that the product $B_{max} f$ is independent of the frequency, if the voltage across the inductor remains constant.

14.   Suppose you need a 1μH inductor with low losses. You are given a choice between two toroidal cored inductors, one with a small core and the other with a larger one. Exactly the same material is used for both cores. Which inductor would you choose?

15.   Is it true that larger air-cored inductors have higher Q?

16.   Design a single-layer air-cored inductor to have 10μH inductance with an unloaded Q of 200 at 10 MHz. The resonant frequency must be higher than 30 MHz.

17.   Determine the highest Q attainable by using a single-layer solenoidal coil to obtain 500nH inductance at 50 MHz, if the resonant frequency is to be above 200 MHz and the coil size is to be smaller than 3cm$\times$3cm$\times$3cm.

18.   Design an inductor to have 1μH inductance. The unloaded Q must be higher than 40 at 1.6 MHz. A toroidal core of 4C4 material must be used. The maximum voltage across the inductor will be 20V. Assume the relative permeability to be equal to 100.

The available cores have the following dimensions:

Core 1: A=12.5 μm^3; l=36 mm; Al=0.44 μm^3 (14$\times$9$\times$5 mm^3)
Core 2: A=31.5 μm^3; l=57 mm; Al=1.80 μm^3 (23$\times$14$\times$7 mm^3)
Core 3: A=37.5 μm^3; l=75 mm; Al=2.81 μm^3 (29$\times$19$\times$7.5 mm^3)
Core 4: A=65.0 μm^3; l=92 mm; Al=5.98 μm^3 (36$\times$23$\times$10 mm^3)

Core 5: A=97.5 μm^3; l=92 mm; Al=8.97 μm^3 (36$\times$23$\times$15 mm^3)

19.   If a core consisting of four stacked toroids is to be used, design an inductor to meet the same specifications as the inductor in the previous problem.

20.   How much power can a 2.8mm flexible coaxial cable handle at 100 MHz?

21.   If the input power to the cable in the previous question is 100W, what will the output power be if the cable is 10m long?

22.   Show the series connection for two transmission lines.

References and Additional Reading

1.   Hardy, K.H., *High Frequency Circuit Design,* Reston, VA: Preston Publishing Company, 1979.

2.   Krauss, H.L., W.B. Bostian, and F.H. Raab, *Solid State Radio Engineering,* New York: John Wiley and Sons, 1980.

3.   American Technical Ceramics, *The RF Capacitor Handbook,* 1979.

4.   Skilling, H.H., *Fundamentals of Electric Waves,* New York: John Wiley and Sons, 1948.

5.   Snelling, E.C., *Soft Ferrites: Properties and Applications,* London: Iliffe Books Ltd., 1969.

6.   Howe, H. *Stripline Circuit Design,* Dedham, MA: Artech House, 1974.

7.   Welsby, V.G., *The Theory and Design of Inductance Coils,* London: Macdonald and Co. Ltd., 1960.

8.   Medhurst, R.G., "High Frequency Resistance and Capacity of Single-Layer Solenoids," *Wireless Engineer,* March 1947, p. 35.

9.   Hilbers, A.H., "On the Design of H.F. Wideband Power Transformers (ECO 6907)," Philips C.A.B. Group, Eindhoven, 1969.

10.   Carson, R.S., *High Frequency Amplifiers,* New York: John Wiley and Sons, 1979.

11.   Precision Tube Company, Inc., Coaxitube semi-rigid coaxial cable, North Wales, (n.d.).

12.   Krauss, H.L., and C.W. Allen, "Designing Toroidal Transformers to Optimize Wideband Performance," *Electronics,* August 16, 1973.

13.   March, S., "Microstrip Packaging: Watch The Last Step," *Microwaves,* December 1981.

14.   Pues, H.F., and A.R. van de Capelle, *Electron. Lett.,* Vol. 16, November 6, 1980, pp. 870-872.

15.   Bahl, I.J., and D.K. Trevedi, "A Designer's Guide to Microstrip Line," *Microwaves,* May, 1977.

16.   Pucel, R.A., D.J. Masse, and C.P. Hartwig, "Losses in Microstrip," *IEEE Trans. Microwave Theory Techn.,* Vol. MTT-16, June 1968, pp. 342-350; "Correction to Losses in Microstrip," *Ibid.,* (Corresp.), Vol. MTT-16, December 1968, p. 1064.

17.   Welsh, J.D., and H.J. Pratt, "Losses in Microstrip Transmission Systems for Integrated Microwave Circuits," *NEREM Rec.,* Vol. 8, 1966, pp. 100-101.

CHAPTER 3

NARROWBAND IMPEDANCE MATCHING WITH LC NETWORKS

3.1 INTRODUCTION

Maximum power is transferred between a source and a load when the source and load impedances are conjugately matched (that is, $Z_L = Z_s^*$).

When the power gain of a transistor is low, as is the case with power transistors at higher frequencies, it is important to match the input impedance of the transistor to the source resistance. If this is not done, some or most of the gain will be lost.

Impedance-matching networks are not only used to match load and source impedances for maximum power transfer, but also to transform impedances for other purposes.

As an example of this, it is usually necessary to transform the load of a power amplifier to a value which is a function of the supply voltage and the required power. This impedance is usually different than that of the physical load of the amplifier.

Another example of impedance transformation is the optimization of an amplifier for optimum noise performance. The required source impedance (as viewed from the transistor terminals) is not the conjugate of the transistor input impedance.

Impedance-matching networks are therefore used for transforming impedances to certain required values, which may or may not be the conjugate of the source impedance.

Narrowband impedance-matching is done with two or more components. Where two components are used to bring about a resistance transformation, the matching network is called an L-section.

Three-element matching networks are usually Π- or T-sections. The names are descriptive of the configuration formed by the reactive elements.

The design of L-, T-, and Π-sections will be discussed in this chapter. Transformation of real, as well as reactive, loads will be considered.

When T- and Π-sections are used, it is possible to bring about the required transformation and to control the bandwidth of the network. Although the 3 dB bandwidth of an L-section can be determined easily, it is not a design parameter.

It is sometimes necessary to know the bandwidth resulting from a transforming section more accurately than possible with the approximation method which is usually used. In these cases, as well as in instances where bandwidths other than the 3 dB bandwidth is of interest, the procedure outlined in sec. 3.9 can be used.

It was shown in Ch. 2 that lossless reactive components do not exist. For this reason, all impedance-matching networks will have some insertion loss. These losses can be quite pronounced in circuits with very narrow bandwidths.

A simple procedure for calculating the insertion loss caused by a cascaded LC network will be outlined in sec. 3.8.

Apart from matching and transforming impedances, impedance-matching networks are often also used to reject unwanted signals outside the pass band.

This rejection can often be obtained by using impedance-matching networks with high Q-factors, that is, if the required rejection is not too great.

The rejection obtainable by using parallel and series resonant circuits will be considered in sec. 3.2.

When the required rejection becomes very high, the Q of the components, their temperature stability, and the tuning of these networks can become problems. If the associated insertion losses can be tolerated and the filtering occurs at low power levels, the required rejection can often be obtained by using surface acoustic wave (SAW) devices, ceramic filters, or crystal filters. These components are very stable and can provide extremely sharp rejection.

3.2 - PARALLEL RESONANCE

A parallel resonant circuit is shown in Fig. 3.1. Although it is not an impedance-matching network, it is of interest here because of its frequency response.

The frequency response of this circuit is determined by the zero at the origin, the zero at infinity, and the two poles. That is,

$$V_o(s) = Z(s) I$$

$$= \{1 / [1 / R_L + sC + 1/(sL)]\}I$$

$$= \frac{sLI}{s^2 LC + sL/R_L + 1} \tag{3.1}$$

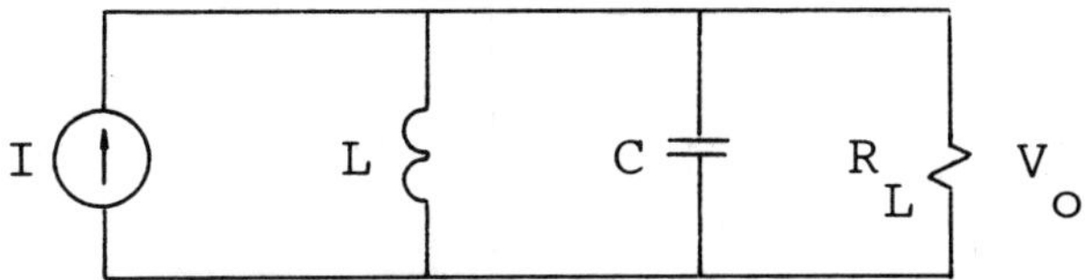

Figure 3.1   A Parallel Resonant Circuit

By inspection of Fig. 3.1, it is obvious that the highest possible output voltage will occur where

$$\omega L = 1/(\omega C)$$

that is, when

$$\omega_0 = 1/\sqrt{LC} \tag{3.2}$$

The 3 dB frequencies of the circuit can be determined by using (3.1) and (3.2). These frequencies occur where

$$|1/R_L + j\omega C + 1/(jwL)| = \sqrt{2}\,|1/R_L + j\omega_0 C + 1/(j\omega_0 L)|$$
$$= \sqrt{2}/R_L \tag{3.3}$$

After some manipulation the solutions of this equation are found to be

$$\omega_{3dB} = \omega_0 \left[1 + 1/(4Q^2)\right]^{1/2} \pm 1/(2RC) \tag{3.4}$$

Therefore, the bandwidth of the circuit is

$$B = \omega_{3dB2} - \omega_{3dB1} = 1/(RC) \text{ (rad/s)} \tag{3.5}$$

It can be seen from (3.4) that the circuit response is not symmetrical around the resonant frequency ω_0. It can, however, be proved easily through multiplying the two solutions given by (3.4), that the resonant frequency is the geometric mean of the two cut-off frequencies, that is

$$\omega_1 = \left[\omega_{3dB1}\,\omega_{3dB2}\right]^{1/2} \tag{3.6}$$

The Q-factor of the circuit is defined as the ratio of the center frequency to the bandwidth, that is

$$Q = \omega_0/B \tag{3.7}$$
$$= \omega_0\,CR \tag{3.8}$$
$$= R/(\omega_0 L) \tag{3.9}$$

A high Q therefore implies a very small relative bandwidth and, in the case of parallel resonance, reactances with low impedance compared to that of the load resistance. The reactances are shown in Fig. 3.2 for a Q of 10.

When the Q of the circuit is high, the arithmetic and the geometric mean of the cut-off frequencies are approximately the same (see (3.4)).

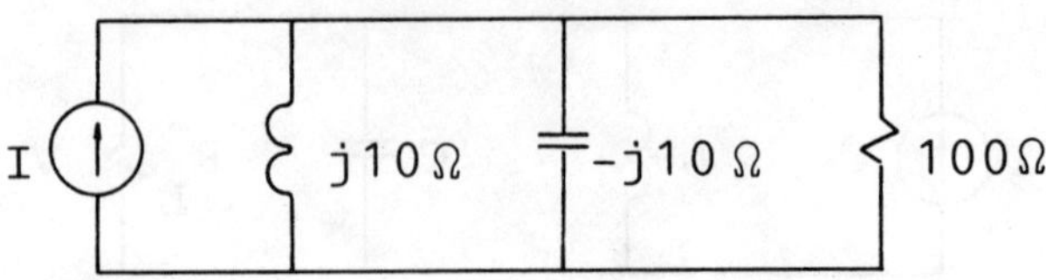

Figure 3.2   A Parallel Resonant Circuit with Q=10

Extremely sharp rejection can be obtained by using a parallel (or series) resonant circuit. Where the components are considered to be ideal, the ratio of the power transmitted to the load at resonance (P_{o-max}) and that at any other frequency ($P_o(f)$) can be calculated by using the equation:

$$\frac{P_{o-max}}{P_o(f)} = [1-2Q^2] + Q^2\left[(\omega/\omega_0)^2 + (\omega_0/\omega)^2\right] \qquad (3.10)$$

As an illustration of the rejection characteristics of the circuit, the attenuation is given in Table 3.1 as a function of the normalized frequency (f/f_0) for different values of the circuit Q. Only the frequencies above resonance are considered because the response is close to symmetrical when the Q is high. Where the response curve levels off to a single-pole response, no more entries were made into the table.

In order to appreciate the rate of rejection a –30 dB quality factory (Q_{-30}) is defined here as

$$Q_{-30} = f_0/B_{-30} \qquad (3.11)$$

where B_{-30} is the –30 dB "bandwidth" of the circuit (in Hz).

The –30 dB Q-factors for the three Q-factors used in Table 3.1 are 0.315 (Q = 10), 3.15 (Q = 100), and 7.90 (Q = 250), respectively.

It follows by observation of the results obtained that the –30 dB Q-factor of a resonant circuit is related to the 3 dB Q-factor in a simple way when the 3 dB Q-factor is greater than 10:

$$Q_{-30} \simeq 0.0315\, Q \qquad (3.12)$$

The normalized –30 dB bandwidth of the circuit is, therefore, given with good approximation by the equation:

$$B_{-30} = 31.75/Q \qquad (3.13)$$

The two normalized –30 dB rejection frequencies are given with good approximation by the equation:

$$f_{-30} = \pm 15.875/Q + \sqrt{\left(\frac{15.875}{Q}\right)^2 + 1} \qquad (3.14)$$

By using (3.13), the Q-factor required for a specified –30 dB bandwidth can be calculated easily.

Table 3.1

**The Frequencies at which the Output Signal of an Ideal Parallel (or Series)
Resonant Circuit Are Attenuated by Specified Amounts for Some Values
of the Circuit Quality Factor**

Attenuation (dB)	The normalized frequencies (f/f_0) at which the specified amount of attenuation occurs for some values of the circuit quality factor		
	$Q = 10$	$Q = 100$	$Q = 250$
0	1.0000	1.0000	1.0000
−3	1.0512	1.005	1.002
−10	1.1615	1.015	1.006
−20	1.615	1.051	1.020
−30	3.46	1.171	1.065
−40	—	1.620	1.22
−50	—	3.46	1.82
−60	—	—	4.24
−70	—	—	—

Example 3.1

As an example of the application of (3.13), the Q-factor necessary to provide
−30 dB rejection at 40 MHz and 60 MHz with a parallel resonant circuit will be
determined.

The resonant frequency of the circuit is

$$f_0 = (40 \times 60)^{1/2} = 48.99 \text{ MHz}$$

The normalized −30 dB bandwidth is

$$B_{-30} = (60\text{-}40)/48.99 = 0.4082$$

The required Q is obtained by using (3.13):

$$Q = 31.75/B_{-30} = 77.76 \tag{3.15}$$

Up to this point the losses in the components of the parallel resonant circuit
have been ignored. When the required Q of the circuit becomes of the same
order as the Q-factors of the components used, they cannot be ignored any
more.

The effective load resistance (R_T) at the resonant frequency, is then given by

$$1/R_T = 1/R_L + 1/(Q_L X_L) + 1/(Q_C X_C) \tag{3.16}$$

where X_L and X_L are the reactance of the inductor and capacitor, respectively,
at the resonant frequency. Q_L and Q_C are the unloaded Q-factors of the

inductor and capacitor, respectively.

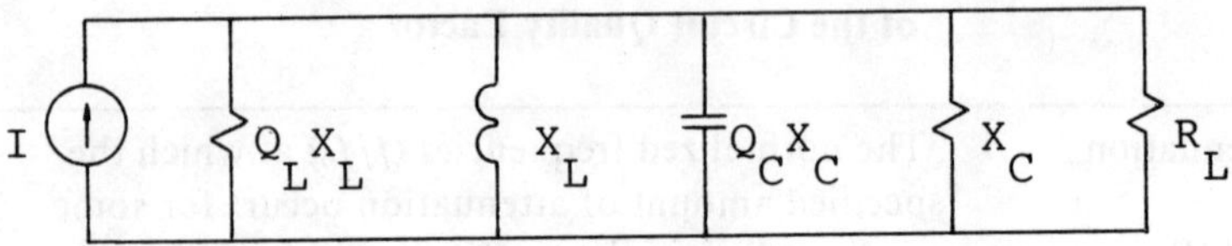

Figure 3.3   A Parallel Resonant Circuit with Lossy Components

At resonance the capacitive and inductive reactances are equal and (3.16) can be changed to

$$X_L / R_T = X_L / R_L + [1/Q_L + 1/Q_C] \tag{3.17}$$

The last term in this equation is defined as the unloaded Q (Q_u) of the circuit:

$$1/Q_u = 1/Q_L + 1/Q_C \tag{3.18}$$

The effective Q of the circuit (Q_{eff}) is, therefore, given by

$$1/Q_{eff} = 1/Q_I + 1/Q_u \tag{3.19}$$

where Q_I is the Q when the components are assumed to be lossless.

The highest Q obtainable with a parallel resonant circuit is limited by component losses and the temperature stability of the components.

3.3 SERIES RESONANCE

The results obtained for a parallel resonant circuit can be applied directly to a series resonant circuit by using the principle of dualism.

According to this principle, for every circuit, there is another circuit for which whatever applies to the current of one circuit, the same applies to the voltage of the other circuit, and *vice versa*.

This equivalent can be obtained by following the procedure illustrated in Fig. 3.4. A node is placed in every loop of the first circuit, as well as in the space outside it. These nodes are then connected by passing from one loop to another through the components of the different loops. Inductors are replaced with capacitors, capacitors with inductors, resistors with conductors, and conductors with resistors. The values assigned to the new components (H, F, Ω, S) are numerically equal to those of the original components.

The output voltage of the parallel resonant circuit in Fig. 3.4 is given by the equation:

$$V_o = I/[1/R+sC+1/(sL)] = 0.2/[0.5+3s+1/(5s)] \tag{3.20}$$

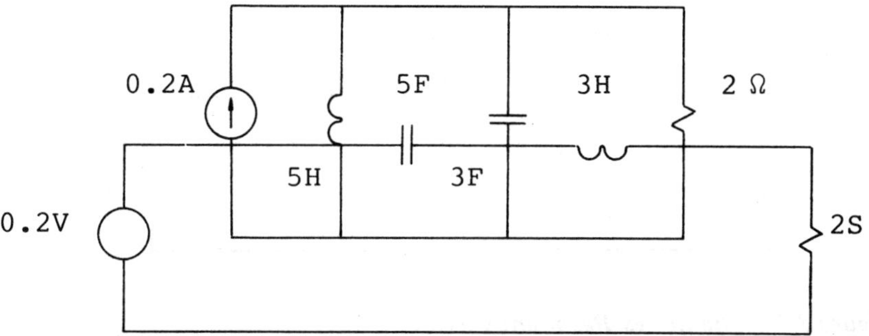

Figure 3.4   The Principle of Dualism Applied to a Parallel Resonant Circuit

The output of the series resonant circuit is obtained by replacing:

V_o with I_o
I with E
R with G
C with L
L with C

Thus, $I_o = E/[1/G + sL + sL + 1/(sC)] = 0.2/[0.5 + 3s + 1/(5s)]$ $\qquad$ (3.21)

It follows by inspection of Fig. 3.4 that the output current of the series resonant circuit is indeed given by this equation.

By applying the principle of dualism to the results deduced in the previous
section, the following equations are found to apply to the series resonant
circuit of Fig. 3.5:

$$Q \quad = \omega_0 L/R = 1/(\omega_0 C R) \qquad (3.22)$$

$$\omega_{3dB} = \omega_0 [1 + 1/(4Q^2)]^{1/2} \pm R/(2L) \qquad (3.23)$$

$$B \quad = R/L \text{ (rad/s)} \qquad (3.24)$$

$$R_T \quad = R_L + X_L/Q_L + X_C/Q_C \qquad (3.25)$$

$$1/Q_{eff} = R_L/X_L + [1/Q_L + 1/Q_C] \qquad (3.26)$$

It follows from (3.26) that similarly to the parallel resonant circuit, the unloaded Q for the series resonant circuit is given by the equation:

$$1/Q_u = 1/Q_L + 1/Q_C \qquad (3.27)$$

The reactances for a series resonant circuit with $Q=10$ are shown in Fig. 3.6 at
the resonant frequency. Note that these reactances are high compared to the
load resistance.

With the same loaded Q, the frequency response of the series resonant circuit
is identical to that of the parallel resonant circuit.

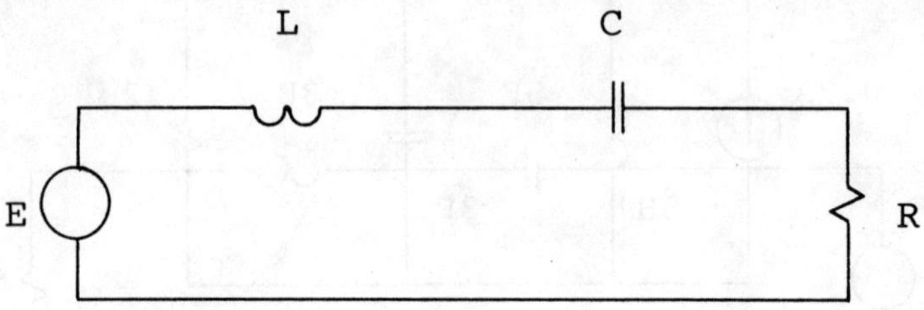

Figure 3.5   The Series Resonant Circuit

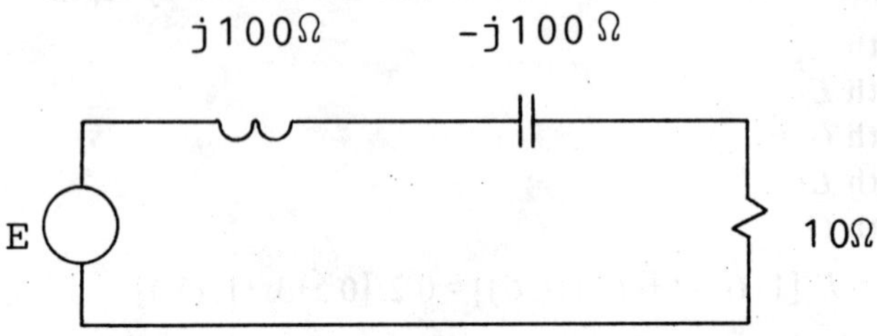

Figure 3.6   A Series Resonant Circuit with Q=10

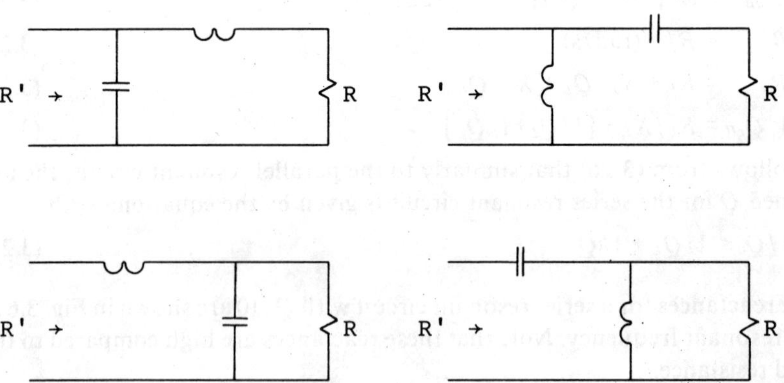

Figure 3.7   The Four Possible Configurations for an L-Section

3.4 L-SECTIONS

An L-section is a two-element matching network. The four possible configurations are shown in Fig. 3.7.

Depending on the position of the first component (as viewed from the load), an up or downward transformation of the load resistance can be obtained by using an L-section.

When the first reactive component is a series component, the transformation is upward, and when it is a parallel element the transformation is downward.

The second element in the L-section is used for removing the residual reactance caused by the transformation element (that is, the first element). This second element is the compensating element.

The basic principle used in narrowband impedance-matching is that the resistance of a reactive impedance is not the same when the impedance is considered a series or parallel combination of the resistance and reactance. This is illustrated in the following example.

Example 3.2

The transformation and compensation properties of an L-section will be illustrated here by using the L-section shown in Fig. 3.7a, as an example. The resistance and reactances at the frequency where the transformation is to be done are taken as

$$R \quad = 1\,\Omega$$
$$\omega L \quad = 1\,\Omega$$
$$1/\omega C = 2\,\Omega$$

if the first element is a series element, and

$$Q_1 = R/X_1 \qquad\qquad (3.30)$$

if the first element (as viewed from the load) is a parallel element.

In the last two equations, R is the load resistance to be transformed and X_1 is the required reactance of the first element of the L-section.

It can be shown easily that when an L-section is used, the resistance changes with a factor

$$D_1 = 1 + Q_1^2 \qquad\qquad (3.28)$$

where

$$Q_1 = X_1/R \qquad\qquad (3.29)$$

L-section:

Resistance to be transformed:

Addition of the transforming element:

Parallel equivalent of the series combination ($Y = 1/Z$):

Cancellation of the residual reactance:

Transformed resistance:

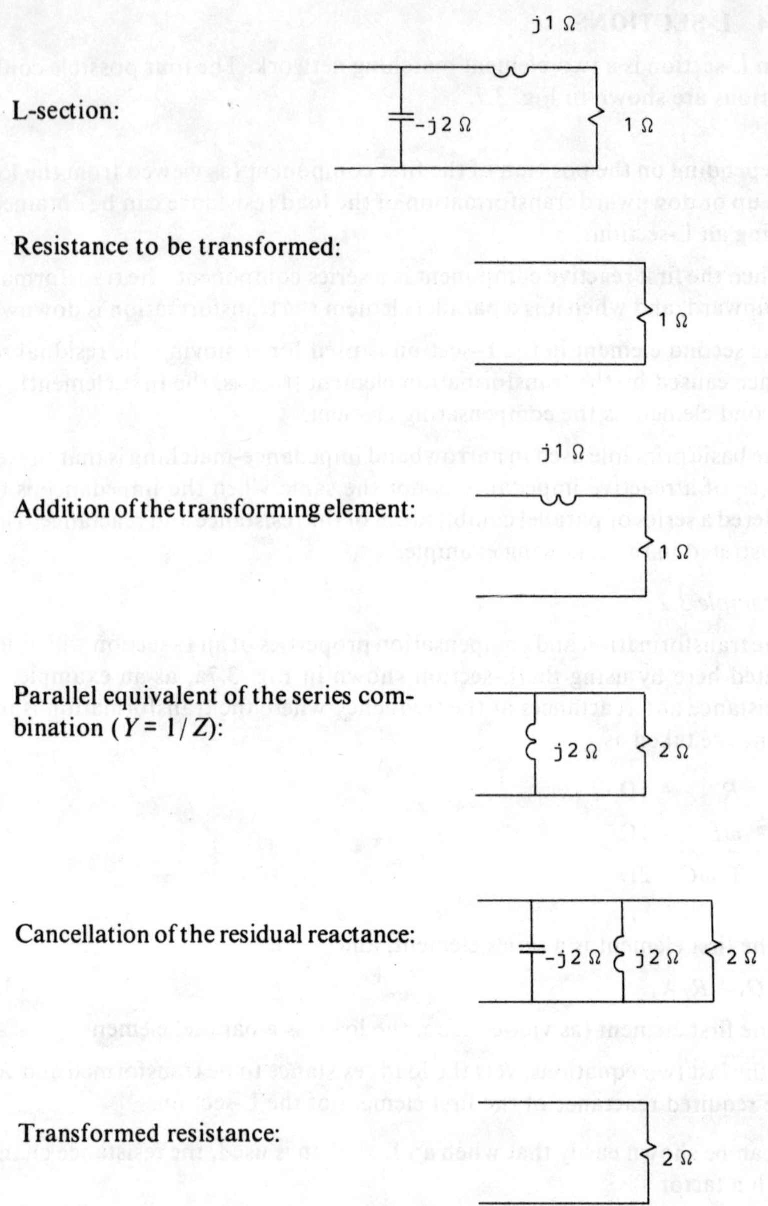

Figure 3.8   Illustration of the Transformation and Compensation Properties of an L-Section.

The ratios defined in (3.29) and (3.30) are similar in form to the Q-factors of the series or parallel resonant circuits, respectively. These ratios are defined as Q-factors of transformation.

Where the first reactive element is a series element, the load resistance is transformed upward, and when it is a parallel element it will be transformed downward.

The reactance changes by a factor

$$E_1 = 1+1/Q_1^2 \tag{3.31}$$

in the transformation step.

As is the case with the resistance, the reactance increases if the first element is a series element and decreases if it is a parallel element.

It is useful from a programming viewpoint to assign a sign to the transformation Q. A position sign is assigned to it if the first element is an inductor in the series case or a capacitor in the parallel case.

The second element in the L-section is used to achieve the desired reactance level. If a purely resistive input impedance is required, the reactance of this element is given by

$$X_2 = -X_1 \, (1+1/Q_1^2) = -R'/Q_1 \tag{3.32}$$

if the first element is a series element, and by

$$X_2 = -X_1/(1+1/Q_1^2) = -R' \, Q_1 \tag{3.33}$$

if the first element is a parallel element.

R' is the transformed value of the load resistance R. Thus,

$$R' = R \, D_1 \tag{3.34}$$

The last equality in (3.32) and (3.33) can be verified easily by using the relationship $Z=1/Y$ and $Y=1/Z$, respectively.

The formulas relevant to the design of an L-section are summarized in Table 3.2.

When the transformation Q is high, the frequency response of an L-section will be similar to that of a simple series or parallel resonant circuit, that is, near the resonant frequency.

The Q of the L-section is approximately equal to 0.5 times the Q of the transforming section (the transformation Q).

If a more accurate value for the Q of the circuit is required, the procedure outlined in sec. 3.9 can be followed.

Example 3.3

An L-section will be designed to transform a load of $50\,\Omega$ to $250\,\Omega$ at 50 MHz, as an example of the application of the theory discussed above.

Table 3.2

Formulas Relevant to the Design of L-Sections

Downward transformations

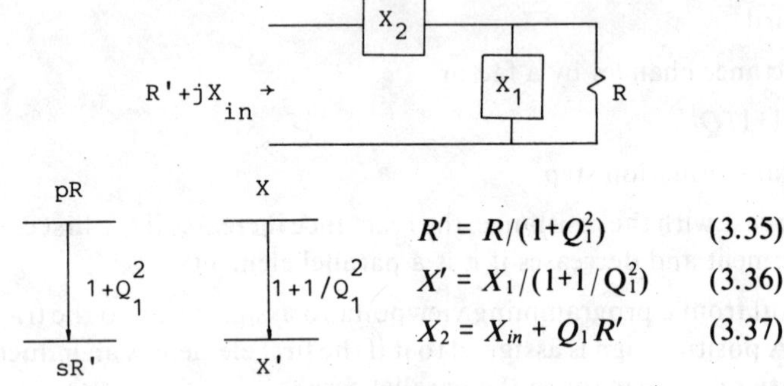

$$R' = R/(1+Q_1^2) \qquad (3.35)$$
$$X' = X_1/(1+1/Q_1^2) \qquad (3.36)$$
$$X_2 = X_{in} + Q_1 R' \qquad (3.37)$$

Upward transformations

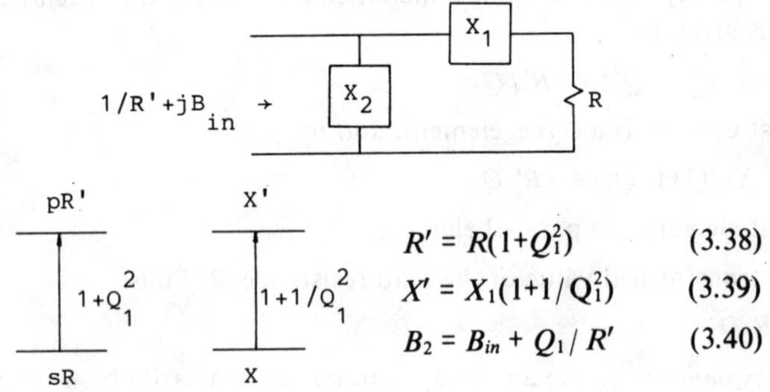

$$R' = R(1+Q_1^2) \qquad (3.38)$$
$$X' = X_1(1+1/Q_1^2) \qquad (3.39)$$
$$B_2 = B_{in} + Q_1/R' \qquad (3.40)$$

Since the transformation is upward, the first element of the L-section must be a series element. The following diagram applies:

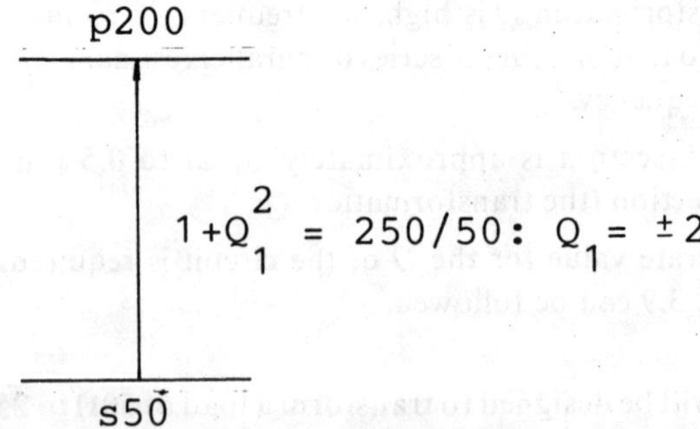

Figure 3.9   The Transformation Diagram Relevant to Example 3.3

Since the transformation Q must be equal to 2, a series inductor or capacitor with reactance equal to

$$X = Q \cdot R = 2 \cdot 50 = 100 \ \Omega$$

must be used.

If the first element is chosen to be an inductor, the required inductance is

$$L = 100/(2\pi 50 \cdot 10^6) = 0.318 \mu H$$

The parallel equivalent of the transforming section is the required 250Ω in parallel with a reactance

$$X' = R'/Q_1 = 250/2 = 125\Omega$$

It is useful to note that the Q-factors of the series combination and its equivalent parallel combination must be equal.

The capacitance required to remove the reactiveness of the input admittance of the resistor and inductor combination is

$$C = 1/[125 \ (2\pi 50 \cdot 10^6)] = 25pF$$

The designed network is shown in Fig. 3.10.

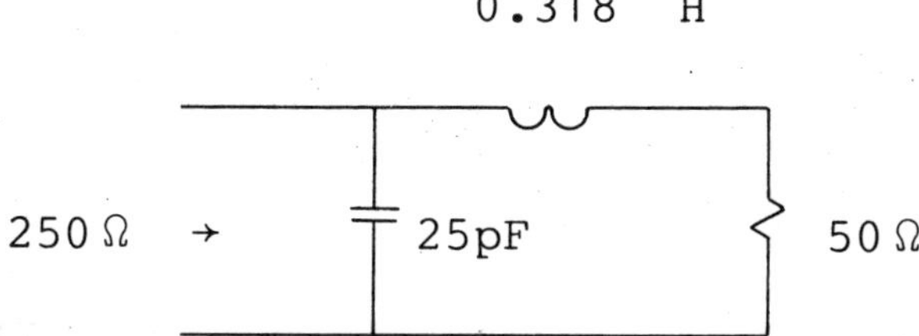

Figure 3.10   An L-Section for Matching a Load of 50Ω to 250Ω at 50 MHz

3.5   Π AND T SECTIONS

Π- and T-sections are three-element matching networks. A Π-section has two parallel elements and the T-section has two series elements as shown in Fig. 3-11.

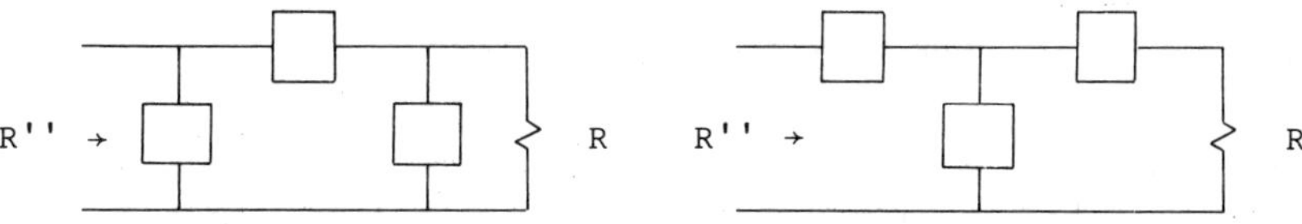

Figure 3.11   Topology for [a] Π-Section and [b] T-Section

The first two elements in these sections are transforming elements. One of these elements causes the resistance to increase, while the other decreases it.

The reactance level is set by the last element in the section (the compensating element).

Because the resistance is transformed twice, there are two transformation Q-factors for these sections. The highest transformation Q can be chosen to have any value higher than that necessary for an equivalent L-section.

The bandwidths of Π- and T-sections are also determined by the transformation Q-factors. Where the two Q-factors are different, the Q of the network will be approximately equal to one-half of the highest transformation Q.

Because the highest transformation Q is adjustable, Π- and T-sections can be designed to have specific bandwidths.

3.5.1   The Π-Section

The resistance transformations caused by a Π-section are illustrated in Fig. 3.12. The resistance is first transformed downward by a factor $1+Q_1^2$, where Q_1 is the first transformation Q formed by the load resistance and the first element of the network, and the resistance is then transformed upward again by a factor $1+Q_2^2$, where Q_2 is the second transformation Q.

The second transformation Q is equal to the ratio of the effective reactance in series with the transformed resistance (R') and the transformed resistance itself.

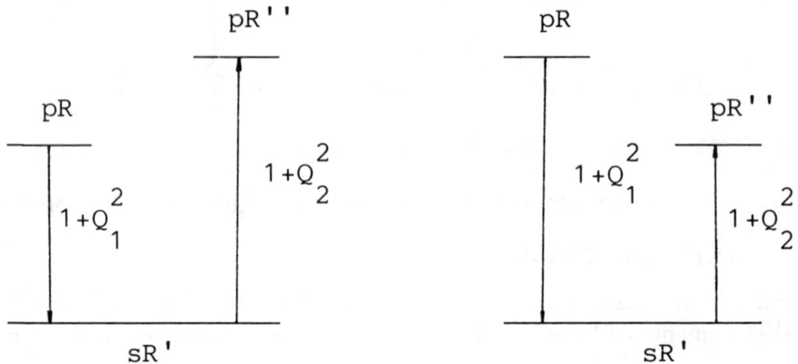

Figure 3.12   [a] Upward Transformation of the Load Resistance with a Π-Section; [b] Downward Transformation of the Load Resistance with a Π-Section

The transformed resistance will be lower than the load resistance when the first transformation Q is higher than the second. An upward transformation requires the second transformation Q to be higher than the first.

The value of the highest transformation Q is determined by the required bandwidth of the network. The Q of the network is approximately equal to one-half of the highest transformation Q when the transformation Q-factors are sufficiently different.

The transformation of a 10 Ω load to 50 Ω by using a Π-section is illustrated in detail in Fig. 3.13.

The designed Π-section:

Load resistance to be transformed:

First transformation of the load resistance ($R \rightarrow R'$):

Changing the transformation Q for the second transformation:

Second transformation of the load resistance ($R' \rightarrow R''$):

Cancellation of the residual reactance:

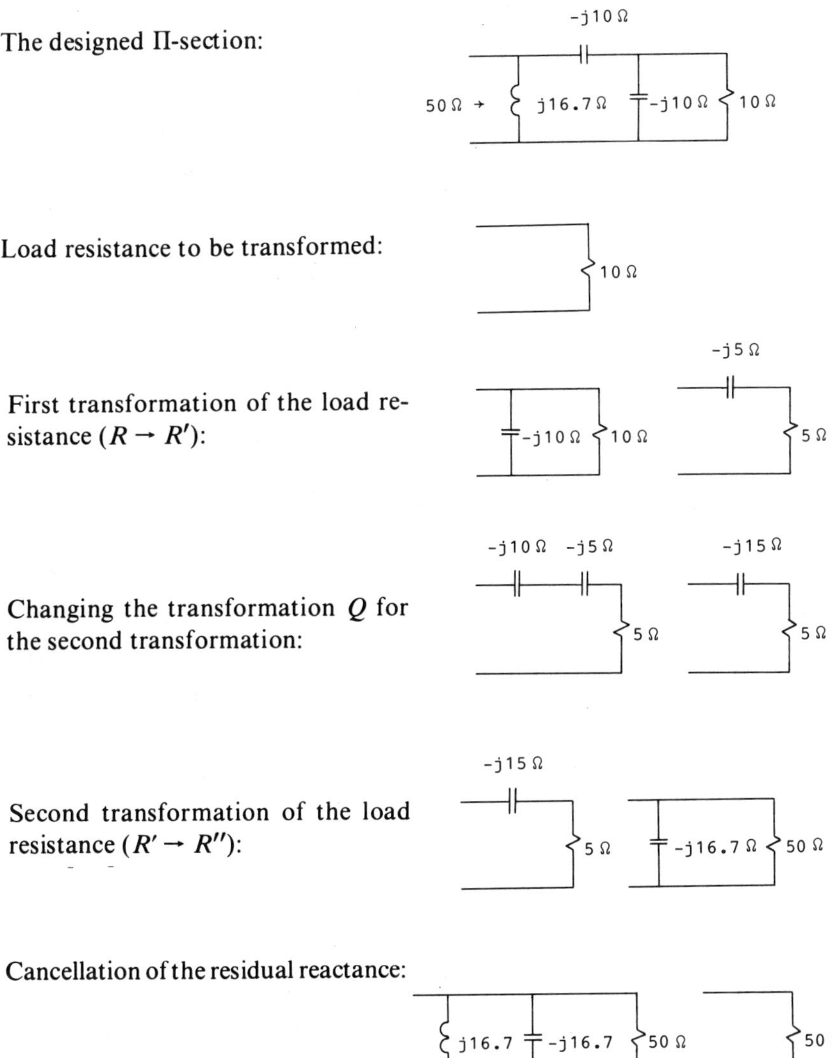

Figure 3.13   The transformation of a 10Ω Load to 50Ω with a Π-section

The formulas relevant to the design of a Π-section are summarized in Table 3.3.

Table 3.3

Formulas for Designing a Π-Section

Decreasing the load resistance

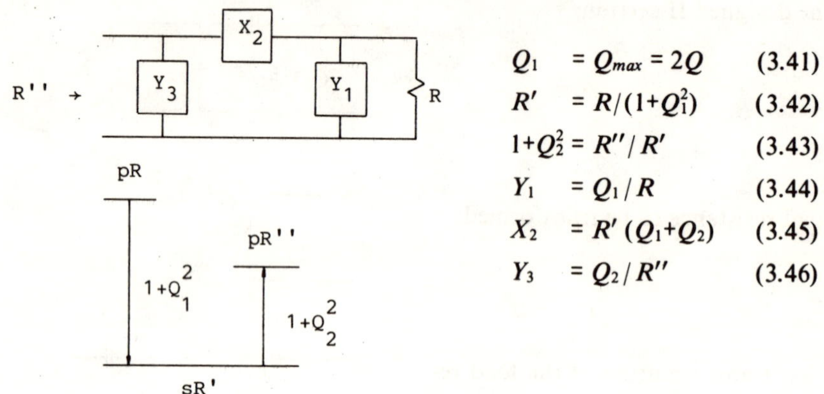

$$Q_1 = Q_{max} = 2Q \qquad (3.41)$$
$$R' = R/(1+Q_1^2) \qquad (3.42)$$
$$1+Q_2^2 = R''/R' \qquad (3.43)$$
$$Y_1 = Q_1/R \qquad (3.44)$$
$$X_2 = R'(Q_1+Q_2) \qquad (3.45)$$
$$Y_3 = Q_2/R'' \qquad (3.46)$$

Increasing the load resistance

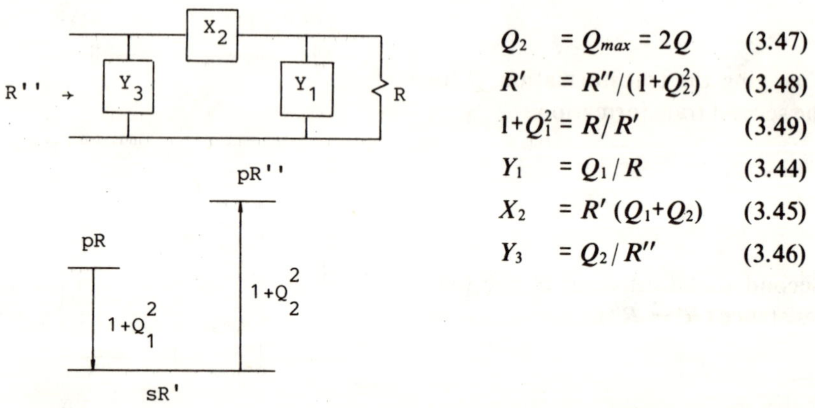

$$Q_2 = Q_{max} = 2Q \qquad (3.47)$$
$$R' = R''/(1+Q_2^2) \qquad (3.48)$$
$$1+Q_1^2 = R/R' \qquad (3.49)$$
$$Y_1 = Q_1/R \qquad (3.44)$$
$$X_2 = R'(Q_1+Q_2) \qquad (3.45)$$
$$Y_3 = Q_2/R'' \qquad (3.46)$$

Example 3.4

A matching network for transforming 50Ω to 12.5Ω will be designed. The maximum transformation Q is to be 5.

Because the transformation is downward, the first transformation Q will be the highest.)

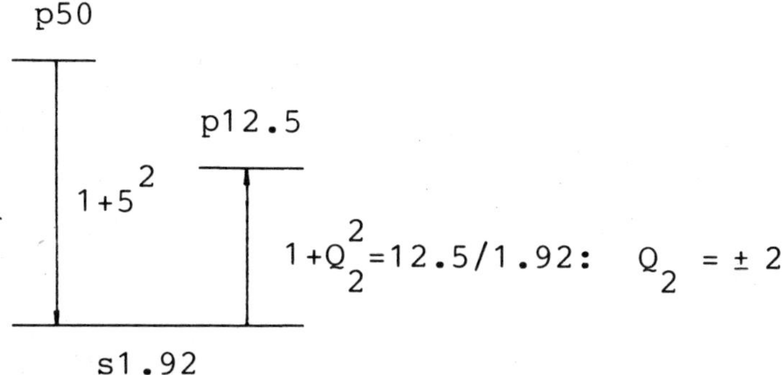

Figure 3.14   The Transformation Diagram Corresponding to Example 3.4

The next step is to choose the network topology to be used. The network is arbitrarily assumed to have an inductor as the first element. The other components are chosen to be capacitors. (It is not possible to choose both of the first two components to be inductors or capacitors. If this is done the second transformation Q will be the highest.)

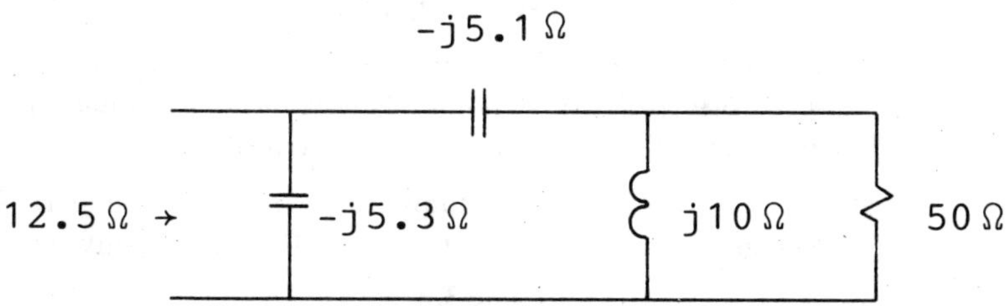

Figure 3.15   A Π-Section for Matching 50Ω to 12.5Ω

Because the first transformation Q is

$Q_1 = -5,$

the reactance of the inductor must be

$X_1 = 50/5 = 10\Omega.$

The second component must change the transformation Q to 2.35. The Q .must be positive (inductive) if the last component is to be a capacitor:

$X_2 = R'\,(Q_1+Q_2) = 1.92\,(2.35+(-5)) = -5.1\Omega$

The reactance of the last component is

$X_3 = -R''/Q_2 = -12.5/2.35 = -5.3\Omega$

The designed network is shown in Fig. 3.15.

3.5.2   The T-Section

The dual of a Π-section is a T-section. Therefore, the formulas for designing a Π-section can be used to design a T-section.

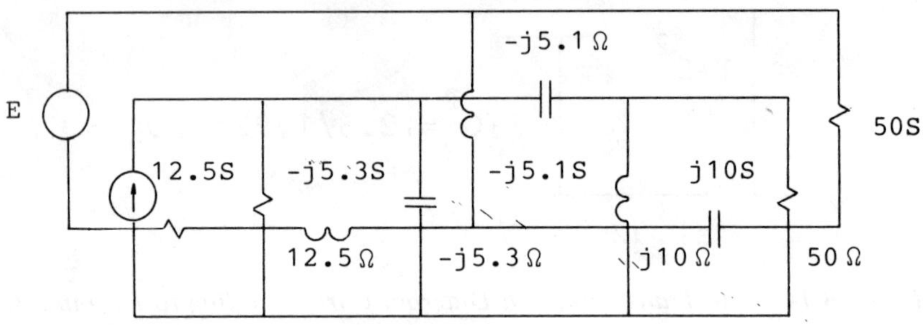

Figure 3.16   The Dual of a Π-Section

In order to do this, it is necessary to replace the resistance and reactance in these formulas with conductance and susceptance, respectively.

If a T-section is therefore to be designed to match a load of 50Ω to a 10Ω source, the specifications for the Π-section become $1/50\Omega$ and $1/10\Omega$,

The reactance results of the Π-section apply directly to the T-section, if they are interpreted to be susceptances. If the results for the Π-section are $j10\Omega$, $-j5\Omega$, $j3\Omega$, they become $j10S$, $-j5S$, $j3S$ for the T-section.

This information is very useful for developing a program to design Π- and T-sections. The program can be written to design Π-sections only, and by

Table 3.4

Formulas for Designing T-Sections

Decreasing the load resistance

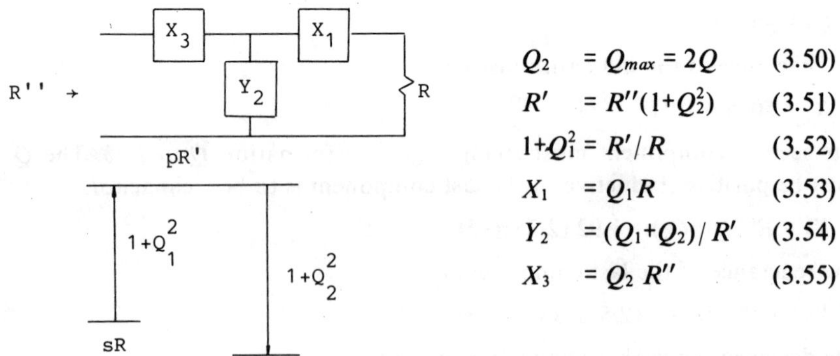

$$Q_2 = Q_{max} = 2Q \qquad (3.50)$$

$$R' = R''(1+Q_2^2) \qquad (3.51)$$

$$1+Q_1^2 = R'/R \qquad (3.52)$$

$$X_1 = Q_1 R \qquad (3.53)$$

$$Y_2 = (Q_1+Q_2)/R' \qquad (3.54)$$

$$X_3 = Q_2 R'' \qquad (3.55)$$

Increasing the load resistance

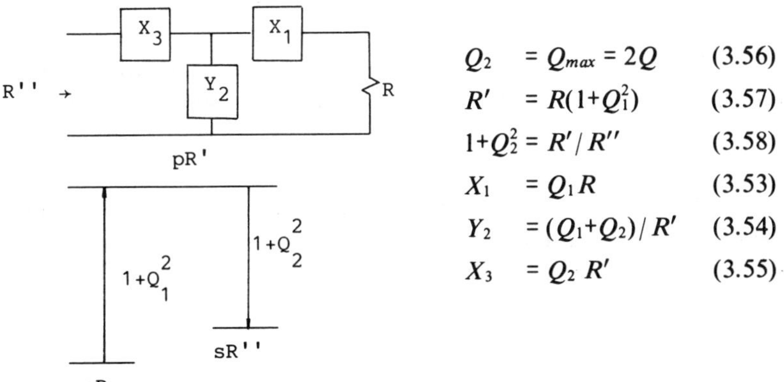

$$Q_2 = Q_{max} = 2Q \qquad (3.56)$$
$$R' = R(1+Q_1^2) \qquad (3.57)$$
$$1+Q_2^2 = R'/R'' \qquad (3.58)$$
$$X_1 = Q_1 R \qquad (3.53)$$
$$Y_2 = (Q_1+Q_2)/R' \qquad (3.54)$$
$$X_3 = Q_2 R' \qquad (3.55)$$

entering the specifications correctly it also can be used to design T-sections.

When the design is not done by computer, it is better to follow the procedure outlined in Table 3.4.

3.6 THE DESIGN OF Π-SECTIONS AND T-SECTIONS WITH COMPLEX TERMINATIONS

The procedures outlined in the previous sections can be extended easily to the

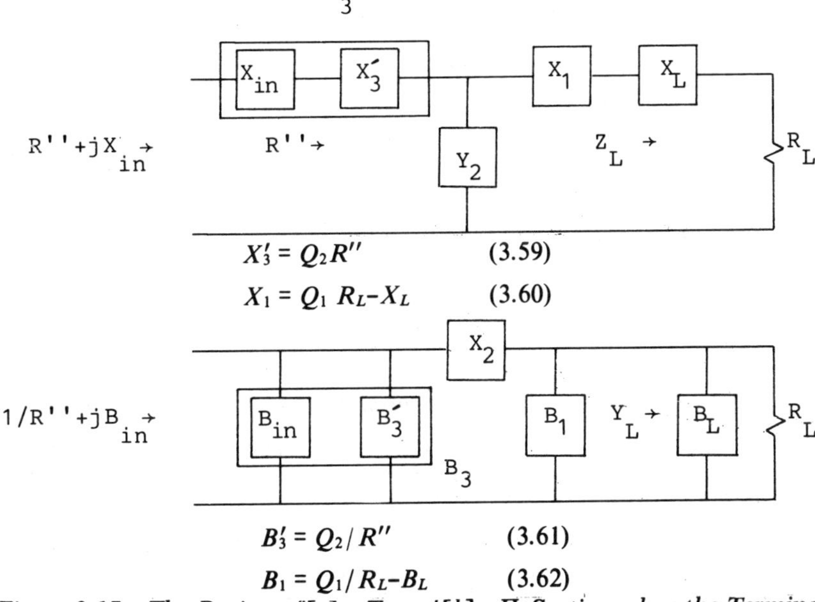

$$X_3' = Q_2 R'' \qquad (3.59)$$
$$X_1 = Q_1 R_L - X_L \qquad (3.60)$$

$$B_3' = Q_2/R'' \qquad (3.61)$$
$$B_1 = Q_1/R_L - B_L \qquad (3.62)$$

Figure 3.17 *The Design of [a] a T- and [b] a Π-Section when the Terminations Are Complex*

general case where the load and source impedances are complex. The approach is illustrated in Fig. 3.17.

The reactive parts of the load and source impedance (T-section) or admittance (Π-section) are ignored initially and the network is designed to match the relevant resistances to each other. The first and last components are then changed to take into account the imaginary parts of the load and source impedance or admittance.

Because a T-section transforms a series load resistance to a new series value, the load and the required input impedance must be specified in series form, that is, as impedances. The specifications for a Π-section must be of parallel form.

Example 3.5

A T-section for matching a $10+j10\,\Omega$ load to $50+j40\,\Omega$ (see Fig. 3.19) with a maximum transformation Q equal to 5 will be designed.

These specifications are in impedance form, as required for a T-section.

The transformation diagram for this problem is shown in Fig. 3.18. Because the transformation is upward, the first transformation Q must be the highest in this case. The second transformation Q must be equal to 2.05.

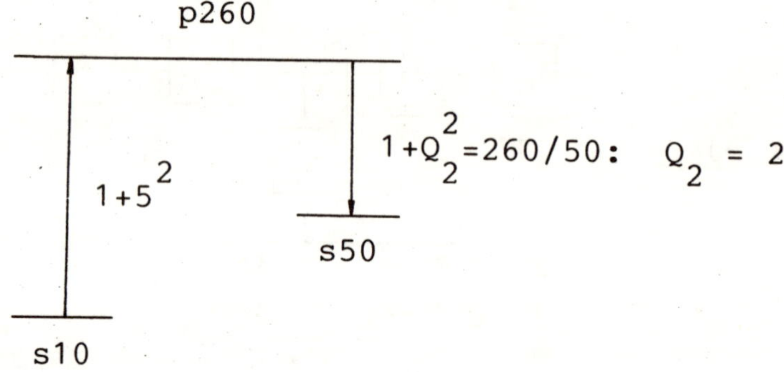

Figure 3.18   The Transformation Diagram for the T-Section of Example 3.5

With the Q-factors known, the next step is to choose a topology. If the network shown in Fig. 3.19 is chosen, the bandwidth of the circuit can be calculated as was done before. Since the Q-factors of the load and source impedances are low compared to the maximum transformation Q, predictable results will also be obtained with other topologies. The only major difference will be in the rate at which the slope outside the pass band levels off because of the higher number of poles.

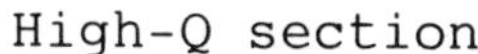

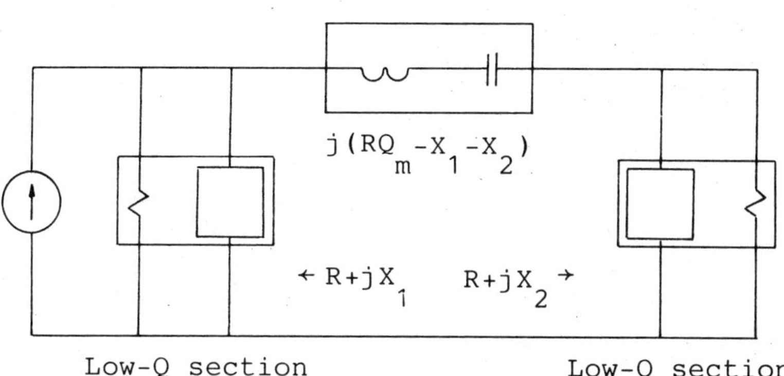

Figure 3.19   A T-Section for Transforming a 10+j10Ω Load to 50+j40Ω.

The component values of the network chosen can now be determined by using the relevant transformation Q-factors.

In order to have a transformation Q of 5 at the load, a reactance of $+j40\Omega$ must be added to the existing $+j10\Omega$, that is,

$$X_1 = (Q_1\ 10) - 10 = 40\Omega$$

After the first transformation, the transformation Q is still equal to 5. In order to change it to 2.05, a capacitor with susceptance

$$Y_2 = (Q_1 + Q_2)/R' = (5+2.05)/260 = 27.1 \text{ mS}$$

must be used, that is, both Q_1 and Q_2 must be positive.

It is not possible to use a capacitor with susceptance

$$Y = (Q_1 - |Q_2|)/R'$$

in this case, since the last component of the network was chosen to be an inductor.

The last component of the T-section must remove the residual reactance and change it to the required level of $+j40\Omega$. In order to do this, an inductor with reactance

$$X_3 = Q_2\ R'' + 40 = 2.05\ (50) + 40 = 142.5\Omega$$

is required.

The designed network is shown in Fig. 3.19.

3.7   FOUR-ELEMENT MATCHING NETWORKS

When four elements are used, the control over the frequency response of an impedance matching network increases. Bandwidths can be obtained which are both lower or higher than that of an L-section.

A possible approach for designing a network having a very high Q, and which can be tuned very easily, is shown in Fig. 3.20.

The two low-Q sections transform the load and source impedances to have the same value ($R<R_1$, R_2 with the network as shown in Fig. 3.20) and the high-Q section provides the required rejection.

Strictly speaking, only one downward transforming section is required in this high-Q network. However, where two downward transforming sections are used, it is often possible to decrease the insertion losses of the circuit. This follows because it is often possible to use higher Q components in the circuit when this is done.

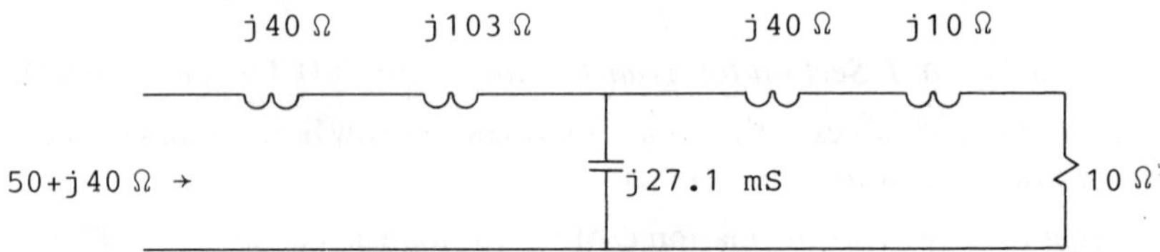

Figure 3.20   A High-Q, Easily Tunable, 4-Element Impedance-Matching Network

Wider bandwidths than those obtainable with L-sections can be achieved by following the approach illustrated in Fig. 3.21 (R_2 is assumed to be smaller than R_1).

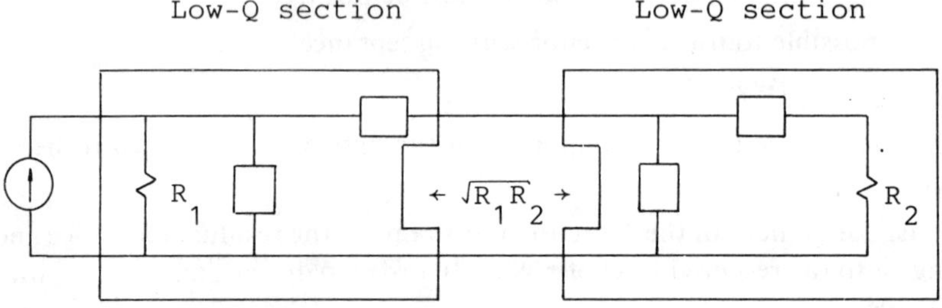

Figure 3.21   A Wideband, 4-Element, Impedance-Matching Network

3.8   CALCULATION OF THE INSERTION LOSSES OF LC IMPEDANCE-MATCHING NETWORKS

It was shown in Ch. 2 that the ideal component does not exist. For this reason, all practical circuits will have some insertion losses. If the losses are to be kept low, the component Q-factors must be significantly higher than those of the circuit.

The insertion losses of any cascaded LC network can be computed very rapidly by following the procedure outlined below:

1)   Model each reactive component in the network as an ideal component with a resistor in series or in parallel with it, depending on whether it is a series or shunt element, respectively.

The value of this resistance can be determined from the expected Q-factor of the component.

Q-factors for capacitors and magnetic-core inductors can usually be found by using the data given by the manufacturer, while those of air-cored solenoidal coils can be determined by following the procedure outlined in sec. 2.3.6.

Where the Q-factor changes significantly with frequency it should be taken into account.

2)   Assume that the power dissipated in the load in equal to 1W.

3)   If the first component (as viewed from the load) of the network is a series element, calculate the power dissipated in it by using the equation:

$$P_D = (R_Q/R_L)\, P_L \tag{3.63}$$

where R_Q is the associated series resistance of the first element and R_L the (effective) load resistance. P_L is the power dissipated in the load.

If the first component is a parallel element, calculate the power dissipated in it by using the equation:

$$P_D = (G_Q/G_L)\, P_L \tag{3.64}$$

where G_Q is the associated parallel conductance of the first component and G_L the (effective) conductance of the load. P_L is the power dissipated in the load.

4)   Add the power dissipated in the first component to that dissipated in the load:

$$P_T = P_L + P_D \tag{3.65}$$

5)   Consider the first component to be part of the load and calculate the new (effective) load admittance or impedance.

6)   Repeat steps 3 to 5 until the power entering the matching network (P_T) and the effective input impedance of the network (Z_{IN}) are known.

7)   Calculate the transducer power gain of the network (G_T) by using the equation:

$$G_T = [1-|S_s|^2]\,\frac{P_L}{P_T} \tag{3.66}$$

$$= \left[1 - \left|\frac{Z_{IN}-Z_S^*}{Z_{IN}+Z_S}\right|^2\right]\Big/ P_T \tag{3.67}$$

$$= \frac{4R_{IN}R_S}{[R_{IN}+R_S]^2 + [X_{IN}+X_S]^2} \Big/ P_T \tag{3.68}$$

where S_s is the input reflection parameter with Z_s as normalizing impedance,

$$Z_{IN} = R_{IN} + jX_{IN} \tag{3.69}$$

and

$$Z_S = R_S + jX_S \tag{3.70}$$

where Z_s is the internal impedance of the source driving the network.

Example 3.6

As an example of the application of this procedure, the insertion loss of the Π-section designed in *Example 3.4* will be calculated at the center frequency. The unloaded Q-factors of the capacitors are assumed to be 500, while that of the inductor is taken as 100.
The conductance associated with the inductor is 1mS, the series resistance associated with the first capacitor is 0.01Ω, and the conductance associated with the second capacitor is 0.38mS.

The power dissipated in the inductor is 50mW. The power entering the last section of the network is therefore 1.05W. The input impedance of this section is

$$Z = 2.0+j9.6\Omega$$

The power dissipated in the first capacitor is 5mW. The power entering the network at this point is therefore 1.055W. The input admittance at this point is

$$Y = (84-j186)\text{mS}$$

The power dissipated in the last capacitor is also 5mW. The total power entering the network is therefore 1.06W. The input impedance is

$$Z_{IN} = 11.9-j0.45\Omega$$

The transducer power gain of the network was calculated to be 0.94, that is, an insertion loss of 0.3 dB.

3.9   CALCULATION OF THE BANDWIDTH OF CASCADED LC IMPEDANCE MATCHING NETWORKS

The bandwidth of a network can be found iteratively if its transducer power gain is determined as a function of frequency. The transducer power gain of any cascaded LC network can be found with minimum effort by following the procedure outlined in the previous section.

Because the cut-off frequencies (3 dB) of L-, T-, and Π-sections are known with good approximation ($f_{-3dB} = f_0 \pm f_0 / Q_{max}$), the exact bandwidths of these circuits can be determined quickly by following this procedure.

Example 3.7

By following this procedure, the 3 dB cut-off frequencies of the network in *Example 3.6* are found to be 83 MHz and 130 MHz, that is, if the center frequency is selected as 100 MHz. The exact Q of the circuit is, therefore, 2.3 instead of the expected 2.5.

If bandwidths other than the 3 dB bandwidth are required, they can be found easily by following the same procedure.

Questions and Problems

1. What is the definition of the Q-factor of a circuit?

2. (a) Find the –30 dB frequencies of a parallel resonant circuit with $Q=300$ at 10.7 MHz. (b) A parallel resonant circuit is to be used to obtain 30 dB rejection at 5 MHz and 7 MHz (center frequency 5.92 MHz). What is the Q required to meet these specifications?

3. The Q of a parallel resonant circuit is 50 when ideal components are used. What will it be if an inductor and capacitor with unloaded Q-factors of 150 and 250, respectively, are used?

4. Differentiate between the Q of a circuit and a transformation Q.

5. Is it true that the load will always be transformed downward by an L-section if the first element of the section is in parallel with the load?

6. What is the relationship between the transformation Q of an L-section and the Q of the network?

7. Design an L-section to transform a 50Ω load to 12.5Ω at 50 MHz. Use the information in Ch. 2 to determine the unloaded Q of the inductor in the design, that is, if it is to be a single-layer air-cored coil (dimensions 2cm×2cm×2cm, realistic wire thickness). Do this for both of the topologies that can be used.

8. What is the relationship between the transformation Q of a T- or Π-section and the Q of the network?

9. Design a Π-section to transform a load of 500Ω to 50Ω. The bandwidth of the network must be 2 MHz with the resonant frequency at 10 MHz. A capacitor must be used as the firt element and the other two elements must be inductors.

10. Design a T-section to achieve the transformation in the previous problem.

11. Design a low-pass L-section to transform the input impedance of a 200W push-pull amplifier to 12.5Ω. The input impedance of the amplifier is $2-j2\Omega$ and the frequency of interest 30 MHz.

12.   (a) Design a T-section to transform a load of $3-j6\Omega$ to $50+j25\Omega$. The Q of the circuit must be approximately 8; f_0 = 100 MHz. (b) Design a Π-section to meet the specifications in (a).

13.   (a) Use the material in sec. 2.3.6 of Ch. 2 to determine the range of inductance values for which an unloaded Q of 250 can be obtained at 50 MHz with standard wire thicknesses and coil sizes below $2cm\times2cm\times2cm$. (b) Design a four-element matching network to match a load of 200Ω to a source with 50Ω internal resistance. The Q of the circuit must be 25 and the resonant frequency 50 MHz. (c) Determine the $-30\,dB$ frequencies of the circuit. (d) Calculate the insertion losses of the circuit at the resonant frequency.

14.   Transform a load of 10Ω to 50Ω. The bandwidth is to be as large as possible. Use a four-element matching network; f_0 = 20 MHz.

15.   Determine realistic values for the unloaded Q-factors of the components in *Problems 3.9* and *3.10*. Compare the insertion losses of the two networks.

16.   Calculate the insertion loss of the network designed in *Problem 3.11* if the unloaded Q-factors of the capacitor and inductor used in the network are 150 and 50, respectively.

17.   Calculate the exact 3 dB bandwidths of the circuits designed in *Problems 3.9, 3.10, 3.12, 3.13,* and *3.14.* Compare the expected and actual bandwidths.

Reference

RCA Corporation (Solid State Division, Somerville, NJ), RF Power Transistor Manual, 1971.

CHAPTER 4
COUPLED COILS AND TRANSFORMERS

4.1 INTRODUCTION

When the parasitics can be ignored, transformers are ideally suited for impedance scaling.

A practical transformer differs from the ideal in that it has leakage flux, finite magnetizing inductance, losses, and parasitic capacitance, all of which degrade its performance. Several equivalent circuits for practical transformers will be presented.

The wideband performance of a transformer is mainly determined by the coupling factor, as will be shown. Very wide bandwidths are possible if stacked toroids or balun cores are used.

Although the finite magnetizing inductance and the leakage inductance are undesirable in a wideband transformer, they are put to good use in narrowband matching networks. Several narrowband matching networks using transformers will be examined in detail.

Because it is important to adjust the coupling factor of a transformer to the required value when narrowband matching networks are used, its measurement will also be discussed in this chapter.

4.2 THE IDEAL TRANSFORMER

The equivalent circuit of the ideal transformer is shown in Fig. 4.1. The number of turns on the primary and secondary sides of the transformer are, respectively, n_1 and n_2.

The ideal transformer has the following characteristics:

1) The magnetic flux in the two windings is the same. There is, therefore, no leakage flux.

The voltage induced by the changing flux in each winding is given by Faraday's law:

$$V = n_x \, \partial \Phi / \partial t \tag{4.1}$$

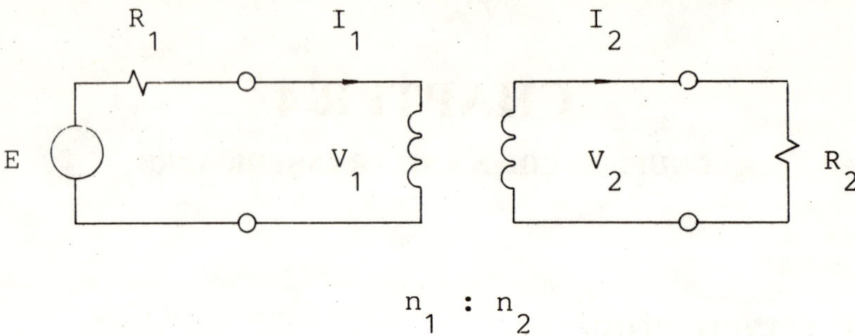

Figure 4.1   The Ideal Transformer

where n_x is the number of turns in the winding under consideration.

Because the flux coupling each winding is the same, the ratio of the primary to secondary voltage of the transformer is

$$V_1/V_2 = n_1/n_2 \tag{4.2}$$

This relationship is more significant if it is written in the form

$$V_1/n_1 = V_2/n_2 \tag{4.3}$$

The voltage per turn is, therefore, the same for both sides of the transformer.

2)   The primary current necessary to establish the flux in the ideal transformer is negligible. The input impedance of the transformer with the load open-circuited is, therefore, infinite.

3)   There are no losses in the ideal transformer.

The average power dissipated in the load is, therefore, exactly the same as the average power entering the transformer.

Because the ideal transformer has no reactive components, the instantaneous power dissipated in the load is also equal to the instantaneous power entering the transformer, that is

$$v_1 i_1 = v_2 i_2 \tag{4.4}$$

By using (4.2) to replace the voltages in this equation, the relationship between the primary and secondary currents is found to be

$$n_1 I_1 = n_2 I_2 \tag{4.5}$$

The demagnetizing force (magnetomotive force) of the current in the secondary winding is, therefore, balanced by that of the current in the primary winding.

By using this equation and (4.3), the relationship between the primary and secondary impedances is found to be

$$Z_1 = V_1/I_1 = [n_1/n_2]^2 \, Z_2 \tag{4.6}$$

The impedance ratio is, therefore, only a function of the turn ratio of the transformer.

4)   The permeability of the ideal transformer is independent of the flux density in the core. This implies that the ideal transformer is a perfectly linear device.

From an impedance-matching viewpoint, the ideal transformer is very useful in that it can be used to scale impedances by any factor.

4.3   EQUIVALENT CIRCUITS FOR THE PRACTICAL TRANS-FORMER

The practical transformer deviates from the ideal in the following ways:

1)   There is some leakage flux and, therefore, leakage inductance.

2)   The magnetizing inductance is finite.

3)   There are losses in the windings of the transformer (copper losses), as well as in the core (hysteresis and eddy current losses).

4)   The relative permeability of the magnetic material changes with signal level and dc current (saturation) as well as with frequency and temperature.

5)   Apart from the effect of the leakage inductance, the high frequency response is degraded by the presence of parasitic capacitance between the windings and turns of each winding.

A circuit model for the practical transformer, ignoring the capacitance and nonlinearities, is shown in Fig. 4.2. The two dots indicate the sides of the two windings which have the same voltage polarity.

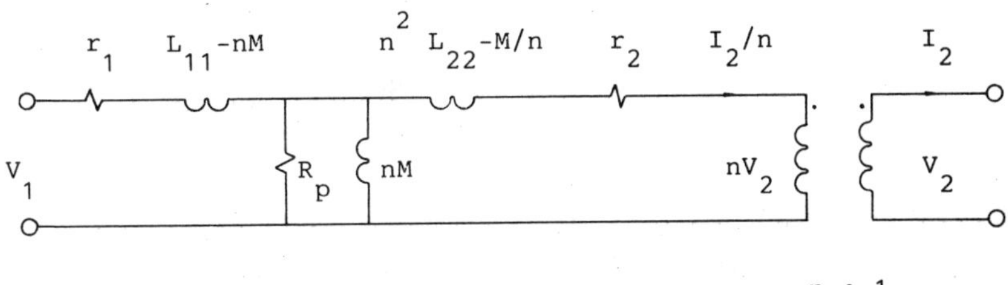

Figure 4.2   An Equivalent Circuit for the Practical Transformer

The two series inductances represent the leakage flux, the series inductance to the left, together with the shunt inductance are the magnetizing inductance, the resistances r_1 and r_2 represent the copper losses, and $\mathbf{R}_p$ the losses in the magnetic material.

The mutual inductance M can be determined from the magnetizing inductances L_{11} and L_{22} by using the relationship

$$M = k \, (L_{11}L_{22})^{1/2} \tag{4.7}$$

where k is the coupling factor of the transformer.

The symbol n can have any arbitrary value, but is usually chosen to be equal to the turn ratio of the two windings of the transformer. However, a better choice for it is

$$n = \frac{1}{k} \sqrt{\frac{L_{11}}{L_{22}}} \tag{4.8}$$

If the losses in the magnetic material can be ignored, the equivalent circuit for the two coupled coils (Fig. 4.3a) can be used for the transformer.

The circuit shown in Fig. 4.3b is equivalent to the coupled coil circuit [3]. That it is equivalent can be established by setting up the Z-parameter matrices for the two circuits.

The transformation ratio shown in Fig. 4.3b is that for the impedances.

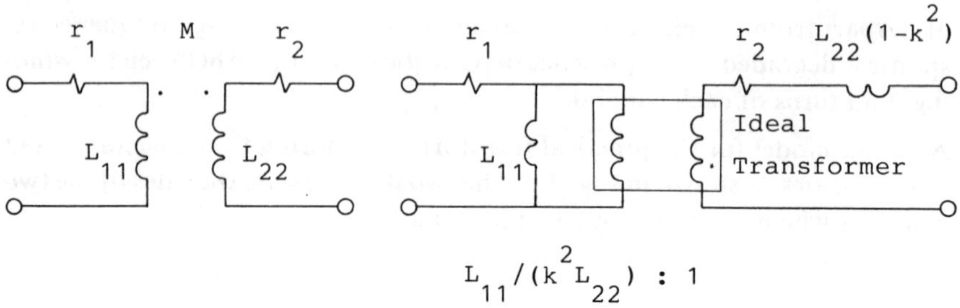

Figure 4.3   Two Equivalent Circuits for the Practical Transformer [R_p neglected]

The following characteristics of the transformer are immediately evident from the equivalent circuit in Fig. 4.3b:

1)   The load impedance is transformed to the primary side of the transformer as

$$Z'_L = [L_{11}/(k^2 L_{22})] \; [Z_L + r_2] \tag{4.9}$$

that is, as long as

$$[\omega L_{22}(1-k^2)]^2 \ll |Z_L + r_2|^2 \tag{4.10}$$

This equation can be changed to a more useful form by substituting L_{22} by using (4.9):

$$\{\omega L_{11}[(1-k^2)/k^2]\}^2 \ll |Z'_L|^2$$

leading to/

$$[1/k^2-1] \ll |Z'_L/(\omega L_{11})|^2 \tag{4.11}$$

This inequality will always apply at low frequencies and if the coupling between the windings is good, it will also apply at higher frequencies.

It follows from (4.9) that the impedance transformation factor at low frequencies is always

$$n^2 = L_{11}/(k^2 L_{22}) \tag{4.12}$$

2)   The high frequency response of the transformer is limited by the leakage inductance $\omega L_{22}(1-k^2)$.

3)   The low frequency responses of the transformer is limited by the magnetizing inductance L_{11}.

4)   The input inductance at low frequencies is the magnetizing inductance L_{11}.

If the frequency is low enough for inequality (4.11) to apply, the equivalent circuit of Fig. 4.3b can be simplified to that shown in Fig. 4.4.

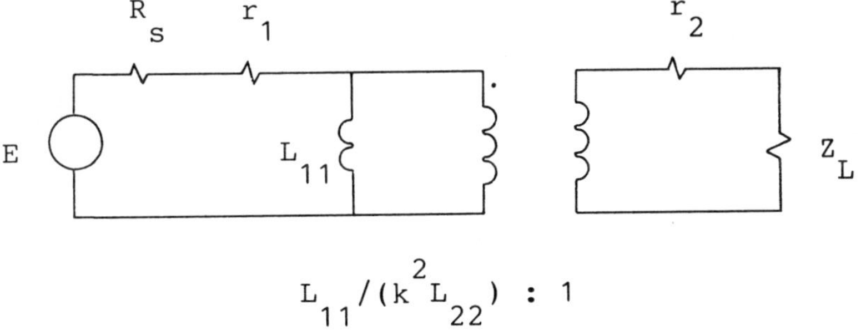

Figure 4.4   An equivalent Circuit for the Practical Transformer at Low Frequencies

By using this equivalent circuit, the low frequency cut-off frequency is found to be

$$\omega_{L3dB} = R_s/(2L_{11}) \tag{4.13}$$

This equation applies if r_1 and r_2 can be ignored, and if the load resistance R_L is transformed to be equal to the source resistance R_s.

Therefore, the required primary inductance can be determined by using the specifications for the cut-off frequency and the source resistance.
The required secondary inductance is given by the equation

$$L_{22} = \frac{R_L}{R_s} \cdot \frac{L_{11}}{k^2} \tag{4.14}$$

This inductance is clearly a function of the coupling factor, which is usually not known at the design stage.

The easiest way to overcome this problem is to ensure that the coupling factor is close to unity, if possible.

Very good coupling can usually be obtained by using materials with high relative permeabilities. The coupling between the windings of the transformer is also better if balun or stacked toroidal cores instead of a single toroidal core are used.

4.4   WIDEBAND IMPEDANCE MATCHING WITH TRANSFORMERS

Impedance matching over very wide bandwidths (decades) can be done by using transformers, if the coupling between the windings is good, and the parasitic capacitance between the windings and turns of each winding is small.

If the parasitic capacitance can be ignored, the 3 dB bandwidth of a transformer can be determined by finding the poles of the equivalent circuit in Fig. 4.5.

This equivalent circuit can be derived from the one in Fig. 4.2 by setting n equal to one.

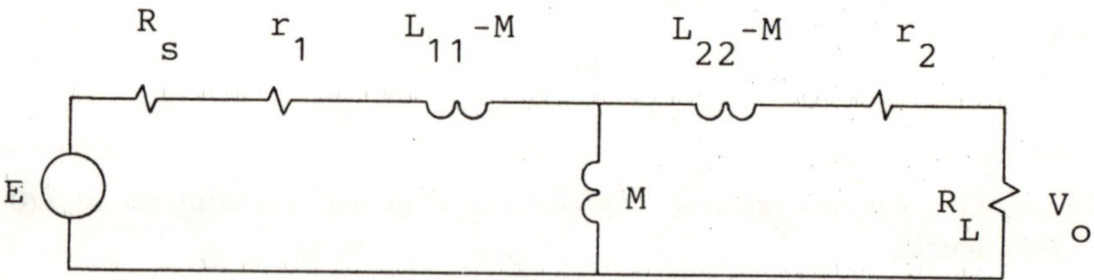

Figure 4.5   A T-Section Equivalent for the Transformer

This circuit has two poles, a zero at the origin and one at infinity.

The poles can be found by setting the impedances in either of the two loops of the equivalent circuit equal to zero. It this is done for the first loop, the following equation is obtained:

$$R_{s+1} + s(L_{11}-M) + sM \parallel [sL_{22} -sM + R_{L+2}] = 0 \tag{4.15}$$

This equation simplifies to

$$s^2 L_{11} L_{22}(1-k^2) + s[L_{22} R_{s+1} + L_{11} R_{L+2}] + R_{s+1} R_{L+2} = 0 \tag{4.16}$$

The poles, and therefore the cut-off frequencies, of the transformer can be obtained by solving this equation.

When the load and source resistance are matched, as in generally the case, the equation

$$R_{L+2} = (k^2 L_{22}/L_{11}) R_{s+1} \tag{4.17}$$

applies.

Because the coupling factor is usually unknown at the outset of the problem and the coupling factor is usually close to unity at low frequencies, the transformer is normally designed for

$$R_{L+2} = (L_{22}/L_{11}) R_{s+1} \tag{4.18}$$

If this value for R_{L+2} is substituted in (4.16), it simplifies to

$$s^2 L_{11} L_{22}(1-k^2) + s2L_{22} R_{s+1} + (L_{22}/L_{11}) R_{s+1} R_{s+1} = 0 \tag{4.19}$$

This equation can be written as

$$[sL_{11}(1+k) + R_{s+1}] [sL_{11} (1-k) + R_{s+1}] = 0 \tag{4.20}$$

The cut-off frequencies are, therefore, given by the equations:

$$\omega_L = R_{s+1}/[L_{11}(1+k)] \tag{4.21}$$

$$\omega_H = R_{s+1}/[L_{11}(1-k)] \tag{4.22}$$

that is when (4.18) applies.

It is clear from these equations that the low cut-off frequency of the transformer is determined by the effective resistance in parallel with the magnetizing inductance and that the high cut-off frequency is a strong function of the coupling factor.

The relative bandwidth of the transformer with the load and source impedances matched is given by

$$\omega_H/\omega_L = [1+k]/[1-k] \tag{4.23}$$

The relative bandwidth of the transformer is, therefore, only a function of the coupling factor, that is, when the load and source impedances are matched and the parasitic capacitance can be ignored.

Because the coupling factor is frequency dependent and close to unity at the lower frequencies, (4.23) can be simplified to

$$\omega_H/\omega_L = 2/[1-k] \tag{4.24}$$

Example 4.1

The primary inductance, secondary inductance, and coupling factor required to transform a load of 50 Ω to 300 Ω with 3 dB cut-off frequencies at 1 MHz and 20 MHz will be determined as an example.

Assuming that the copper losses in the windings and the hysteresis losses in the magnetic core can be ignored, the required inductance ratio of the transformer can be obtained by using (4.18):

$$L_{11}/L_{22} = R_s/R_L = 50/300 = 0.1667$$

The required magnetizing inductance L_{11} can be obtained by using (4.21):

$$L_{11} = 300/[2\pi 1 \cdot 10^6 (1 + 1)] = 23.9 \ \mu H.$$

The secondary inductance is therefore

$$L_{22} = L_{11}/6 = 4.0 \ \mu H.$$

The required coupling factor at 20 MHz can be obtained by using (4.24):

$$1-k = 2/[20/1] = 0.1$$

$$k = 0.9$$

The coupling factor required to obtain a relative bandwidth of 20 is, therefore, 0.9

4.5 THE SINGLE-TUNED TRANSFORMER

The single-tuned transformer shown in Fig. 4.6 can be used to step the load impedance up or down and to obtain a frequency response identical to that of a parallel resonant circuit.

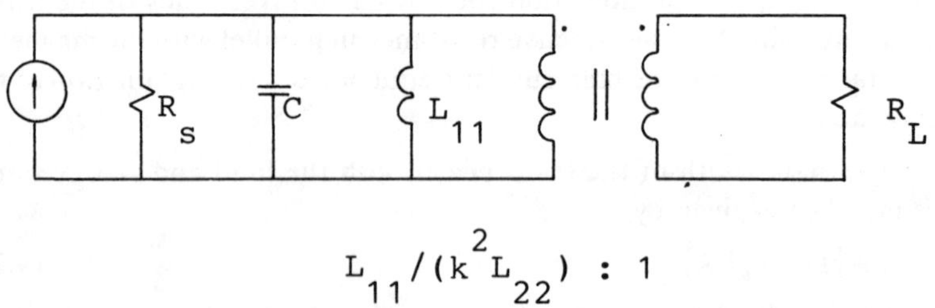

Figure 4.6   The Single-Tuned Transformer

If coupling between the windings of the transformer is good, the leakage inductance of the transformer can be ignored.

The required magnetizing inductance of the transformer can be found in terms of the Q specification, and the source and load resistance, by using the equation:

$$\omega_0 L_{11} = (R_s \| R'_L) / Q = R_s / (2Q) \tag{4.25}$$

The capacitance necessary to provide resonance is

$$C = 1 / (\omega_0^2 L_{11}) \tag{4.26}$$

Because the coupling is assumed to be good, the required secondary inductance is given by the equation:

$$L_{22} = R_L / R_s \cdot L_{11} / k^2 \simeq R_L / R_s \cdot L_{11} \tag{4.27}$$

The next step in the design of the single-tuned transformer is to select a suitable magnetic material and to determine the type and size of the core required (see Ch. 2).

If necessary, the required number of turns around the core can be found by measuring the inductance of a few turns of wire around the core to be used. Because the inductance is proportional to the square of the number of turns, the required number of turns can be found easily.

4.6 THE TAPPED COIL

The tapped coil (Fig. 4.7) is a very useful narrowband matching network. Independent adjustment of the center frequency and the transformation ratio of the transformer are possible. It is also very easy to built once it has been designed.

The frequency response of this transformer is usually similar to that of the single-tuned transformer.

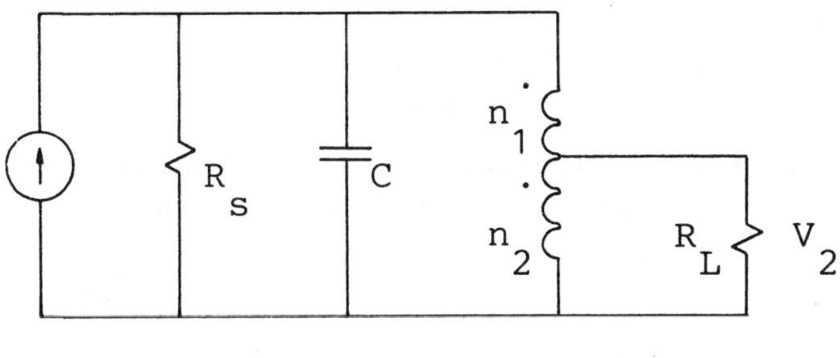

Figure 4.7   The Tapped Coil Resonant Circuit

Analysis of the tapped coil is possible by considering it to be two coupled coils: The inductance of the upper section of the coil is given by

$$L_1 = rn_1^2\, 10^{-6} \left/ \left[22.9\, \frac{l_T}{r}\, \frac{n_1}{n_T} + 25.4 \right] \right.$$

(4.28)

and that of the lower section by

$$L_2 = rn_2^2\, 10^{-6} \left/ \left[22.9\, \frac{l_T}{r}\, \frac{n_2}{n_T} + 25.4 \right] \right.$$

(4.29)

The inductance of the coil as a whole is given by the equation:

$$L_T = rn_T^2\, 10^{-6} \left/ \left[22.9\, \frac{l_T}{r} + 25.4 \right] \right.$$

(4.30)

If viewed in terms of its component parts, the total inductance can also be written as

$$L_T = L_1 + L_2 + 2M$$

(4.31)

By using this equation together with the fact that

$$M = k\, (L_1 L_2)^{1/2}$$

(4.7)

the coupling factor is found to be

$$k = [L_T - L_1 - L_2] / [2(L_1 L_2)^{1/2}]$$

(4.32)

By substituting (4.28) through (4.31) into (4.32) it can be deduced that

1. The coupling factor is dependent on the length-to-radius ratio of the coil, as well as the relative position of the tap-point.

2. The coupling factor is independent of the total number of turns (n_T).

The coupling factor of the tapped coil is given in Table 4.1 as a function of the relative position of the tap-point for several l/r ratios. The coupling factors are not shown for relative positions of greater than 0.5, since it is the mirror image (arithmetic) of the lower values (0.8 corresponds to 0.2).

TABLE 4.1

The Coupling Factor of the Tapped Coil as a Function of the l/r Ratio of the Coil and the Relative Position of the Tap-Point

n_1/n_T	$l/r=1$	$l/r=1.5$	$l/r=2$	$l/r=3$	$l/r=4$	$l/r=5$
0.1	0.543	0.449	0.386	0.304	0.253	0.218
0.2	0.535	0.438	0.372	0.288	0.235	0.200
0.3	0.530	0.431	0.363	0.277	0.225	0.189
0.4	0.529	0.426	0.358	0.272	0.219	0.183
0.5	0.526	0.425	0.357	0.270	0.217	0.182

It can be seen by inspection of Table 4.1 that the coupling factor is not a strong function of the relative position of the tap-point.

The input admittance of the coupled coil can be found by using the equation:

$$\begin{bmatrix} j\omega L_{1+2+2M} & -j\omega L_{2+M} \\ -j\omega L_{2+M} & j\omega L_2 + R_L \end{bmatrix} \begin{bmatrix} I_1 \\ I_2 \end{bmatrix} = \begin{bmatrix} E \\ 0 \end{bmatrix} \tag{4.33}$$

By using Cramer's rule, the input admittance of the tapped coil is found to be

$$Y_{in} = I_1/E$$

$$= \begin{bmatrix} 1 & -j\omega L_{2+M} \\ 0 & j\omega L_2 + R_L \end{bmatrix} \Big/ \Delta$$

$$= (R_L + j\omega L_2)/[j\omega L_{1+2+2M}(R_L + j\omega L_2) - j\omega L_{2+M} \cdot j\omega L_{2+M}]$$

With some manipulation this equation can be changed to

$$Y_{in} = \frac{L_2 L_T R_L + L_1 L_2 (k^2-1) R_L}{\omega^2 L_1^2 L_2^2 (k^2-1)^2 + L_T^2 \cdot R_L^2} - j\, \frac{L_T R_L^2 - \omega^2 L_1 L_2 (k^2-1)}{\omega^3 L_1^2 L_2^2 (k^2-1)^2 + \omega L_T^2 R_L^2} \tag{4.34}$$

If the coupling is perfect, this equation reduces to

$$Y_{in} = \frac{L_2}{L_T} \Big/ R_L - j/(\omega L_T) \tag{4.35}$$

as expected.

When the coupling is not unity, the resistive part of the input admittance will be frequency independent, if

$$\omega^2 L_1^2 L_2^2 (k^2-1)^2 \ll L_T^2 R_L^2$$

that is, if

$$[\omega L_1 L_2 (1-k^2)]^2 \ll (L_T R_L)^2 \tag{4.36}$$

When this is true, the input conductance will be

$$G'_L \doteq \frac{L_2 L_T R_L + L_1 L_2 (k^2-1) R_L}{L_T^2 R_L^2}$$

$$= \frac{L_2}{L_T} G_L - \frac{L_1 L_2 (1-k^2)}{L_T^2} G_L \tag{4.37}$$

It can be seen by inspection of (4.34) that the actual conductance of the coil will be 10% higher than the value predicted by this equation, if

$$[\omega L_1 L_2 (k^2-1)]^2 = 0.1 (L_T R_L)^2$$

For the parallel input inductance to be approximately L_T, it is necessary that

(4.36) as well as the following equation apply:

$$\omega^2 L_1 L_2 (1-k^2) \ll L_T R_L^2 \tag{4.38}$$

The equivalent circuit of the tapped coil can therefore be simplified to the circuit shown in Fig. 4.8, if these two equations apply.

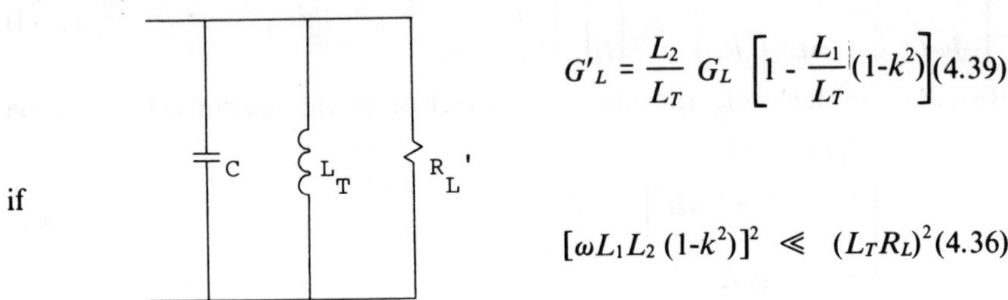

$$G'_L = \frac{L_2}{L_T} G_L \left[1 - \frac{L_1}{L_T}(1-k^2) \right] \tag{4.39}$$

if

$$[\omega L_1 L_2 (1-k^2)]^2 \ll (L_T R_L)^2 \tag{4.36}$$

Figure 4.8   A Simplified Equivalent Circuit for a Tapped Coil Resonant Circuit

It should be noted that from a practical viewpoint, it is more important that (4.36) rather than (4.38) applies, for it is easy to change the resonant frequency of the circuit to the desired frequency if the inductance is not exactly equal to L_T. This can be done by stretching or compressing the coil slightly. (The coupling factor is only a weak function of the l/r ratio of the coil).

The transformation factor of the load (as given by (4.39)) is a function of the inductance L_T, the ratio L_1/L_T, L_2/L_T and the coupling factor k. The last three variables are all determined by the relative position of the tap-point and the l_T/r ratio of the coil. The transformation factor itself is, therefore, only a function of the l/r ratio of the coil and the relative position of the tap-point.

The transformation factor for different l/r ratios and .positions of the tap-point was determined and is tabulated in Table 4.2. This table is only applicable when inequality (4.36) applies. When it does not, the transformation factor can be determined by using (4.28) to (4.30), (4.32) and (4.34).

It can be seen by inspection of Table 4.2, that the transformation factor of the tapped coil is only a weak function of the l/r ratio of the coil. It is also evident that the transformation factor becomes very sensitive when the coil is tapped close to its ground point.

A tapped coil resonant circuit can be designed to match a source to a load with a specified Q by following the procedure outlined below.

Procedure for Designing a Tapped-Coil Resonant Circuit

1)   Assume that the input inductance of the tapped coil will be equal to the inductance (L_T) of the coil itself (refer to Fig. 4.7 and 4.8).

Find the required inductance by using the equation:

$$\omega L_T = [R' \| R_s]/Q = R_s/[2Q] \tag{4.25}$$

This equation will only apply if the unloaded Q-factor of the inductor and capacitor used are high compared to the specified circuit Q.

2) Design the inductor (L_T) to have the required unloaded Q as described in Ch. 2.

Check if the self-capacitance of the designed coil is lower than the capacitance required for resonance.

At this stage L_T, l_T, r, and n_T are known.

3) The next step is to determine the relative position of the tap-point. If the inequality

$$[\omega L_1 L_2 (1-k^2)]^2 \ll (L_T R_L)^2 \tag{4.36}$$

applies, Table 4.2 can be used for this purpose. If it does not apply, then it will be easier to determine the tap-point practically.

Example 4.2

A tapped coil resonant circuit will be designed to meet the following specifications, if possible: $f_0 = 10$ MHz, $R_L = 50\Omega$, $R_s = 1000\Omega$, $Q = 20$, and $IL < 10\%$.

TABLE 4.2
The Transformation Factor of the Tapped Coil as a Function of the l_T/r Ratio and the Relative Position of the Tap-point from the Upper End of the Coil.

n_1/n_T	R'_L/R_L		
	$l_T/r=2$	$l_T/r=4$	$l_T/r=6$
0.10	1.18	1.18	1.18
0.20	1.46	1.46	1.48
0.30	1.92	1.92	1.94
0.40	2.66	2.68	2.70
0.50	4.00	4.00	4.00
0.55	5.08	5.07	5.05
0.60	6.64	6.62	6.56
0.65	9.04	8.99	8.86
0.70	12.90	12.80	12.60
0.75	19.70	19.60	19.10
0.80	33.00	33.00	32.10
0.85	64.00	64.70	62.80
0.90	160.00	166.00	162.00
0.95	732.00	505.00	808.00

In order to have less than 10% losses, the loss resistance must be (approximately) higher than 10 times the transformed load resistance, that is

$$R_{loss} = 10 \cdot 1 \cdot 10^3 = 10k\Omega$$

This implies that the unloaded Q of the inductor must be at least

$$Q_u = R_{loss}/(\omega L) = 10k/25 = 400.$$

It is assumed here that the losses in the capacitor can be ignored. The required Q is, however, high in this design, and it will be necessary to use a very good capacitor and even a parallel combination of smaller values to obtain a Q which is high compared to that of the inductor.

Assuming the l_T/r ratio of the coil to be equal to 2 and a d/c ratio of 0.55, the required radius of the coil is found to be

$$r = Q_u/[k(f)^{1/2}] = 1.26 \text{ cm} \tag{2.25}$$

The required length of the coil is 2.53 cm ($l_T/r=2$).

The number of turns can be found by using (2.27):

$$N = [L(22.9l/r + 25.4)/r]^{1/2} = 4.75 \tag{2.27}$$

The required wire thickness is

$$d = l/D \cdot d/c \cdot 2r/(N\text{-}1) = 0.37 \text{ cm} \tag{2.28}$$

Because the thickness of No. 12 SWG wire is 0.264 cm, required wire thickness is unrealistic and a Q of 400 can, therefore, not be obtained with standard available wire.

If the radius of the coil is decreased to

$$r_2 = 1.01 \text{ cm}$$

the required wire thickness will be 0.26 cm.

The ratio of the former radius (r_1) to that of the required radius was obtained iteratively by using the equation:

$$\frac{r_1}{r_2}\left[\sqrt{\frac{r_1}{r_2}} - 1/N_1\right] = \frac{d_1}{d_2}\left[1 - 1/N_1\right] \tag{4.40}$$

This equation can be derived easily from (2.27) and (2.28).

The number of turns required is now

$$N_2 = [r_1/r_2]^{1/2} \cdot N_1 = 5.3 \tag{4.41}$$

Because of the reduction in radius the unloaded Q of the inductor will decrease to 317. The insertion loss will, therefore, increase to approximately 13%.

The parasitic capacitance of the coil (0.51pF) is much smaller than the capacitance required to provide resonance (637pF).

With the coil designed and realizable, the tap-point can be determined. This can be done by using Table 4.2, that is, if inequality (4.36) applies.

It follows by inspection of the column for $l_T/r=2$ that the coil must be tapped where

$$N_1/N_T = 0.75$$

that is, where

$$N_1 = 4.0.$$

In order to establish whether the inequality does apply, it is necessary to calculate the values of the inductances L_1 and L_2, and to determine the coupling factor of the tapped coil:

$$L_1 = rN_1^2 \, 10^{-6} / \left[22.9 \, \frac{l_T}{r} \, \frac{N_1}{N_T} + 25.4 \right] \tag{4.28}$$
$$= 0.269 \, \mu H$$

By using (4.29), the value of L_2 is found to be

$$L_2 = 46.6 \text{ nH}$$

The value of the coupling factor can be determined from Table 3.1 (or by using (4.32)). Its approximate value is 0.37.

Because

$$[\omega L_1 L_2 \, (1-k^2) \,]^2 = (462.1)10^{-15}$$

and

$$[L_T R_L]^2 = (400.0)10^{-12}$$

inequality (4.36) does apply and the tap-point as determined from Table 4.2 will be accurate.

4.7 THE PARALLEL DOUBLE-TUNED TRANSFORMER

The narrowband circuits discussed up to this point match the load conjugately to the source at a single frequency only. With the double-tuned transformer shown in Fig. 4.9, it is possible to match the source to the load at two different frequencies. If these frequencies are close enough to each other, the ripple in the pass band can be designed to be small.

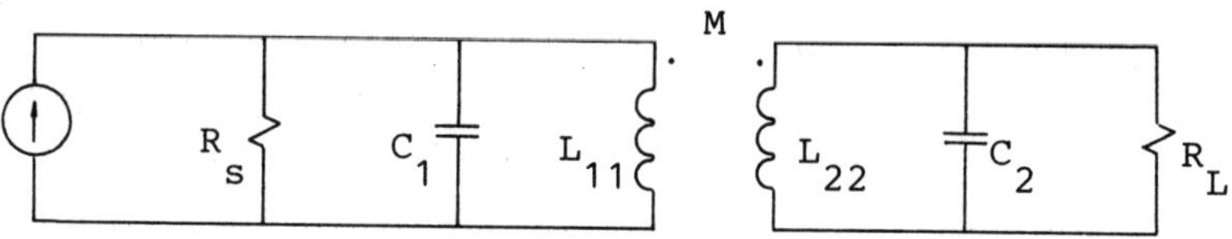

Figure 4.9   The Parallel Double-Tuned Transformer

The shape of the frequency response of the double-tuned transformer is shown in Fig. 4.10 for different values of the coupling factor. When the coupling factor is smaller than a certain critical value (k_c), the source cannot be matched to the load.

The rejection characteristics of the double-tuned transformer are superior to those of a simple parallel resonant circuit. The difference becomes very pronounced when the circuit Q becomes high.

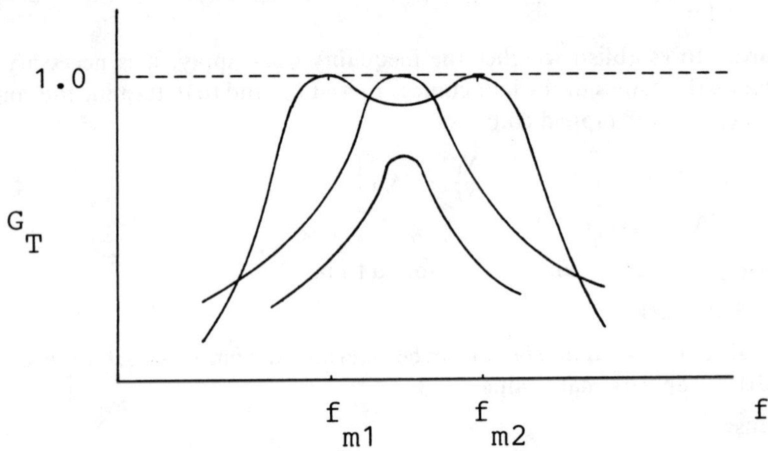

Figure 4.10   Three Typical Responses of the Double-Tuned Transformer

The transformer can be analyzed by using the equivalent circuits shown in Fig. 4.11.

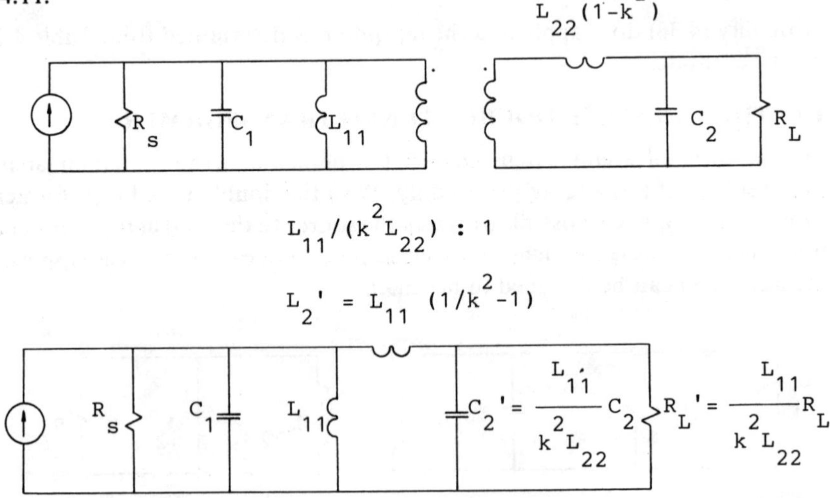

Figure 4.11   Two Equivalent Circuits for the Double-Tuned Transformer

It can be seen that it is possible for circuit to match the load to the source at two different frequencies, if it is split up as shown in Fig. 4.12 (or Fig. 4.13).

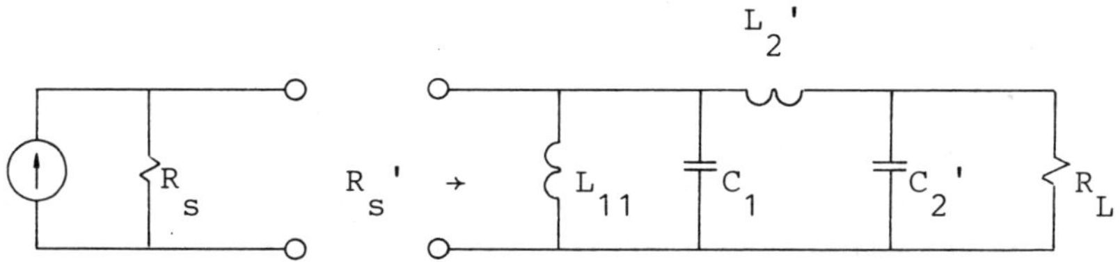

Figure 4.12   An Equivalent Circuit for Determining the Frequencies where Maximum Power Is Transferred

In Fig. 4.12, L_2' and C_2' are transforming elements, while L_{11} and C_1 are compensating elements.

The two transforming elements can match the load resistance to R_s at two different frequencies, while the two compensating elements can be designed to cancel the residual reactances at these frequencies.

Equations for the ripple in the pass band of the double-tuned transformer (see Fig. 4.10) can be established by breaking up the equivalent circuit as shown in Fig. 4.13.

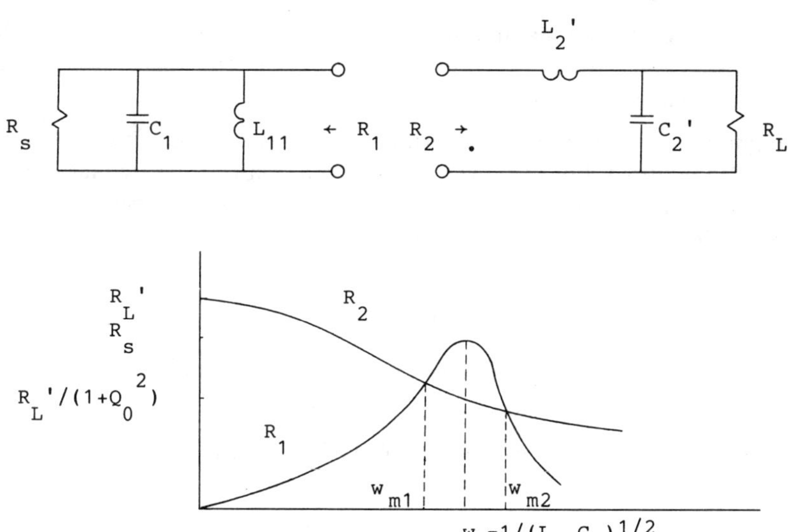

Figure 4.13   The Series Resistance of the Two Impedance Matching Sections as a Function of Frequency

The capacitor and inductor of the left-hand matching section cause the series resistance R_1 to decrease at high and low frequencies, respectively. R_1 is equal to the source resistance R_s at the resonant frequency of the inductor and capacitor.

The capacitor in the right-hand section causes the series resistance R_s to decrease monotonically with increasing frequency.

If R'_L is small enough there will be two frequencies where the series resistance of the left-hand section is equal to that of the right-hand section.

Because these resistances are not equal in the pass band, there must inevitably be some ripple in the pass band.

The size of the ripple is a function of the ratio of the source resistance R_s to the series resistance of the right-hand section at the resonant frequency $\omega_O = 1/\sqrt{L_{11}C_1}$. The reactance of the two sections can be ignored when this calculation is done.

The double-tuned transformer can be designed to match the load to the source at two specified frequencies with a specified amount of ripple in the pass band if the theory presented in Ch. 3 is applied to the equivalent circuits shown in Fig. 4.12 and 4.13.

The transformer can be designed as described below.

Design Procedure for the Double-Tuned Transformer

Specifications: The two maximum power transfer frequencies f_{m1} and f_{m2}, the load resistance R_L, the source resistance R_s, and the transducer power gain (G_T) at the center frequency.

As might be expected, the minimum ripple in the pass band is a function of the bandwidth required. For two given maximum power transfer frequencies f_{m1} and f_{m2}, the minimum value of the transducer power gain in the pass band, will always be lower than

$$G_{T,\ min} = \frac{4f_{m2}/f_{m1}}{(f_{m2}/f_{m1})^2 + 2f_{m2}/f_{m1} + 1} \tag{4.42}$$

The first step in the design process, therefore is, to establish whether the specified $G_{T,\ min}$ can be realized.

2)   Calculate the resistance ratio r $(r = R_s/[R'_L(1 + Q_0^2)])$ for the required pass band ripple:

$$r = \frac{1 + |1 - G_T|^{1/2}}{1 - |1 - G_T|^{1/2}} \tag{4.43}$$

3)   Calculate the value of Q_2 (that is, the second transformation Q of the

transforming section in Fig. 4.12) at the two desired maximum power transfer frequencies:

$$Q^2_{2-m1} = r\, f_{m1}/f_{m2} - 1 \tag{4.44}$$

$$Q^2_{2-m2} = r\, f_{m2}/f_{m1} - 1 \tag{4.45}$$

4) Solve the following two equations for the values of L'_2 *and* C'_2:

$$-\omega_{m1}L'_2 + (1/C'_2)/\omega_{m1} = |\,Q_{2-m1}\,|\; R_s/[\; 1+Q^2_{2-m1}\;] \tag{4.46}$$

$$-\omega_{m2}L'_2 + (1/C'_2)/\omega_{m2} = |\,Q_{2-m2}\,|\; R_s/[\; 1+Q^2_{2-m2}\;] \tag{4.47}$$

5) Calculate the value of R'_L:

$$R'_L = (1+Q^2_{2-m1})/(\omega^2_{m1} C'^2_2 R_s) \tag{4.48}$$

6) Calculate the value of the input susceptance (B_{m1}, B_{m2}) of the right-hand section in Fig. 4.13 at the maximum power transfer frequencies f_{m1} and f_{m2}.

These susceptances are given by the equations:

$$B_{m1} = Imag\;\cfrac{1}{j\omega_{m1}\,L'_2 + \cfrac{1}{G'_2 + j\omega_{m1}\,C'_2}} \tag{4.49}$$

$$B_{m2} = Imag\;\cfrac{1}{j\omega_{m2}\,L'_2 + \cfrac{1}{G'_2 + j\omega_{m2}\,C'_2}} \tag{4.50}$$

7) Solve the following two equations for the values of L_{11} and C_1 :

$$(1/L_{11})/\omega_{m1} - \omega_{m1}\,C_1 = |\,B_{m1}\,| \tag{4.51}$$

$$(1/L_{11})/\omega_{m2} - \omega_{m2}\,C_1 = -|\,B_{m2}\,| \tag{4.52}$$

8) Calculate the values of k, L_{22}, and C_2 by using the following equations:

$$k \;\; = 1/[1+L'_2/L_{11}]^{1/2} \tag{4.53}$$

$$L_{22} = [L_{11}/k^2]\, R_L/R'_L \tag{4.54}$$

$$C_2 \;\; = [L_{11}/(k^2 L_{22})]C'_2 \tag{4.55}$$

9) If the components of the transformer turn out to be unrealizable, it is often possible to obtain a practical circuit by impedance scaling of the results. The required transformation of the load and source resistances can then be obtained by using L-sections.

The alternative is to design a series double-tuned transformer.

10) Check the insertion loss and the frequency response by following the procedures outlined in sec. 3.8 and 3.9.

Example 4.3

The procedure outlined above was followed to design a double-tuned transformer with perfect matching at 9 MHz and 11 MHz, and pass-band ripple of less than 0.5 dB ($G_T = 10^{-0.5/10} = 0.89$). The load resistance is 50Ω and the source resistance is 500Ω.

The results were the following:

1. $G_{T,\ min}$ = 0.99

2. r = 1.9925

3. $Q_{2-9\ MHz}^2$ = 0.630;   ($Q_{2-9\ MHz}$ = 0.793)

   $Q_{2-11\ MHz}^2$ = 1.435;   ($Q_{2-11\ MHz}$ = 1.198)

4. L_2' = 19.5μH

   C_2' = 13.2pF

5. R_L' = 5.9kΩ

6. B_{m1} = 1.49mS

   B_{m2} = −2.36mS

7. C_1 = 157pF

   L_{11} = 1.71μH

8. k = 0.284

   L_{22} = 0.18μH

   C_2 = 1.55nF

9. The transducer power gain of the double-tuned transformer is given in Table 4.3 as a function of frequency. (The components were assumed to be lossless).

Table 4.3

The Transducer Power Gain of the Double-Tuned Transformer as a Function of Frequency

Frequency (MHz)	G_T (dB)
4.00	−23.00
5.00	−19.00
7.50	−5.50
8.00	−2.40
9.00	−0.06
9.95	−0.53
11.00	−0.05
12.00	−3.70
15.00	−19.00
16.00	−23.00

Example 4.4

As an illustration of the extremely good rejection characteristics of the double-tuned transformer, the −30 dB quality factor of a double-tuned transformer with maximum power transfer frequencies at 5.97 MHz and 6.03 MHz (0.1 dB ripple) was determined.

The −30 dB cut-off frequencies for the designed transformer are approximately 5.915 MHz and 6.082 MHz. The resulting −30 dB Q-factor, therefore, is

$$Q_{-30} = 36$$

If the same results were to be obtained with a simple parallel resonant circuit, the required 3 dB Q-factor would have been 1026! The highest transformation Q in the double-tuned transformer is 60.

The −3 dB Q-factor for the double-tuned transformer is approximately equal to 7. The ratio of the −3 dB and −30 dB Q-factors of the transformer, therefore is

$$Q_{-30} / Q = 36/7 = 5.1.$$

This ratio is 145 times that of the equivalent simple parallel resonant circuit.

4.8   THE SERIES DOUBLE-TUNED TRANSFORMER

Although it is not the dual of the parallel double-tuned transformer, the response of the series double-tuned transformer shown in Fig. 4.14 is similar to that of the parallel double-tuned transformer.

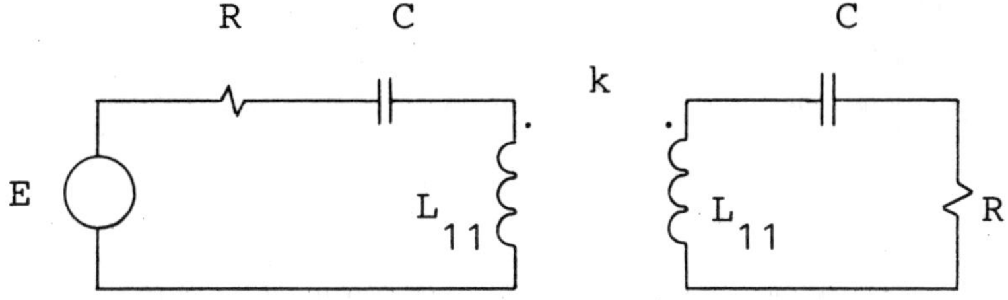

Figure 4.14   The Series Double-Tuned Transformer

The series circuit will have less insertion losses than the equivalent parallel circuit when the load and source resistances are low.

Impedance transformation can be obtained with the circuit shown in Fig. 4.14 by scaling the components on each side of the transformer to the required levels while the coupling factor remains unchanged.

The series tuned transformer can be designed to match the source conjugately to the load at two specified frequencies by following the procedure outlined below.

Design Procedure for the Series Double-Tuned Transformer

Specifications: The two maximum power transfer frequencies f_{m1} and f_{m2}, the load and source resistances, and the transducer power gain (G_T) at the center frequency.

1) Calculate the value of the product kQ:

$$kQ = 1/\sqrt{G_T} + \sqrt{\frac{1}{G_T} - 1} \tag{4.56}$$

k is the required coupling factor and Q is the quality factor of each of the two sides of the transformer when the coupling between them is equal to zero.

2) Determine the coupling factor required by solving the following equation:

$$k^4 + k^2 [(kQ)^4 M^2 - 4 (kQ)^2] + [4 - M^2] (kQ)^4 = 0 \tag{4.57}$$

where

$$M = \frac{f_{m1}^2 + f_{m2}^2}{f_{m1} f_{m2}} \tag{4.58}$$

3) Determine the required uncoupled Q-factor and the circuit components:

$$Q \;\; = (kQ)/k \tag{4.59}$$

$$f_0 \;\; = [f_{m1} f_{m2}(1 - k^2)^{1/2}]^{1/2} \tag{4.60}$$

$$C_1 \;\; = 1/[\omega_0 R_s Q] \tag{4.61}$$

$$L_{11} = Q R_s / \omega_0 \tag{4.62}$$

$$L_{22} = Q R_L / \omega_0 \tag{4.63}$$

$$C_2 \;\; = 1/ [\omega_0 R_L Q] \tag{4.64}$$

If some of these components are unrealizable, the specifications can be changed to be more realistic, or, in some cases, the components can be scaled to more realistic values. In the latter case, the source and load terminations can be transformed to the required values by using L-sections.

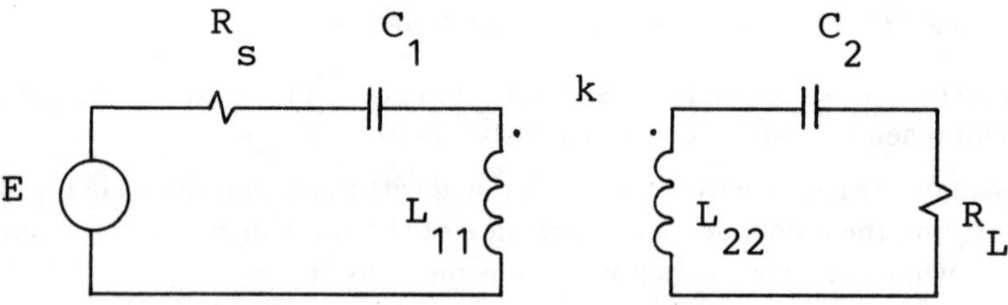

Figure 4.15 *The Equivalent Circuit of the Series Double-Tuned Transformer When the Load and Source Resistance Are Different*

4) The insertion loss and frequency response of the transformer can be determined by following the procedure outlined in sec. 4.8 and 4.9.

The 3 dB bandwidth of the series double-tuned transformer can be calculated by using the equation:

$$B = \sqrt{b^2 + 2b - 1}\ \ f_0/Q \qquad\qquad (4.65)$$

where

$$b = k/k_c \qquad\qquad (4.66)$$

and

$$k_c = 1/Q \qquad\qquad (4.67)$$

Example 4.5

As an example of the application of the procedure outlined above, a transformer was designed to have maximum power transfer frequencies at 27 MHz and 28 MHz. The other specifications were $G_T = 0.89$, $R_s = 50\Omega$ and $R_L = 20\Omega$.

The results are:

1. $kQ = 1.41156$

2. $M = 2.00132$

   $k^4 + k^2\ 7.93123 - 0.02101 = 0$

   $k = 0.05146$

3. $Q = 27.43$

   $f_0 = 27.48\ \text{MHz}$

   $C_1 = 4.2\text{pF}$

   $L_{11} = 7.94\mu\text{H}$

   $L_{22} = 3.18\mu\text{H}$

   $C_2 = 10.6\text{pF}$

4. Assuming the transformer to be lossless, the −3 dB bandwidth of the transformer is

   $B = 1.97\ \text{MHz}$

The −3 dB and −30 dB Q-factors of the transformer are

   $Q = 13.98$

   $Q_{-30} = 2.81$

The ratio of the two Q-factors is

   $Q_{-30}/Q = 0.2$

4.9 MEASUREMENT OF THE COUPLING FACTOR OF A TRANS-FORMER

Three ways to determine the coupling factor of a transformer will be discussed.

In the first method, the open- and short-circuit input inductances of the transformer are measured. This method can only be used when the losses in the transformer can be ignored.

In the second method, the open-circuit voltage gain of the transformer is measured. An oscilloscope or voltmeter with a high input impedance, as compared to the leakage reactance of the transformer at the measuring frequency, is required if reliable results are to be obtained with this method.

The Z-parameters of the transformer are used in the last method. It is usually more convenient to measure the S-parameters of the transformer. These parameters can be converted easily to Z-parameters by using (1.101).

4.9.1 Measurement of the Coupling Factor by Short-Circuiting the Secondary Winding of the Transformer .

If the resistive losses and the parasitic capacitance of the transformer can be ignored, the equivalent circuit of Fig. 4.16 can be used for the transformer.

If the secondary winding of the transformer is short-circuited, the input admittance of the transformer is given by the equation:

$$(1/(\omega L_T)) = 1/(\omega L_{11}) + 1/[L_{11}/(k^2 L_{22}) \cdot \omega L_{22} (1-k^2)] = 1/(\omega L_{11})$$
$$+ 1/[\omega L_{11} (1/k^2-1)] \tag{4.68}$$

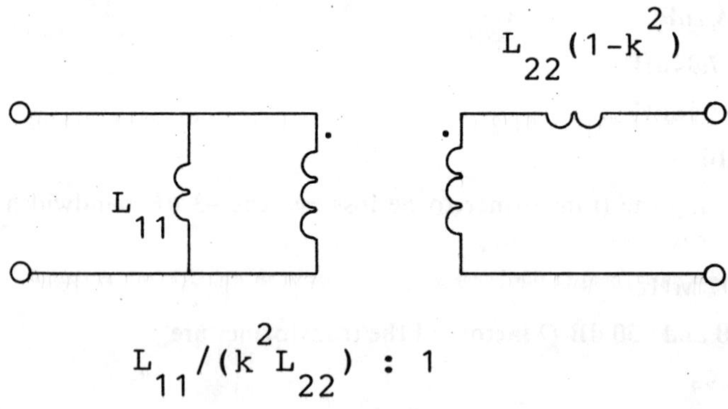

Figure 4.16   *The Equivalent Circuit of the Transformer When The Losses Can Be Ignored*

This equation can be rewritten to obtain the value of the coupling factor as a function of the other variables:

$$k = [1 - L_T / L_{11}]^{1/2} \tag{4.69}$$

It is therefore possible to determine the coupling factor of the transformer by measuring the open- and short-circuit impedances of the transformer (that is, if the losses in the transformer can be ignored).

4.9.2 Measurement of the Coupling Factor by Measuring the Open-Circuited Voltage Gain of the Transformer

When the equivalent circuit of Fig. 4.17 applies, the open-circuit voltage gain of the transformer is given by the equation:

$$V_2 = \pm \left[j\omega k \, (L_{11} L_{22})^{1/2} \right] V_1 / (r_1 + j\omega L_{11}) \tag{4.70}$$

From this equation the coupling factor can be obtained as

$$k = \left[(r_1^2 + \omega^2 L_{11}^2) / (\omega^2 L_{11} L_{22}) \right]^{1/2} \cdot V_2 / V_1 \tag{4.71}$$

Therefore, the coupling factor can be determined by measuring the open-circuit voltage gain of the transformer. We can determine r_1 and the inductances L_{11} and L_{22} by measuring the open-circuited primary and secondary input impedances of the transformer.

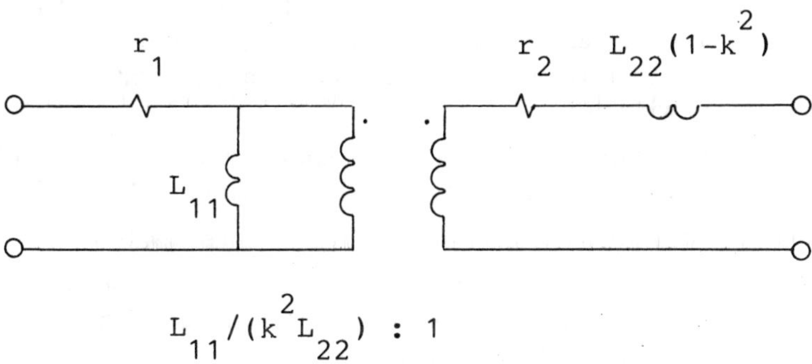

Figure 4.17   The Equivalent Circuit of the Transformer with the Parasitic Capacitance and Core Losses Ignored

It is important for the input impedance of the voltmeter or oscilloscope used to measure the transformer voltage gain to be much higher than the leakage reactance $\omega L_{22} (1 - k^2)$. Because of this, it is sensible to define the low-impedance side of the transformer as the secondary side.

4.9.3 Measurement of the Coupling Factor by Measuring the S-Parameters of the Transformer

If the equipment is available, the transformer S-parameters can be measured

     Coupled Coils and Transformers

easily. These parameters can be converted to Z-parameters by using standard conversion formulas.

If the equivalent circuit of Fig. 4.17 applies, the transformer Z-parameters are given by the equation:

$$\overline{z} = \begin{bmatrix} r_1 + j\omega L_{11} & \pm j\omega M \\ \pm j\omega M & r_2 + j\omega L_{22} \end{bmatrix} \tag{4.72}$$

With the Z-parameters known, it is a simple matter to determine the copper losses as well as the primary and secondary inductances.

The coupling factor can be determined from the mutual inductance M and the magnetizing inductances L_{11} and L_{22} by using (4.7).

Questions and Problems

1. Prove that the load of an ideal transformer is transformed by a factor equal to the square of the turn ratio of the transformer.

2. Prove that $L_{11}/L_{22} = (n_1/n_2)^2$ if the primary and secondary windings of a transformer are wound around a highly permeable toroidal core.

3. The low cut-off frequency (3 dB) of a transformer is to be 2 MHz. The load and source resistances are 400Ω and 50Ω, respectively. What is the value of the magnetizing reactance required? If the coupling factor can be assumed to be equal to one, what is the required inductance of the secondary winding?

4. What is the minimum coupling factor required to match a load of 100Ω to 10Ω over the frequency range 2-30 MHz (that is, 3 dB bandwidth)?

5. Prove that the two circuits shown in Fig. 4.3 are equivalent.

6. Explain why it is that the coupling factor of a transformer with a core of stacked toroids (or a balun core) is higher than that of an equivalent transformer with a single toroidal core.

7. Design a single-tuned transformer to match a load of 150Ω to a source with 50Ω internal resistance. The Q of the circuit must be 10 and $f_0 = 100$ MHz.

8. Design a tapped-coil resonant circuit to transform a load of 1000Ω to 50Ω. The required Q of the circuit is 25 and $f_0 = 10$ MHz. The insertion losses of the circuit must be as low as practically possible.

9. Design the filter shown in Fig. 4.18 to have the highest Q practically possible. The resonant frequency must be 10 MHz and the space available for the inductor is 4cm$\times$4cm$\times$4cm. High Q porcelain capacitors may be used. (Note that the Q-factors of these capacitors are not necessarily negligible.)

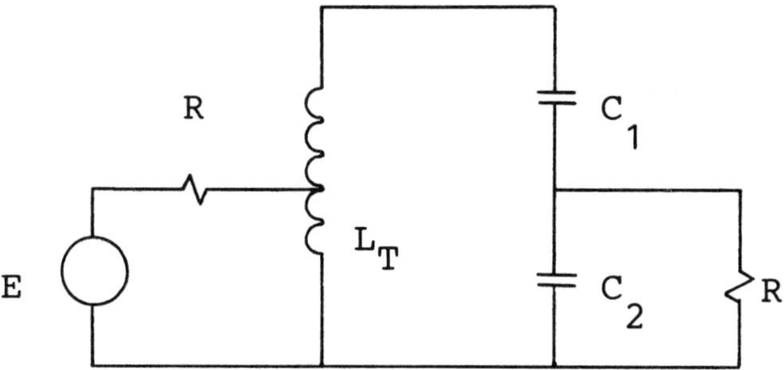

Fig. 4.18   A High Q Tapped-Coil Resonant Circuit

10. Design a double-tuned transformer to match a load of 250Ω conjugately to a source of 1000Ω at 8.8 MHz and 9.6 MHz. The maximum ripple in the pass band must be 1 dB. Determine the –3 dB and –30 dB frequencies of the transformer iteratively.

11. Design the double-tuned transformer to match a load of 50Ω to a source with 20Ω internal resistance. The maximum power transfer frequencies must be 89 MHz and 94 MHz. The maximum ripple in the pass band must be less than 0.25 dB.

    Determine the –3 dB and –30 dB frequencies of the designed transformer.

12. What is the smallest ripple in the pass band of the parallel double-tuned transformer when the ratio of the two maximum power transfer frequencies is

    (a)   1.1,
    (b)   1.25,
    (c)   1.50,
    (d)   1.75,
    (e)   2.00?

13. Determine the insertion losses of the transformers designed in *Problems 4.10* and *4.11*.

14. The transformer in *Problem 4.10* is to be replaced with three discrete inductors (refer to Fig. 4.5). Calculate the values of the inductances required.

15. Find the dual of the series and parallel double-tuned circuits.

16.   Use the knowledge gained in the previous problem and the standard equations available for transforming a Π-section to an equivalent T-section to design the network shown in Fig. 4.19 so as to have maximum power transfer frequencies at 9.75 MHz and 10.25 MHz. $R_L = 250\,\Omega$; $R_s = 500\,\Omega$.

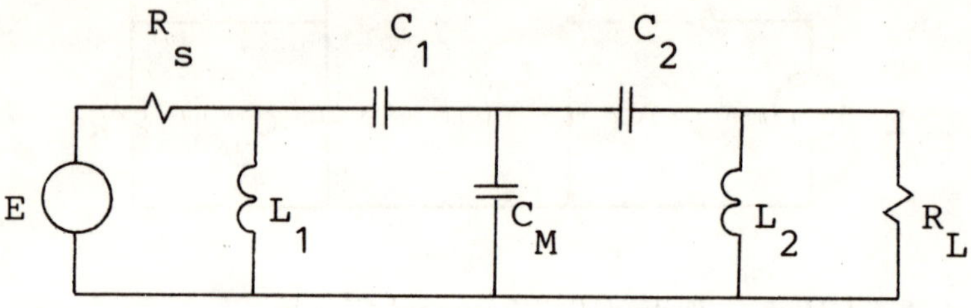

Figure 4.19   Two Coupled Resonant Circuits

17.   The Z-parameters of a transformer are

$$\begin{bmatrix} 10+j50 & -j69 \\ -j69 & 20+j150 \end{bmatrix} \text{ at 25 MHz}$$

Determine the values of L_{11}, L_{22}, r_1, r_2, and k.

18.   The circuit shown in Fig. 4.20 can be used as the output circuit of a vacuum-tube amplifier. The series capacitor is used to tune the circuit. The parallel capacitor is the output capacitance of the tube.

The output power of the amplifier is to be as close as possible to 350 W over the frequency range 115-150 MHz. The load resistance is 50 Ω, and the power supply voltage 2 kV. Assuming that the saturation effects and losses in L_{11} can be ignored, find the optimum values for L_{11}, L_{22}, and k. Determine the tuning range of the capacitor C.

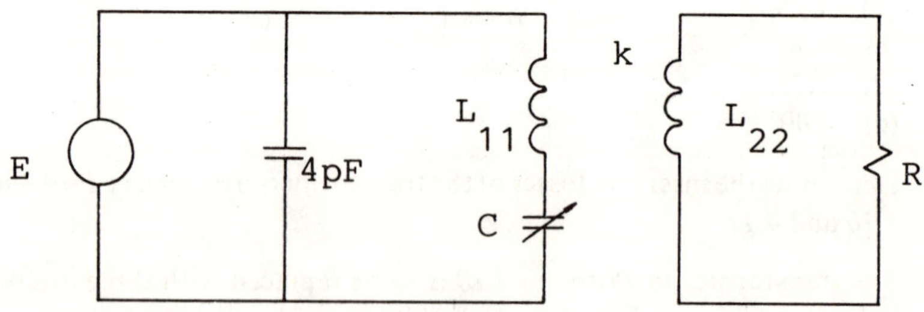

Figure 4.20   An Output Matching Network for a Vacuum-Tube Power Amplifier

References

1. Skilling, H.H., *Electrical Engineering Circuits,* John Wiley and Sons, 1965.

2. Krauss, H.L., W.B. Bostian, and F.H. Raab, *Solid State Radio Engineering,* John Wiley and Sons 1980.

3. Van der Walt, P.W., "A Simple Procedure for Designing Impedance Matching Networks with Loosely-Coupled Transformers," Research Note, University of Stellenbosch, South Africa.

CHAPTER 5

TRANSMISSION-LINE TRANSFORMERS

5.1 INTRODUCTION

The high frequency response of a magnetically coupled transformer is limited by the leakage inductance and the parasitic capacitance of the transformer.

The leakage inductance can be decreased significantly if a balun or stacked core, instead of a toroidal core, is used. It is, however, more difficult to decrease the parasitic capacitance between the two windings and the turns of each winding.

If the screen and center conductor of a coaxial cable are used as the primary and secondary windings of a 1:1 transformer, connected as shown in Fig. 5.1b, a 1:4 impedance transformation can be obtained and the parasitic capacitance between the windings can be controlled.

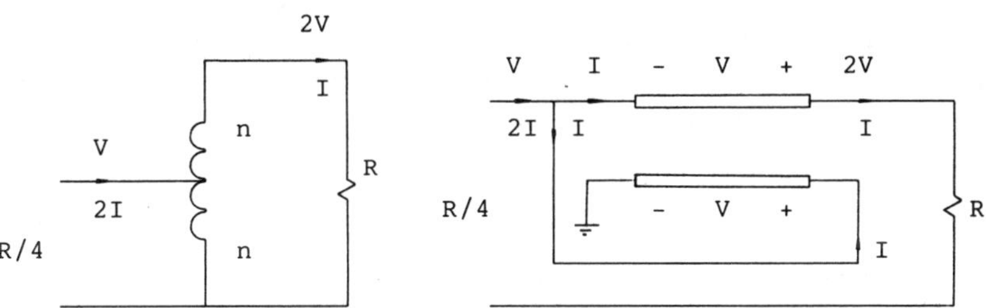

Figure 5.1   [a] A Conventional Auto-Transformer and [b] a 1:4 Transmission-Line Transformer

One would expect the performance of this transmission-line transformer to be optimum when the capacitance between the windings is low, that is, when the characteristic impedance of the line is high.

Fortunately, it turns out that this is not true and there is an optimum characteristic impedance for the line. The high frequency performance of the transformer is therefore improved by the transmission-line effect.

Seeing that the transmission-line plays a dominant role at high frequencies, the transformer cannot be considered as a conventional transformer with good coupling at high frequencies. In fact, it turns out that the magnetic coupling between the windings can be removed totally (by not using a magnetic core and straightening out the line) and the high frequency performance of the transformer will not be influenced at all (that is, if the losses in the magnetic material were negligible).

That this is possible, can be appreciated by assuming the currents in the transmission line to be balanced and the line to be short enough for the phase difference between the voltages across the line and the currents in the line to be small. This is illustrated in Fig. 5.2.

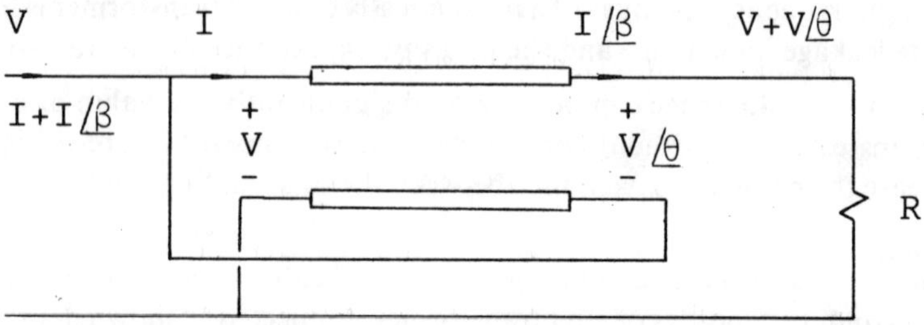

Figure 5.2   The Voltages Across and the Currents in a Transmission-Line Transformer at High Frequencies

It has been assumed that the currents in the transmission line are balanced. If the currents were perfectly balanced, it can be seen by inspecting Fig. 5.3a, that the output voltage would have been zero, which is not the case. The currents in the line must be unbalanced for the output voltage to be non-zero.

Because

$$V_0 = 2sLlI_{1u} \tag{5.1}$$

the unbalanced current is very small at high frequencies.

In (5.1) L is the inductance per unit length, l the length of the line, and I_{1u} the unbalanced component of the current in each of the two conductors of the line.

Although the high frequency performance of the transmission-line transformer is not affected by the removal of the magnetic core, the low frequency response is seriously degraded when this is done.

The reason for this is the increase in the unbalanced current. This current increases approximately inversely with frequency.

When magnetic material is used, the inductance in (5.1) increases and, consequently, the unbalanced current will be small. The frequency response of the transformer will then be very good at high as well as low frequencies.

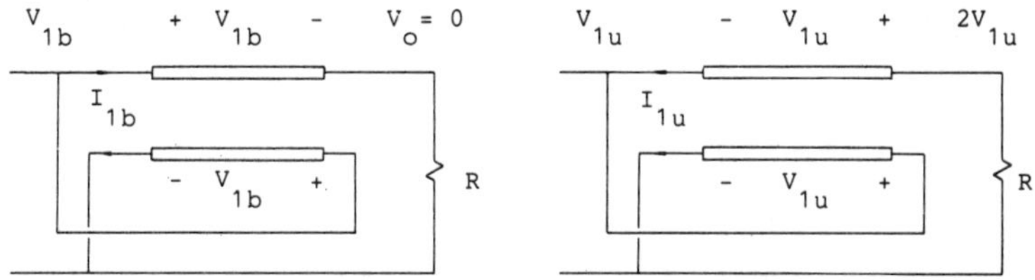

Figure 5.3   The Output Voltage of the 1:4 Transmission Line Transformer as a Function of the Balanced and Unbalanced Currents in the Line

The bandwidths of transmission-line transformers are significantly better than those of conventional transformers.

The analyses of the 1:4 transmission-line transformer and other transmission-line transformers will be discussed in detail in sec. 5.3.

It will be shown that the basic component of a transmission-line transformer is an unbalanced transmission line with increased inductance for the unbalanced currents in the line. This simplifies to a balanced transmission line at high frequencies and a 1:1 transformer with magnetizing inductance L_{11} at low frequencies.

Transmission-line transformers are often used to transform resistances. When only one transmission-line is used, the only transformation ratios that can be obtained are 1:1 and 1:4.

When more than one line is used, it is also possible to realize transformers with other transformation ratios. The number of lines required to do this might be prohibitive, however.

Apart from impedance-matching, transmission-line transformers are also used to perform various combining and splitting functions. The configurations of some of these transformers will be presented in the next section.

The design of transmission-line transformers consists mainly of designing the transformer to meet the low-frequency specifications and compensation at low or high frequencies to extend the bandwidth if necessary. The various steps in designing these transformers will be discussed in detail in section 5.4.

5.2   TRANSMISSION-LINE TRANSFORMER CONFIGURATIONS

Transmission-line transformers are used to change resistance levels in impedance-matching networks and amplifiers, as well as to perform certain splitting and combining functions.

The transformation ratios obtainable with these transformers are limited to those shown in Table 5.1, that is, if less than five transmission lines are used.

The number of lines in practical application is limited by the available space.

To ensure good low frequency response, each line must be wound around a magnetic core.

More than one line can sometimes be wound around the same core. The polarity of the voltage induced by the flux in the core and the relative size of the voltage must be taken into account when this is done.

Table 5.1

Transformation Ratios Obtainable with Transmission-line Transformers as a Function of the Number of Lines Used

Number of Lines	1	2	3	4
Transformation ratios obtainable	$1:1.00\ (1/1)^2$	$1:2.25\ (2/3)^2$	$1:1.78\ (3/4)^2$	$1:1.56\ (4/5)^2$
	$1:4.00\ (1/2)^2$	$1:4.00\ (1/2)^2$	$1:2.78\ (3/5)^2$	$1:1.96\ (5/7)^2$
	—	$1:9.00\ (1/3)^2$	$1:6.25\ (2/5)^2$	$1:2.56\ (5/8)^2$
	—	—	$1:16.00\ (1/4)^2$	$1:3.06\ (4/7)^2$
	—	—	—	$1:5.44\ (3/7)^2)$
	—	—	—	$1:7.11\ (3/8)^2$
	—	—	—	$1:12.30\ (2/7)^2$
	—	—	—	$1:25.00\ (1/5)^2$

The configuration corresponding to a particular transformation ratio can be found by using the technique illustrated in Fig. 5.4 [3].

When this technique is applied, the ratio of the input to output voltage changes from

$$x/y \text{ to } [x/(x+y)]$$

The impedance transformation ratio is changed from

$$(x/y)^2 \text{ to } [x/(x+y)]^2$$

In the simplest case,

$$x = 1 = y$$

and the application of this technique results in the configuration for the 1:4 transmission-line transformer.

If the technique is now applied to the 1:4 transformer, the 1:9 transformer shown in Fig. 5.5 is obtained.

By considering the high impedance side of the 1:4 transformer to be the input, the 1:2.25 transformer shown in Fig. 5.5b results.

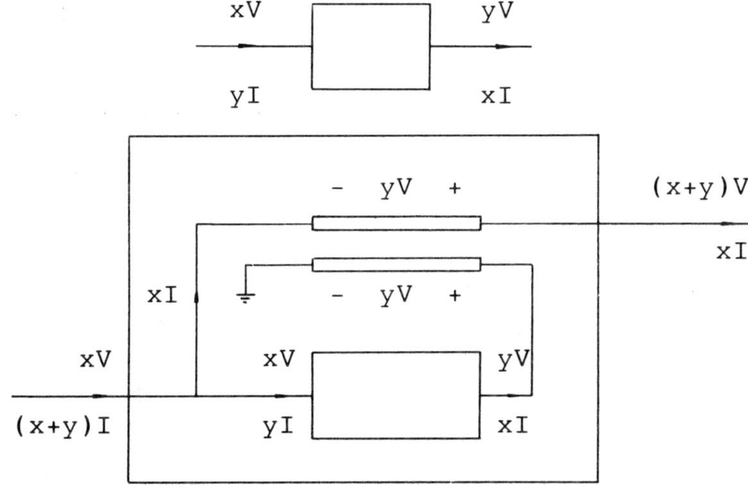

Figure 5.4 *The Influence of Adding an Extra Line on the Impedance Transformation Ratio of a Transmission-Line Transformer*

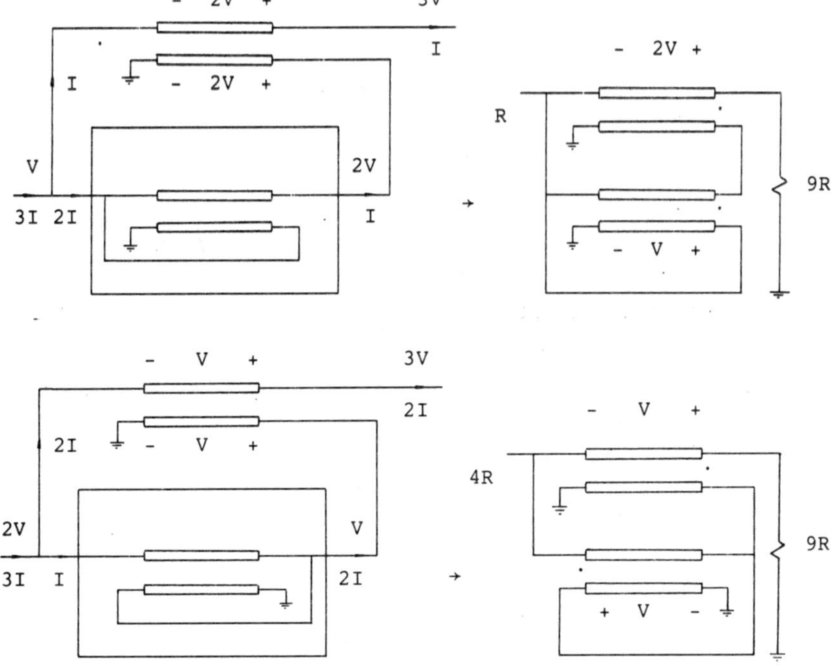

Figure 5.5 *Derivation of the Configurations for the 1:9 and 4:9 Transmission-Line Transformers*

The configurations for the $(3/4)^2$, $(1/4)^2$ and the $(2/5)^2$, $(3/5)^2$ transformers can now be found by applying the technique to the 1:9, 9:1 and 4:9, 9:4 transformers, respectively.

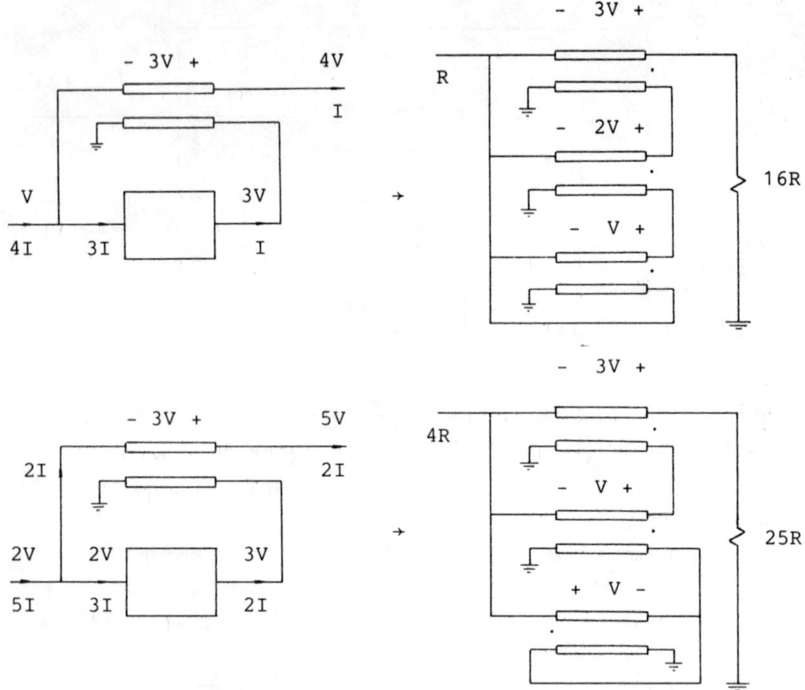

Figure 5.6   *The Configurations of the 1:16 and 4:25 Transmission-Line Transformers*

The transformers most often used in power amplifiers are the 1:4 and 1:9 transmission-line transformers. The high cut-off frequency of the 1:4 transformer can be increased considerably if two lines instead of one are used as shown in Fig. 5.7 [5].

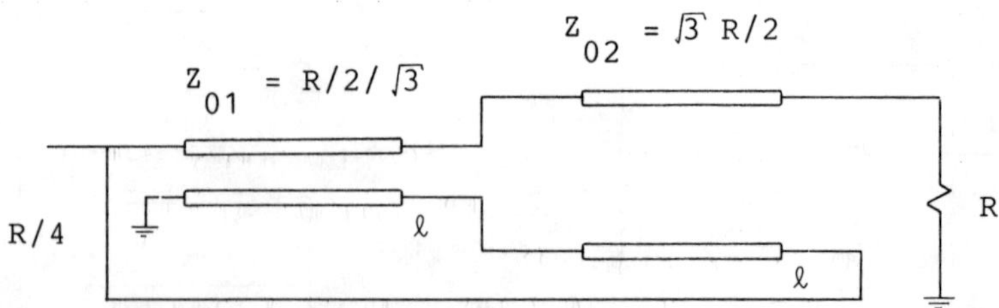

Figure 5.7   *The Configuration of a 1:4 Transmission-Line Transformer which Has No High Cut-Off Frequency [theoretically]*

Impedance transformations between balanced sources and a balanced load are often required. The configurations for the balanced 1:4 and 1:9 transmission-line transformers are shown in Figs. 5.8a and 5.8b, respectively.

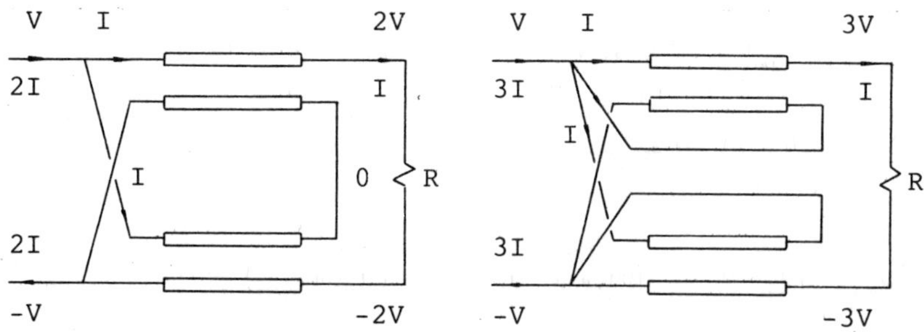

Figure 5.8 The Configurations for the Balanced [a] 1:4 and [b] 1:9 Transmission-Line Transformers

When either the load or the source is unbalanced, the 1:1 transformer shown in Fig. 5.9 can be used to provide the required unbalanced-to-balanced transformation.

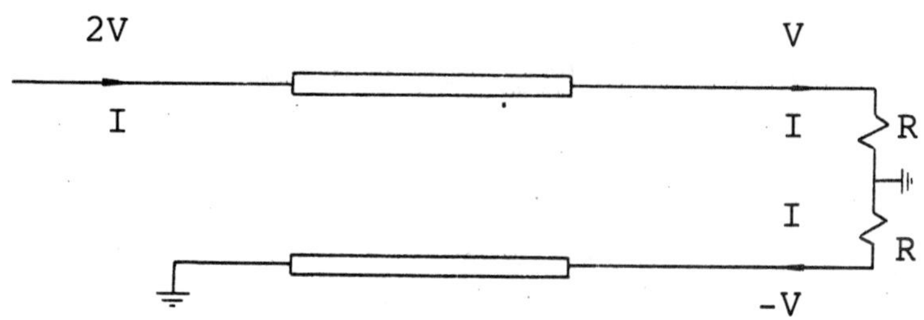

Figure 5.9 The Unbalanced-to-Balanced 1:1 Transmission-Line Tranformer

The frequency response of the 1:1 transformer is exactly the same as that of a transmission line terminated in the same load, that is, at high frequencies.

Because of the symmetry, the high frequency response of the 1:4 balanced transformer is identical to that of the 1:1 transformer. The equivalence is illustrated in Fig. 5.10.

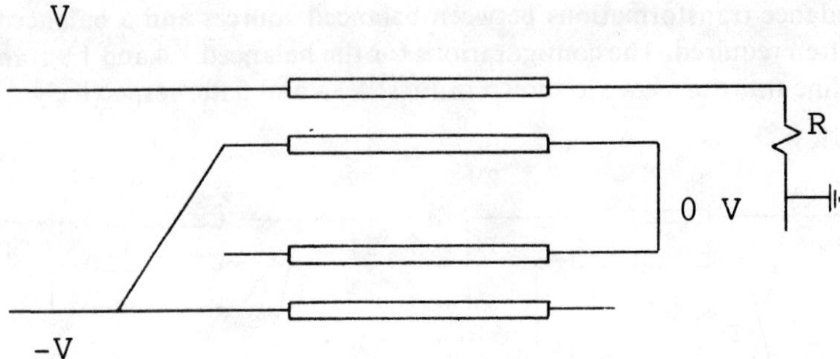

Figure 5.10   Illustration of the Equivalence between the 1:4 Balanced and 1:1 Balanced-to-Unbalanced Transmission-Line Transformers

A 4:1 unbalanced-to-balanced transformation can be obtained by combining the 1:1 and 1:4 balanced transformers as shown in Fig. 5.11a, or by using the transformers shown in Fig. 5.11b. The latter transformer is less sensitive to non-optimum characteristic impedances than the former, although it has a lower cut-off frequency when the optimum characteristic impedance are used.

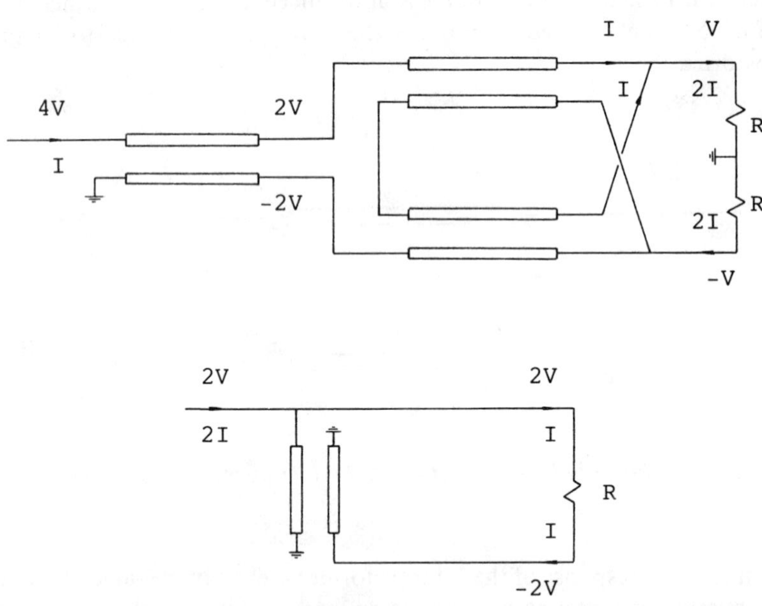

Figure 5.11   [a] 1:1 and 4:1 Transformers Combined to Give an Unbalanced-to-Balanced Transformation; [b] 1:4 Unbalanced Transmission-Line Transformer

The output currents of the two transistors in a push-pull class B amplifier are often (at lower frequencies where the conduction angle is 180°) combined by using either of the 1:4 or 1:9 transformers shown in Fig. 5.12.

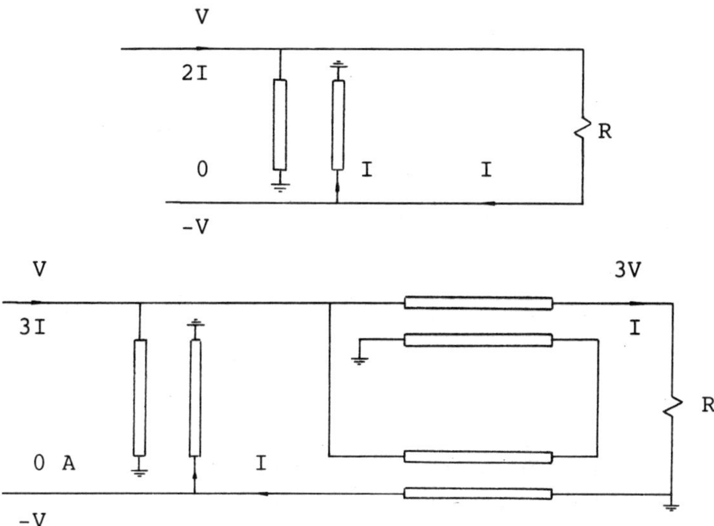

Figure 5.12   1:4 and 1:9 Transformers for Combining the Currents of the Transistors in a Class B Amplifier into a Single Load

Although they are used for different purposes, it can be seen that the configurations of the 1:4 transformer shown in this figure are similar to that of the 1:4 unbalanced-to-balanced transformer shown in Fig. 5.11b.

By redefining the reference plane of the 1:4 transformer in Fig. 5.12, as shown in Fig. 5.13, it becomes clear that the frequency responses of the unbalanced 1:4 transformer and the 1:4 transformers shown in Figs. 5.12 and 5.13 are identical.

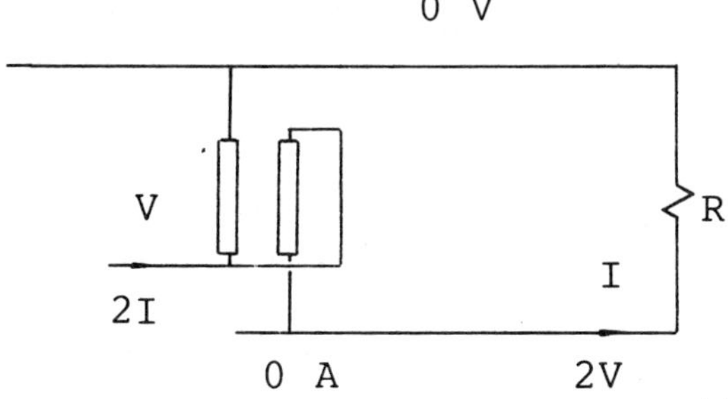

Figure 5.13   The Configuration of the 1:4 Transformer from Fig. 5.12 if the Reference Plane [Ground] Is Redefined as Shown

The combiner shown in Fig. 5.14a is often used to combine two in-phase signals at radio frequencies.

As indicated in the figure, the voltage drop across the 1:1 transformer used in the combiner is equal to zero when the two input signals are equal in amplitude and in phase.

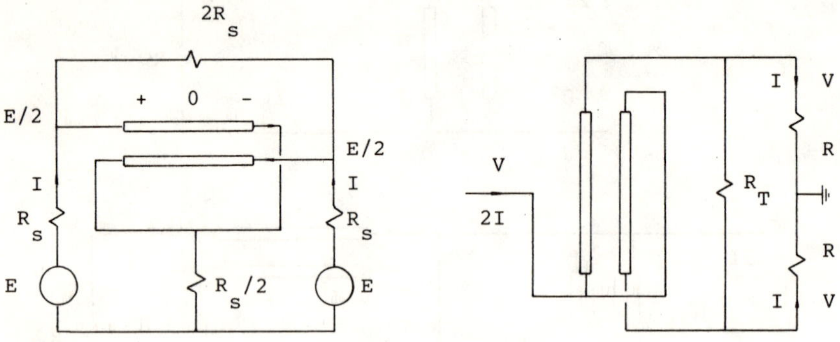

Figure 5.14 [a] *A Transformer for Combining Two In-Phase Signals into the Same Load;* [b] *An In-Phase Power Splitter*

When the signals are unbalanced, the two sources will be isolated from each other by the transformer. As an example of this, if $E_2 = 0$, no current will flow in the resistance R_{s2}, as is shown in Fig. 5.15.

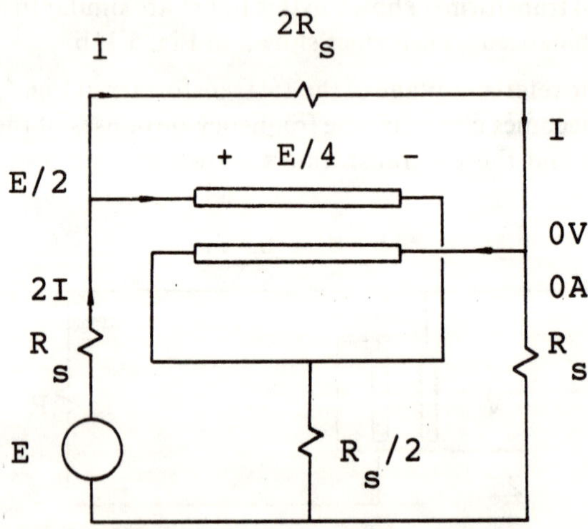

Figure 5.15 *Illustration of the Isolating Action of the Hybrid Transformer Shown in Fig. 5.14a*

At low frequencies, the isolating characteristic of this transformer is a function of the magnetizing inductance of the 1:1 transformer, that is,

$$S = [4\omega L_{11}/(R_s/2)]^2 + 1 \tag{5.2}$$

where S is the ratio of the power dissipated in the load $(R_{s1}/2)$ and the source resistance R_{s2}, when $E_2 = 0$.

The isolation at high frequencies is a function of the electrical length of the line and the characteristic impedance.

The combiner can be changed to a splitter by connecting it as shown in Fig. 5.14b.

As in the case of the combiner, the voltage drop across the 1:1 transformer will be zero as long as the loads are balanced, and the transformer will cause the power to be distributed more evenly between the loads when they are unbalanced.

5.3 THE ANALYSIS OF TRANSMISSION-LINE TRANSFORMERS

The basic building block of transmission-line transformers is an unbalanced transmission line wound around magnetic material, or in the form of a solenoidal coil (semi-rigid coaxial cable can be used for this purpose).

The current in the two line conductors are not equal, but are related by the equation:

$$I_2(x) = I_1(x) + I_0(x) \tag{5.3}$$

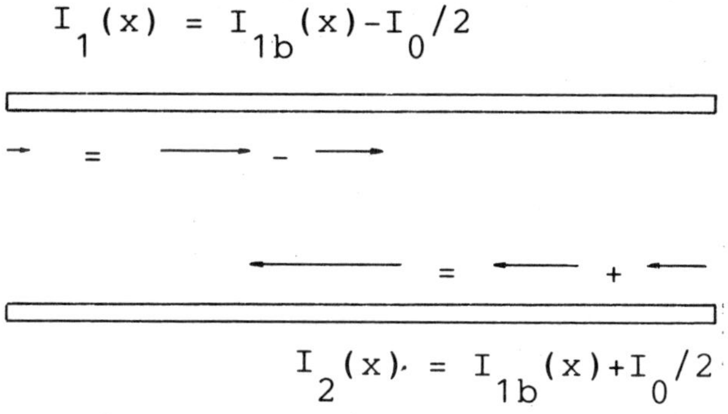

$$I_1(x) = I_{1b}(x) - I_0/2$$

$$I_2(x) = I_{1b}(x) + I_0/2$$

Figure 5.16   The Balanced and Unbalanced Components of the Current in a Transmission Line

Because the effective current entering the line at any point $(I_{eff} = I_1(x) - I_2(x) = -I_0(x))$ must be equal to the current flowing out of the line at any other point farther along the line, the unbalanced current ($I_o(x)$) is independent of the distance (x) along the line and (5.3) simplifies to

$$I_2(x) = I_1(x) + I_0 \qquad (5.4)$$

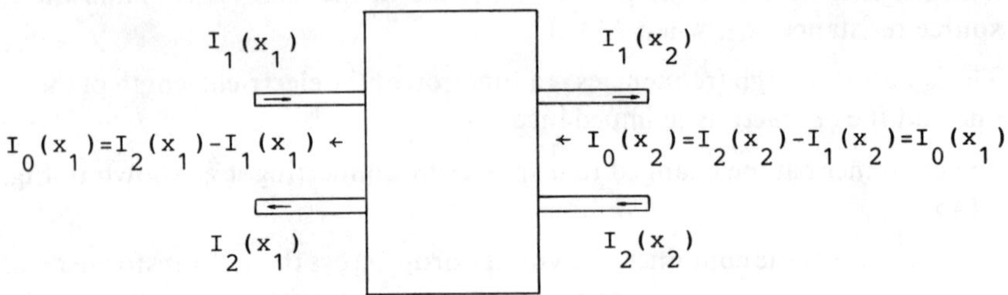

Figure 5.17   The Unbalanced Current on a Transmission Line as a Function of the Distance along the Line

The effect of the magnetic material (or solenoidal coil) is to increase the impedance associated with the unbalanced currents in the line.

Because there is not external magnetic field associated with the balanced currents ($\oint H \cdot dl = I_{1b} - I_{1b} = 0$), these currents are not influenced by the material used or the form in which the line is wound.

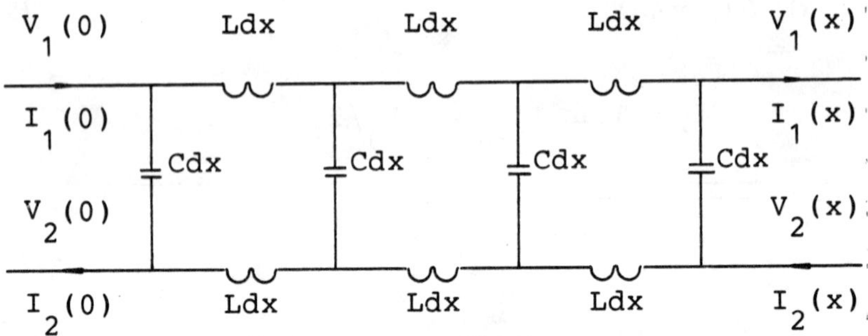

Figure 5.18   The Unbalanced Transmission Line

If the influence of the magnetic core or coil form on the unbalanced currents is ignored initially, equations for the voltage on and current in the unbalanced transmission line can be derived in a similar way to the case of the balanced transmission line. The results of the derivation are the equations:

$$I_1(x) = -I_0/2 + Ae^{-\Gamma x} + Be^{\Gamma x} \qquad (5.5)$$

$$I_2(x) = I_0/2 + Ae^{-\Gamma x} + Be^{\Gamma x} \qquad (5.6)$$

$$V_1(x) = V_1(0) - Z_0/2 \cdot (A\text{-}B) + Z_0/2 \cdot [Ae^{-\Gamma x} - Be^{\Gamma x}] + sLxI_0/2 \qquad (5.7)$$

$$V_2(x) = V_2(0) + Z_0/2 \cdot (A\text{-}B) - Z_0/2 \cdot [Ae^{-\Gamma x} - Be^{\Gamma x}] + sLxI_0/2 \qquad (5.8)$$

$$V_{12}(x) = Z_0[Ae^{-\Gamma x} - Be^{\Gamma x}] \qquad (5.9)$$

where

$$\Gamma = (2LC)^{1/2}s \qquad (5.10)$$

$$Z_0 = (2L/C)^{1/2} \qquad (5.11)$$

L and C are the inductance and capacitance of the line per unit length, respectively.

When magnetic material is used (or the line is wound as a coil), the reactance associated with the unbalanced currents ($I_0/2$) must be changed from $sLl(x{=}l)$ to

$$X_u = 2\,sL_{11} \qquad (5.12)$$

where L_{11} is the inductance associated with each line conductor when the other conductor is open-circuited.

The inductance associated with the unbalanced current in each conductor is twice the expected value (L_{11}) because of the excellent coupling between the two conductors of the transmission line. When current is flowing in only one conductor, the voltage across the length of the other conductor will be equal to that of the first, provided that there are no resistive losses in the conductors. The coupling factor is, therefore, very close to unity.

When magnetic material is used or the line is wound as a coil, (5.7) and (5.8) must be changed to

$$V_1(x) = V_1(0) - Z_0/2 \cdot (A\text{-}B) + Z_0/2 \cdot [Ae^{-\Gamma x} - Be^{\Gamma x}]$$
$$+ s[2L_{11}][x/l]\,I_0/2 \qquad (5.7a)$$

and

$$V_2(x) = V_2(0) + Z_0/2 \cdot (A\text{-}B) + Z_0/2 \cdot [Ae^{-\Gamma x} - Be^{\Gamma x}]$$
$$+ s[2L_{11}][x/l]\,I_0/2 \qquad (5.8a)$$

Before (5.5) to (5.9) can be used to determine the voltages on and the currents in any particular transmission-line transformer, the constants A, B, I_0, $V_1(0)$, and $V_2(0)$ must be determined.

These constants can be found by using the boundary conditions for the transformer under consideration.

When the voltages and currents are known, the power gain and the input and output impedances of the transformer can be determined easily.

Because the transmission line can usually be considered lossless, it is sufficient that the input impedance of a transformer is known. (The magnitudes of the input and output reflection coefficients are then equal, and the magnitude of the transducer power gain is only a function of the reflection coefficient.)

The input impedance of a transmission-line transformer is a function of the frequency, the load impedance, and the length and characteristic impedance of the transmission line used. The expression for the input impedance, therefore, is usually quite complex.

Although the complexity is not a problem when a computer program is used to analyze a transmission-line transformer, it is possible to simplify the equation for the input impedance at low and high frequencies by making appropriate assumptions.

At high frequencies, the reactance associated with each conductor is high compared to the characteristic impedance of the line, and the approximation

$$sLl \gg Z_0 \tag{5.13}$$

can be made.

When this approximation is made, the input impedance of the transformer is only a function of the balanced currents in the line. As far as the impedance is concerned, the transmission line can then be considered balanced.

At low frequencies, the line is electrically very short and the approximation

$$e^{\Gamma l} \simeq 1 \tag{5.14}$$

can be made.

When this approximation can be made, the input impedance of the transformer is independent of the length and characteristic impedance of the transmission line. The transmission-line transformer can then be considered as a conventional transformer.

Therefore, the basic building block of a transmission-line transformer reduces to a balanced transmission line, and a conventional 1:1 transformer with magnetizing inductance L_{11} at high and low frequencies, respectively. This is illustrated in Fig. 5.19.

Example 5.1

The input impedance of a 1:4 unbalanced transmission-line transformer will be determined as an example of the application of Eqs. (5.5) to (5.9).

The boundary conditions for the transformer are

$$V_2(0) = 0 \tag{5.15}$$

$$V_2(l) = V_1(0) \tag{5.16}$$

$$V_1(l) = Z_L I_1(l) \tag{5.17}$$

These conditions will be used to find two independent equations for the

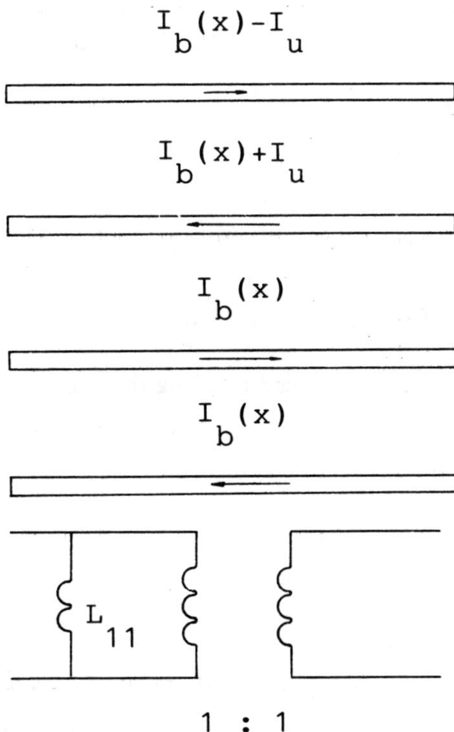

Figure 5.19 [a] *The Basic Building Block of a Transmission-Line Transformer Simplified at* [b] *High and* [c] *Low Frequencies*

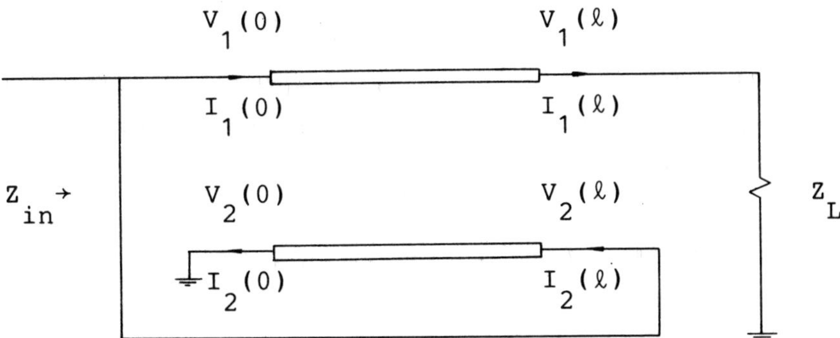

Figure 5.20 *The 1:4 Unbalanced Transmission-Line Transformer*

unbalanced current I_0, in terms of A and B. In this way, the relationship between A and B can be established and the input impedance can be found:

$$V_1(0) = V_{12}(0) = Z_0\left[Ae^{-\Gamma x} - Be^{\Gamma x}\right]$$

$$V_2(l) = 0 + 0.5\,Z_0\left[A-B\right] + sLlI_0/2 - 0.5\,Z_0\left[Ae^{-\Gamma x} - Be^{\Gamma x}\right]$$

Because $V_1(0)$ and $V_2(l)$ are equal, these two equations can be used to obtain an equation for I_0 in terms of A and B. After some manipulation, the following equation is obtained:

$$sLlI_0 = Z_0/(sLl)\cdot[A-B] + Z_0/(sLl)\cdot[Ae^{-\Gamma x} - Be^{\Gamma x}] \qquad (5.18)$$

The second equation is determined by using the constraint imposed by the load:

$$V_1(l) = V_1(0) - 0.5\,Z_0\left[A-B\right] + sLlI_0/2 + 0.5\,Z_0\left[Ae^{-\Gamma l} - Be^{\Gamma l}\right]$$

and

$$Z_LI_1(l) = Z_L\left[-I_0/2 + Ae^{-\Gamma l} + Be^{\Gamma l}\right]$$

By equating these two equations, it follows that

$$[Z_L+sLl]\,I_0 = 2A[Z_0e^{-\Gamma l} - 0.5\,Z_0\,(e^{-\Gamma l}+1)] + 2B[Z_0e^{\Gamma l} + 0.5\,Z_0\,(e^{\Gamma l}+1)] \qquad (5.19)$$

The relationship between A and B can now be determined by using (5.18) and (5.19):

$$\frac{B}{A} = \frac{Z_0\,E2\,[1+Z_L/(sLl)] - 2[Z_L\,e^{-\Gamma l} - Z_0/2\,E2]}{Z_0\,E1\,[1+Z_L/(sLl)] + 2[Z_L\,e^{\Gamma l} + Z_0/2\,E1]} \qquad (5.20)$$

where

$$E1 = 1 + e^{\Gamma l}$$

and

$$E2 = 1 + e^{-\Gamma l}$$

The input impedance of the transformer is given by the equation:

$$Z_{in} = V_1(0)/[I_1(0) + I_2(l)]$$

$$= Z_0\,\frac{1-B/A}{E2 + B/A\cdot E1} \qquad (5.21)$$

If the approximation

$$e^{\Gamma l} \simeq 1$$

is used, the equations for the ratio B/A and the input impedance of the transformer simplifies to

$$\frac{B}{A} = \frac{2Z_0 sLl + Z_L Z_0 - Z_L sLl}{2Z_0 sLl + Z_L Z_0 + Z_L sLl} \tag{5.22}$$

and

$$Z_{in} = Z_0/2 \; \frac{1 - B/A}{1 + B/A} = \frac{Z_L/4 \cdot sLl/2}{Z_L/4 + sLl/2} \tag{5.23}$$

If magnetic material is used, the reactance sLl in these equations must be replaced with $s(2L_{11})$. The input impedance is then

$$Z_{in} = \frac{Z_L/4 \cdot sL_{11}}{Z_L/4 + sL_{11}} \tag{5.24}$$

At high frequencies, the approximation

$$sLl \gg Z_0$$

can be made, and the expression for the ratio B/A simplifics to

$$\frac{B}{A} = \frac{Z_0[1 + e^{-\Gamma l}] - Z_L e^{-\Gamma l}}{Z_0[1 + e^{-\Gamma l}] + Z_L e^{-\Gamma l}} \tag{5.25}$$

The impedance is still given by (5.21).

The transducer power gain for the transformer can be determined by using the equation:

$$G_T = 1 - \left| \frac{Z_{in} - Z_s^*}{Z_{in} + Z_s} \right|^2 \tag{5.26}$$

where Z_s is the impedance of the source driving the transformer.

Example 5.2

The input impedance of the 1:1 balanced-to-unbalanced transmission-line transformer can be determined by using the boundary conditions:

$$V_1(l) = I_1(l) Z_L \tag{5.27}$$

$$V_2(l) = 0 \tag{5.28}$$

$$V_1(0) = -V_2(0) \tag{5.29}$$

By using the first equation, the unbalanced current is found to be

$$I_0/2 = A e^{-\Gamma l}[1 - Z_0/Z_L] + B e^{\Gamma l}[1 + Z_0/Z_L] \tag{5.30}$$

When the second and third equations are used, the second equation necessary for determining the ratio B/A is found to be

$$sLlI_0/2 = Z_0/2 \cdot [A e^{-\Gamma l} - B e^{\Gamma l}] \tag{5.31}$$

The ratio B/A can now be obtained by using these two equations:

$$\frac{B}{A} = -\frac{e^{-\Gamma l}[1 - Z_0/Z_L] - [Z_0/(2sLl)] e^{-\Gamma l}}{e^{\Gamma l}[1 + Z_0/Z_L] + [Z_0/(2sLl)] e^{\Gamma l}} \tag{5.32}$$

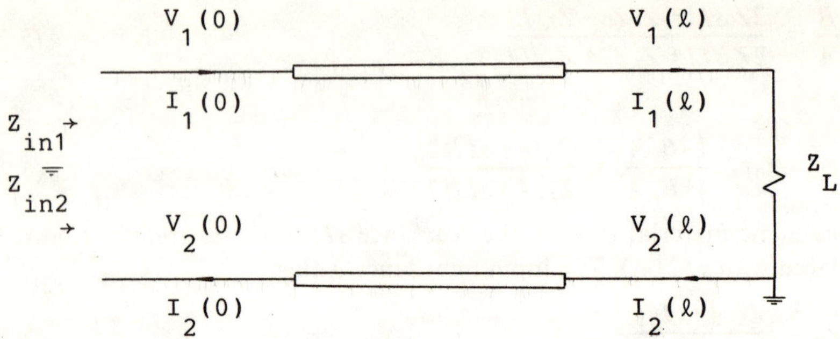

Figure 5.21   The 1:1 Balanced-to-Unbalanced Transmission-Line Transformer

When B/A is known, the input impedances Z_{in1} and Z_{in2} can be determined. These impedances are given by the equations.

$$Z_{in1} = \frac{Z_0\,[1 - B/A]}{-Z_0/(sLl)\,[e^{-\Gamma l} - B/A\,e^{\Gamma l}] + 2[1 + B/A]} \tag{5.33}$$

$$Z_{in2} = \frac{Z_0\,[1 - B/A]}{+Z_0/(sLl)\,[e^{-\Gamma l} - B/A\,e^{\Gamma l}] + 2[1 + B/A]} \tag{5.34}$$

It is clear from the different signs in the denominators of (5.33) and (5.34) that the two input impedances are not equal at low frequencies.

When $sLl \gg Z_0$ the two impedances are approximately equal, independent of the characteristic impedance value of the line.

Furthermore, the input impedance of the transformer is identical to that of a balanced transmission line terminated in the same load impedance (Z_L).

By using this equivalence, it follows that the input impedance of the 1:1 balanced-to-unbalanced transmission-line transformer will be purely resistive at high frequencies if $Z_0 = R_L$.

Because of the symmetry, the same applies to the 1:4 balanced transmission-line transformer.

5.4   THE DESIGN OF TRANSMISSION-LINE TRANSFORMERS

The design of transmission-line transformers consists of the following:

a)   Determining the characteristic impedance and the diameter of the transmission line to be used;

b)   Determining the minimum value of the magnetizing inductance of the transformer at the lowest frequency pass band;

c)   Selecting a suitable material (if needed);

d)   Determining the type and size of the core to be used;

e) Finding the line length and the corresponding high cut-off frequency of the transformer;

f) Compensating the transformer for non-optimum characteristic impedance, if necessary; and

g) Extending the bandwidth by using LC impedance-matching networks, if necessary.

Each of these points will be discussed in detail in the following paragraphs.

5.4.1 Determining the Optimum Characteristic Impedance and Diameter of the Transmission Line to be Used

At high frequencies, the input impedance of a transmission line transformer is a function of the characteristic impedance of the transmission line.

The optimum characteristic impedance can be determined by taking the ratio of the voltage across one end of the transmission line and the current passing through it. The basic building block of the transformer is then considered to be an ideal 1:1 transformer.

The application of this rule is illustrated on the 1:4 transmission-line transformer in Fig. 5.22.

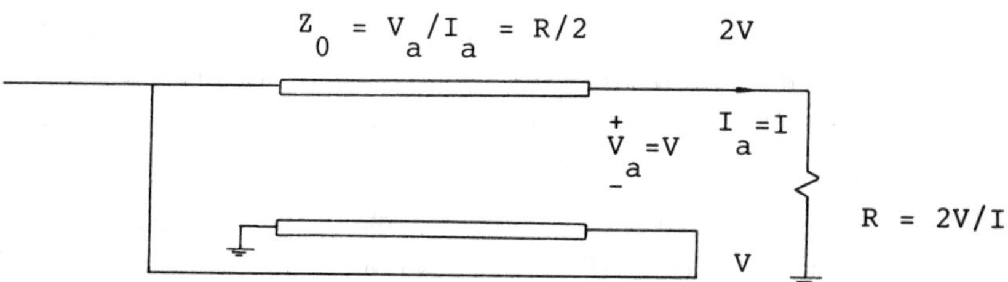

Figure 5.22   Determining the Optimum Characteristic Impedance of an 1:4 Unbalanced Transmission-Line Transformer

If a line with any other characteristic impedance is used, the input reflection coefficient of the transformer will be affected adversely.

The effect of the characteristic impedance on the cut-off frequency of the transformer will be discussed later.

When the optimum characteristic impedance is known, the type of line to be used must be chosen.

Coaxial cables with 25Ω and 50Ω characteristic impedances are freely available. A line with a 12.5Ω characteristic impedance can be obtained by connecting two 25Ω lines in parallel, while 100Ω can be obtained by connecting two 50Ω lines in series.

A wide range of characteristic impedances can be obtained by twisting togeth-er pairs of conductors with various diameters. When very low impedances are required (less than 10Ω), many conductors with smaller diameters can be twisted together.

The characteristic impedances of these twisted lines are influenced by the diameter of the wire used, as well as the number of twists per unit length.

Apart from the characteristic impedance, it is also necessary to decide on the diameter of the cable to be used where applicable. This is determined by the losses that can be tolerated and the power to be transmitted through the line.

The attenuation of bifilar or multifilar transmission lines can become a prob-lem at high frequencies, as mentioned in Ch. 2.

5.4.2 Determining the Minimum Value of the Magnetizing Inductance of the Transformer at the Lowest Frequency in the Pass Band

At low frequencies the transmission-line transformer can be considered as a conventional 1:1 transformer connected in a special way.

When this model is used, the input impedance and power gain *versus* frequen-cy response at low frequencies can be determined by using Kirchhoff's voltage and current laws on the simplified equivalent circuit.

If the load consists of a single resistor, only the input impedance of the transformer needs to be determined. The power dissipated in the load (and therefore the power gain) can be found by using the equation:

$$P_L = V_{cc}^2 \, G_{eff}/2 \tag{5.35}$$

where V_{cc} is the maximum (peak) voltage across the effective parallel input resistance (R_{eff}) of the transformer.

When the input impedance and transfer function are known, the minimum inductance (L_{11}) required to meet the low frequency specifications can be determined.

The minimum value of the magnetizing inductance of the 1:4 unbalanced and 1:1 unbalanced-to-balanced transformers will be determined as examples.

Example 5.3

With the transmission line replaced by a 1:1 transformer with magnetizing inductance L_{11}, the equivalent circuit of the transformer can be simplified as shown in Fig. 5.23.

If the cut-off frequency is to be the 3 dB frequency, it is obvious by inspection of Fig. 5.23c that the required magnetizing inductance L_{11} must be such that

$$\omega L_{11} = R_s/2 \tag{5.36}$$

If this transformer is to be used in a power amplifier, the magnetizing inductance must be high enough for the specified minimum allowable ripple in the pass band to be achieved.

Because the power dissipated in the load is given by (5.35), the output power is directly proportional to the effective parallel input resistance.

The efficiency of the amplifier is decreased if the effective load is reactive, that is, if the output impedance of the transistor is assumed to be purely resistive. In particular, it is decreased by a factor of

$$\eta_r = 1/[1 + (R_{eff}/X_{eff})^2]^{1/2} \tag{5.37}$$

where X_{eff} is the effective parallel input reactance of the transformer.

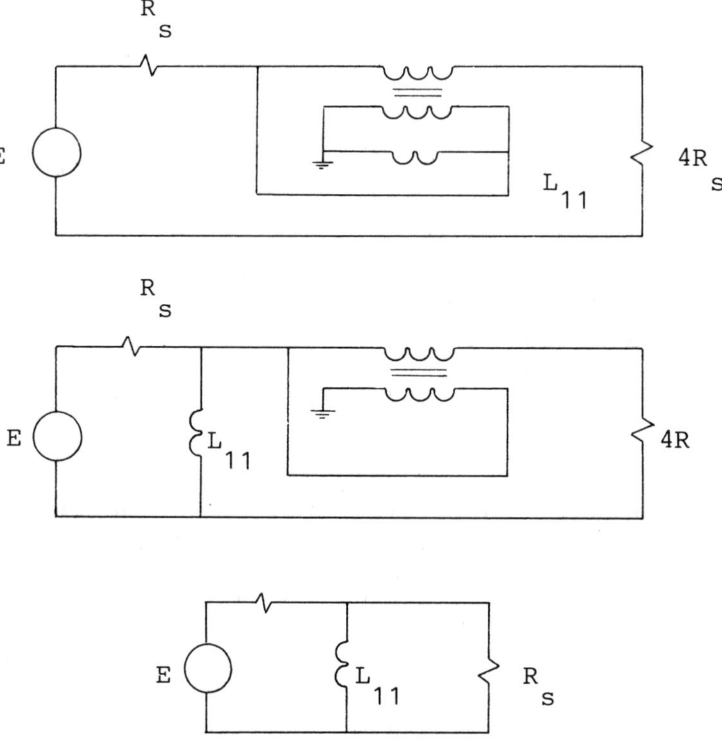

Figure 5.23 *Simplification of the Equivalent Circuit of the 1:4 Unbalanced Transformer at Low Frequencies*

Because R_{eff} is equal to the optimum value (R_s) in this particular problem, the power transmitted through the 1:4 transformer is also equal to the optimum value, that is, at low frequencies.

The relative efficiency is given by

$$\eta_r = 1/[1 + (R_s/(\omega L_{11}))^2]^{1/2}$$

If $\eta_r = 0.95$ is acceptable, the magnetizing inductance must be such that

$$\omega L_{11} = 3\, R_s \qquad\qquad (5.38)$$

(ωL_{11} is often chosen to be equal to 4 R_s).

Example 5.4

The equivalent circuit for the 1:1 unbalanced-to-balanced transformer is shown in Fig. 5.24a.

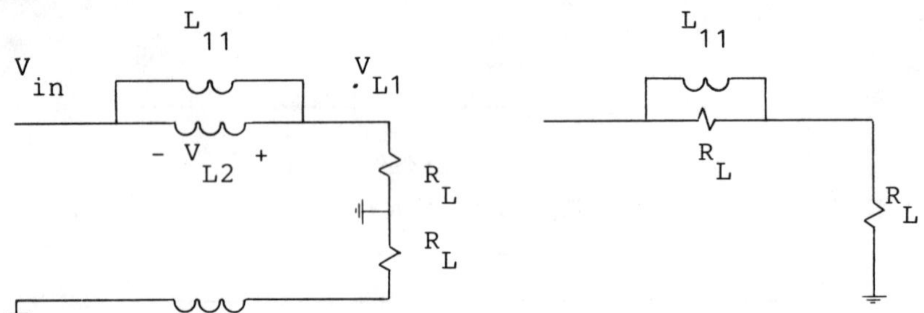

Figure 5.24   The 1:1 Unbalanced-to-Balanced Transmission-Line Transformer at Low Frequencies

By transforming the load on the secondary side of the transformer to the primary side, the equivalent circuit of the 1:1 unbalanced-to-balanced transformer can be simplified to that shown in Fig. 5.23b.

By using this equivalent circuit, the effective input admittance is found to be

$$Y_{in} = \frac{1}{R_L + sL_{11}R_L/[R_L + sL_{11}]}$$

$$= \frac{R_L + 2sL_{11}}{R_L^2 + 2sL_{11}R_L}$$

$$= \frac{1}{2R_L} \cdot \frac{1 + R_L/(sL_{11})}{1 + R_L/(2sL_{11})} \qquad\qquad (5.39)$$

It is clear from this equation that the input resistance will be equal to $2R_L$ if the magnetizing reactance is relatively high.

The relative power dissipation in the two load resistances can be determined by transforming the parallel combination of ωL_{11} and R_L in Fig. 5.24b to the equivalent series impedance shown in Fig. 5.25.

Because the same current flows through the two resistors, the ratio of the power dissipated in each load is equal to the ratio of the resistance of these resistors.

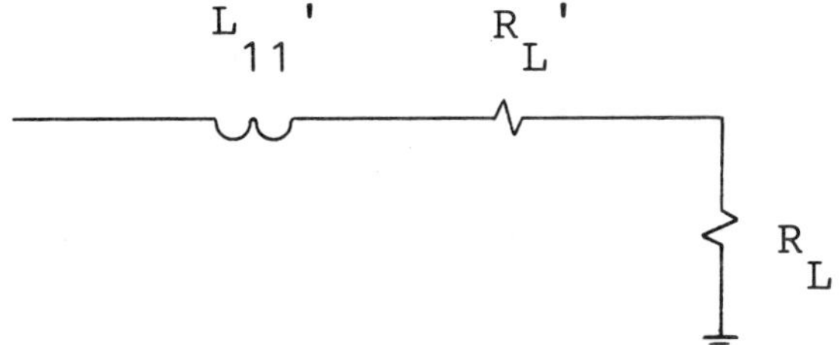

Figure 5.25   The Series Equivalent of the Impedance of the Circuit from Fig. 5.24b

If

$$\omega L_{11} = 4.4\, R_L \qquad (5.40)$$

the power dissipated in the two load resistors will differ by 5%. The input power to the transformer will then be 1% higher than the design value and the relative efficiency will be 0.99.

5.4.3   Determining the Type and Size of the Magnetic Core to be Used

Toroidal cores are often used in transmission-line transformers. The size of the toroidal core is determined by the inductance required, the maximum flux density in the core (and therefore the allowable losses), and the line length required to meet these specifications.

It was shown in Ch. 2 that if the inductance and flux density specifications are to be met simultaneously, a core with

$$Al = \frac{\mu_0 \mu_r}{\omega B_{max}^2}\; \frac{V_{max}^2}{\omega L_{11}} \qquad (2.33)$$

should be used.

It can be shown that the line length will always increase if a core with an *Al*-product larger than that given by this equation is used.

If the core size is decreased, it is possible that the line length might be shorter, at least initially.

Whether it will be shorter is a function of the extent to which the inductance must be increased to meet the loss specification (the flux density in the core will be too high if the inductance is not increased), as well as the dimensions of the core.

If the losses in the material increase sharply when the flux density is increased, the optimum core size will always be that given by (2.33).

It is sometimes possible to reduce the line length necessary to provide the required magnetizing inductance by using a number of smaller toroidal cores instead of only one larger core.

The ratio of the line length for a single core to that of N_c stacked cores is given approximately by the equation:

$$\frac{l_{e1}}{l_{e2}} = \frac{2w_1 + 2h_1 + 4t}{A_1/A_2 \cdot h_2 + l_2/l_1 \cdot (4w_2 + 4t)} \tag{5.41}$$

where

t = The outer diameter of the transmission line used
l_1 = The mean path length of the larger core
l_2 = The mean path length of each of the smaller cores
A_1 = The effective cross-sectional area of the larger core
A_2 = The effective cross-sectional area of each of the smaller cores

and w_1, w_2, h_1, and h_2 are defined in Fig. 5.26

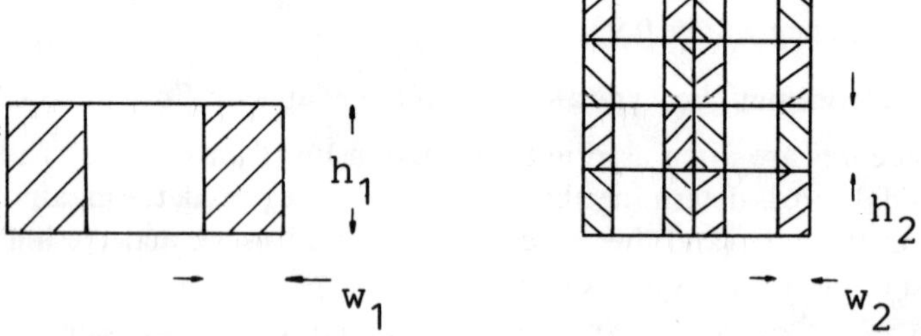

Figure 5.26   Cross-Section of [a] a Single Toroidal Core and [b] a Number of Stacked Toroidal Cores

This equation was derived by assuming that the inductance and flux densities of the two inductors are equal.

In order to have the same flux density, it is necessary that

$$N_1/l_1 = N_2/l_2 \tag{5.42}$$

where

N_1 = The number of turns used with the single core
N_2 = The number of turns used with the stacked core

The inductance of the two inductors will be the same, if

$$N_1^2 A_1/l_1 = N_c N_2^2 A_2/l_2 \tag{5.43}$$

where N_c is the number of cores used in the stacked-core inductor.

By using (5.42), Eq. (5.43) can be changed to

$$A_1 l_1 = N_c \cdot A_2 l_2 \tag{5.44}$$

This equation implies that the effective Al-products of the single-cored and stacked-cored inductors must be the same.

Equations (5.44) and (5.42) can be used to determine the number of cores and the number of turns required, if it turns out to be worthwhile to use a transformer with stacked cores, that is, if the core dimensions are known.

If a core with suitable dimensions (comparable to those of the stacked core) is available, a balun core can also be used to decrease the line length of the transformer.

Example 5.5

The line lengths of a single toroidal core and a stacked core with

$$A_2 = 0.5 A_1$$

$$l_2 = 0.5 l_2$$

$$w = h$$

$$t = w_1 / 3$$

will be compared, as an example of the application of the equations given above.

With $w = h$, (5.40) becomes

$$
\frac{l_{e1}}{l_{e2}} = \frac{4w + 4t}{2w/\sqrt{2} + 1/2 \cdot (4w_2/\sqrt{2} + 4t)}
$$

$$
= \frac{4w + 4t}{\sqrt{2}\,2w + 2t} \tag{5.45}
$$

From this, it follows that the ratio of the line length for the two different cores is

$$
\frac{l_{e1}}{l_{e2}} \simeq \frac{4 + 4/3}{2\sqrt{2} + 2/3} = 1/0.655 = 1.5
$$

The bandwidth of the transformer, therefore, can be increased significantly by using a stacked (or balun) core.

5.4.4 Compensation of Transmission-Line Transformers for Non-Optimum Characteristic Impedances

When a line with the optimum characteristic impedance is not available, it is possible to compensate for the degradation of the high frequency response of the transformer by using compensating inductors or capacitors.

It is usually adequate to use two compensating elements. One element is used in parallel (capacitor) or in series (inductor) with the load to change the input resistance or conductance to the required value at some frequency below the cut-off frequency, while the other is used to remove the reactive part of the input impedance or admittance at the same frequency.

The compensation frequency can be chosen such that the ripple in the output power is equal to the specified value. The optimum compensation frequency can be found by using a computer program.

The compensation of some often used transformers will be considered here as examples.

Example 5.6

The 1:4 unbalanced transmission-line transformer can be compensated as shown in Fig. 5.27, that is, if

$$Z_0 > 1.35 \, Z_{0, \, opt}$$

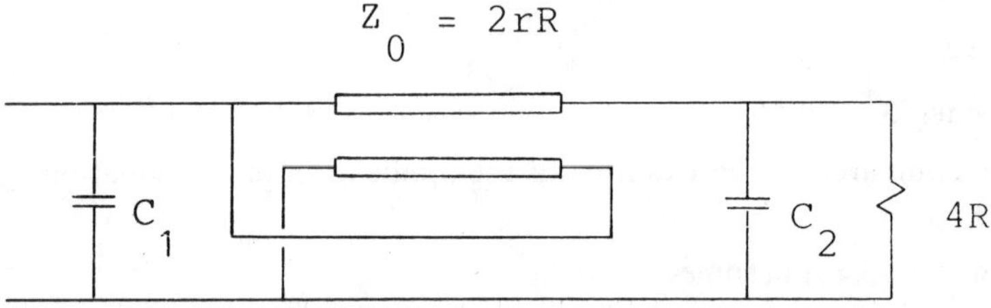

Figure 5.27   Compensation of the Unbalanced 1:4 Transformer when $Z_0 > 1.35 \, Z_{0, \, opt}$

The values of the two compensation capacitors are given by the equations [1]:

$$C_1 = \frac{1 + \cos(\beta l) - \{[1+\cos(\beta l)]^2 - \sin^2(\beta l) \cdot r^2\}^{1/2}}{\omega_{max} \, r R \sin(\beta l)} \tag{5.46}$$

$$C_2 = \frac{2\cos(\beta l) - \{[1+\cos(\beta l)]^2 - \sin^2(\beta l) \cdot r^2\}^{1/2}}{4\omega_{max} \, r R \sin(\beta l)} \tag{5.47}$$

$$r = Z_0/(2R) \tag{5.48}$$

Equations (5.46) and (5.47) can be derived by setting the real part of the input admittance of the transformer terminated in a resistor ($4R$) in parallel with an unknown capacitor (C_2), equal to $1/R$.

C_1 is used to cancel the reactive part of the input admittance.

The compensation frequency ($f_{max} = \omega_{max}/(2\pi)$) is a function of the acceptable variation in the output power and the minimum efficiency required in the passband, and can be found by using a computer program.

The high frequency response of the transformer can be improved considerably by using the compensation technique. This can be appreciated by comparing the electrical lengths of the line, with and without compensation, at the cut-off frequencies of the transformer.

The electrical lengths of the line at the cut-off frequency, with and without compensation, are compared in Table 5.2 for

$$\Delta P_L \leqq 17\%$$

and

$$\eta_r \geq 95\%$$

It can be seen by from Table 5.2 that the high cut-off frequency can be increased by more than an octave above its uncompensated value, independently of the characteristic impedance.

The high frequency response of the transformer can also be improved when

$$Z_0 < 1.35 \, Z_{0, \, opt}$$

Table 5.2

The Electrical Length of the Transmission Line Used in an 1:4 Unbalanced Transmission-Line Transformer, at the Compensation and Cut-Off Frequencies with and without Compensation

$r =$ $Z_0/(2R)$	Line length without compensation (λ)	Line length at compensation frequency (λ)	Line length with compensation (λ)
2.0	0.700	0.144	0.160
2.5	0.050	0.116	0.130
3.0	0.036	0.099	0.110
3.5	0.032	0.085	0.095
4.0	0.027	0.075	0.084
4.5	0.024	0.068	0.075
5.0	0.021	0.061	0.068

This can be done by using an inductor and a capacitor as compensating elements. The inductor is used in series with the high-impedance end of the transformer, while the capacitor is connected in parallel with the low-impedance side, as shown in Fig. 5.28.

When $Z_0 = Z_{0, \, opt}$, the cut-off frequency can be increased by an octave. The required inductance and capacitance can be found by using the following equations:

$$\omega_H L = 0.7558\, Z_0 \tag{5.49}$$

$$\omega_H C = 0.8961\, Y_0 \tag{5.50}$$

$$\omega_H = \omega_{0.2474\lambda} \tag{5.51}$$

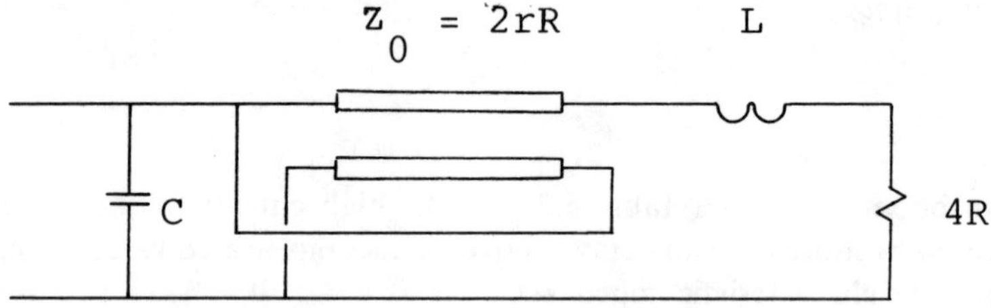

Figure 5.28   Compensation of the Unbalanced 1:4 Transmission-Line Transformer when $Z_0 < 1.35 Z_{0,\,opt}$

Exact equations for the values of the inductance and capacitance can be derived by following the same procedure as before.

Example 5.7

The 1:1 balanced-to-unbalanced (and 1:4 balanced) transmission-line transformer can be compensated for characteristic impedances that are too high, as shown in Fig. 5.29.

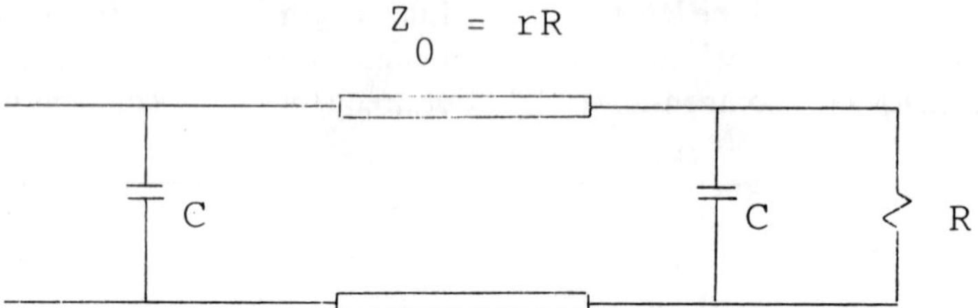

Figure 5.29   Compensation of the 1:1 Balanced-to-Unbalanced Transformer $[r > 1]$

The capacitance of both capacitors are given by the equation

$$C = \frac{1 - [1 - (r^2 - 1)\tan^2(\beta l)]^{1/2}}{\omega_{max}\, r\, R\tan(\beta l)} \tag{5.52}$$

where

$$r = Z_0 / Z_{0,opt} = Z_0 / R \tag{5.53}$$

and ω_{max} is the compensation frequency.

Table 5.3

The Electrical Length of the Transmission Line Used in the 1:1 Balanced-to-Unbalanced Transmission-Line Transformer, at the Compensation and Cut-Off Frequencies, with and without Compensation

$r = Z_0/R$	Line length without compensation λ	Line length at compensation frequency λ	Line length with compensation λ
0.5	0.073	—	—
0.6	0.080	—	—
0.7	0.089	—	—
0.8	0.109	—	—
0.9	0.170	—	—
1.0	∞	—	—
1.1	∞	—	—
1.2	0.119	0.131	0.180
1.3	0.092	0.094	0.155
1.4	0.075	0.075	0.135
1.5	0.065	0.075	0.123
1.6	0.058	0.081	0.116
1.8	0.042	0.081	0.101
2.0	0.041	0.056	0.085
2.5	0.031	0.038	0.066
3.0	0.025	0.044	0.058
3.5	0.021	0.044	0.048
4.0	0.018	0.031	0.043
5.0	0.013	0.031	0.035

The allowable ripple in the output power and the minimum efficiency should be taken into account when the compensation frequency is determined.

The cut-off frequency ($\Delta P_L \geq 17\%$; $\eta_r \geq 95\%$) of the 1:1 transformer, with and without compensation, is shown in Table 5.3 as a function of the characteristic impedance of the line.

It can be seen by inspection of this table, that the cut-off frequency of the 1:1 (and therefore the 1:4 balanced) transmission-line transformer is more sensitive to deviations in the characteristic impedance than the unbalanced 1:4 transformer.

Compensation of the transformer has a significant effect on the cut-off frequency.

Example 5.8

The optimum characteristic impedance of the hybrid transformer shown in Fig. 5.30 cannot be determined by using the rule given in sec. 5.4.1. The reason for this is that the voltage across the end of the line is equal to zero in the balanced case.

It can be shown (by deriving the exact equations for this transformer) that the optimum characteristic impedance for this transformer is the lowest available characteristic impedance.

The hybrid transformer can be compensated with an inductor and a capacitor as shown in Fig. 5.30. The compensation frequency (as well as the cut-off frequency of the transformer, with and without compensation) is given in Table 5.3 as a function of the characteristic impedance.

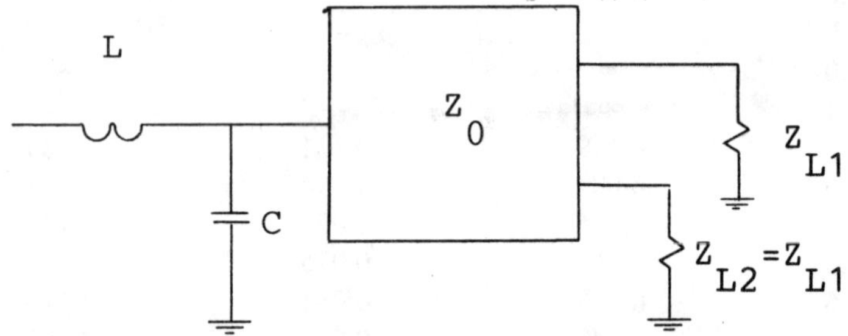

Figure 5.30 Compensation of the Hybrid Divider shown in Fig. 5.14b

The component values are given by the equations:

$$\omega_H L = 0.3015\, Z_{L1}/2 \tag{5.54}$$

$$1/(\omega_H C) = 1.8005\, Z_{L1}/2 \tag{5.55}$$

where the compensation frequency (f_H) can be determined by using Table 5.4.

5.4.5 The Design of Low-Pass LC Networks to Extend the Bandwidth of a Transmission-Line Transformer

The bandwidth of transmission-line transformer can be extended considerably with low or high frequency matching networks.

A network that can be used for low frequency compensation, is shown in Fig. 5.30a. It is sometimes also possible to use its dual, which is shown in Fig. 5.30b.

The optimum reactances of the components of the network are given in Table 5.5 as a function of the allowable ripple in the pass-band. The reactances are normalized for a load resistance of $1\,\Omega$,

Table 5.4

The Electrical Length of the Transmission Line Used in the Hybrid Divider (and Combiner), at the Compensation and Cut-Off Frequencies, with and without Compensation

Z_0/Z_{L1}	Line length without compensation (λ)	Line length at compensation frequency (λ)	Line length with compensation (λ)
0.25	0.212	0.283	0.367
0.50	0.121	0.173	0.303
1.00	0.066	0.093	0.139
2.00	0.029	0.048	0.075
3.00	0.020	0.034	0.057

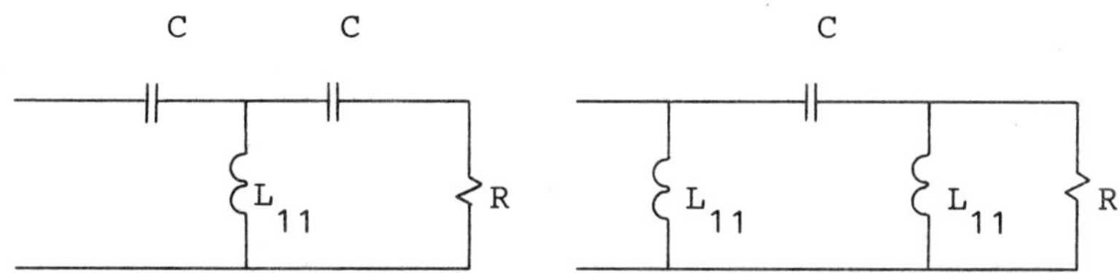

Figure 5.31   Two Low Frequency Impedance-Matching Networks that can be Used to Extend the Bandwidth of a Transformer

Table 5.5

The Normalized Values of the Reactances in Fig. 5.31a at the Lowest Frequency in the Pass-Band as a Function of the Ripple in the Output Power

Ripple (%)	0.5	1.0	2.0	3.0	4.0	5.0
X_C (Ω)	0.2978	0.3647	0.4450	0.5037	0.5475	0.5899
X_L (Ω)	1.8005	1.5204	1.3068	1.2012	1.1399	1.0920
Increase in bandwidth	1.48	1.61	1.73	1.82	1.87	1.90

When this compenstion technique is used, the magnetizing reactance required to meet the low frequency specifications decreases. Because this implies a shorter line, the high cut-off frequency of the transformer will increase.

The exact amount by which the bandwidth will increase is a function of the acceptable ripple in the output power, the magnetic material used, and the losses that can be tolerated.

The new line length can be determined by using the information in sec. 5.4.3.

It should be noted that the maximum voltage across the line is slightly more than that without compensation. The capacitor in series with the load resistance transforms the resistance slightly upward, and the voltage across the line must therefore be higher than that across the load resistance, in order to deliver the same power in the effective resistance in parallel with the magnetizing inductance as in the load. This is illustrated in Fig. 5.32.

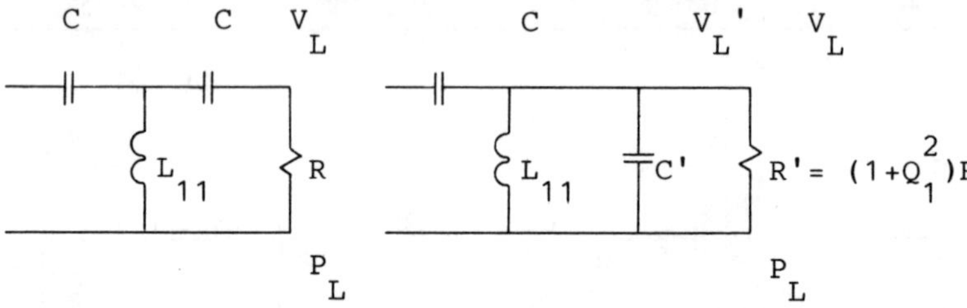

Figure 5.32   Illustration of the Increase in the Voltage across the Magnetizing Inductance of the Transmission-Line Transformer with Low Frequency Compensation

Because of the decrease in the reactance of the magnetizing inductance and the increase in the effective resistance in parallel with it, the unloaded Q of the magnetizing inductance will also change if the losses are to remain the same as without compensation. The maximum allowable flux density in the core will, therefore, also charge (it must be less than before).

Despite the increase in the voltage across the magnetizing inductance and the decrease in the maximum allowable flux density in the core, it is usually worthwhile to use this compensation technique.

If the losses in the core can be ignored and the core size remains the same as before compensation, the improvement in the bandwidth will be that given in Table 5.5, that is, if the transformer is designed to have the same low cut-off frequency as before.

Along with improving the bandwidth, this technique has the added advantage that the required frequency response might be obtained by using an air-cored solenoidal coil instead of magnetic material. When this is done, the increase in the voltage per turn is not a problem and the flux density consideration does not apply.

Example 5.9

As an example of the application of the low frequency compensation technique, the required magnetizing inductance and capacitance for a 1:4 unbalanced transmission-line transformer with

$R_L = 12.5$ and $f_L = 2$ MHz

will be determined. The pass-band ripple is to be 5%.

The compensation networks for $R_L = 1\Omega$ and $R_L = 12.5\Omega$ are shown in Fig. 5.33. The component values in Fig. 5.33a were obtained from Table 5.5.

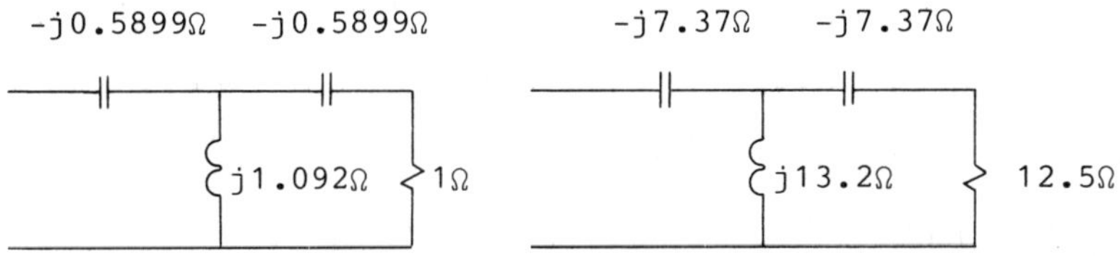

Figure 5.33   The Low Frequency Impedance-Matching Network from Fig. 5.31a, if the Pass-Band Ripple Is 5% and [a] $R_L = 1\Omega$, [b] $R_L = 12.5\Omega$

The positions of the compensating capacitors in the 1:4 transformer are shown in Fig. 5.34a. It can be seen easily that the input impedance of the transformer is the same as that of the equivalent circuit shown in Fig. 5.34b, which is of the same form as the matching network in Fig. 5.33a.

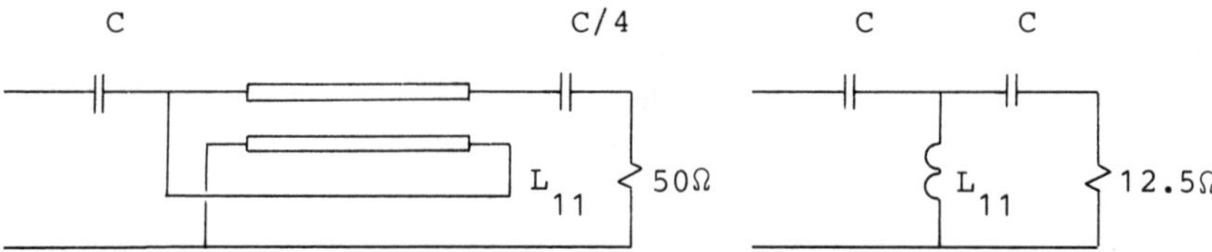

Figure 5.34   The Positions of the Compensating Capacitors in the Unbalanced 1:4 Transmission-Line Transformer and [b] the Equivalent Circuit for the Input Impedance of the Transformer

By using this equivalence, it follows that

C $= 10.8$nF

$C/4 = 2.7$nF

L_{11} $= 1.1 \mu$H

If the required high cut-off frequency of the transformer is low enough, the magnetizing inductance can be realized by using an air-cored solenoidal coil.

Questions and Problems

1. Is the frequency response of a transmission-line transformer dependent on the coupling between the two conductors of the line at (a) high frequencies, (b) low frequencies?

2. Are there any losses in the magnetic core of a transmission-line transformer at high frequencies?

3. Verify the transformation ratios given in Table 5.1 for transmission-line transformers with four lines.

4. Find the configuration of the 1:2.78 transmission-line transformer by using the technique illustrated in Fig. 5.4.

5. Show that the high frequency responses of the balanced-to-unbalanced 1:1 and the balanced 1:4 transmission-line transformers are identical at high frequencies.

6. Prove that the unbalanced current in a transmission-line transformer is not a function of the distance along the line.

7. Determine the optimum characteristic impedances of the transmission-lines used in the transformer of *Problem 5.4.*

8. Determine the optimum characteristic impedances of the transmission lines used in the 1:2.25 transmission-line transformer. Determine the relative lengths of the two lines required if only one core is to be used.

9. What is the optimum characteristic impedance for the hybrid divider of Fig. 5.14b?

10. Find the minimum value of the magnetizing inductance of the transmission lines used in the unbalanced-to-balanced 1:4 transmission-line transformer (Fig. 5.11a). The input reactance of the transformer must be equal to four times the resistance in parallel with it.

11. Determine the minimum value of the magnetizing inductance of a unbalanced-to-balanced 1:1 transmission-line transformer if the output power in the two different loads are to be matched within 2%. $R_{L1} = 12.5\Omega = R_{L2}$.

12. Determine the optimum core size for a 12.5:50Ω balanced transmission-line transformer. $P_L = 100W$; $f = 1.6$ MHz. A toroidal core of 4C4 material is to be used; $\mu_r = 120$.

13. Can the bandwidth of the transformer in *Problem 5.12* be increased if stacked toroidal cores instead of a single toroidal core are used?

14. Is it true that the bandwidth of a transmission-line transformer will usually be higher if a balun core instead of a toroidal core is used?

15. A 50Ω coaxial cable is used in a $12.5{:}50\Omega$ transmission-line transformer. If the line length required to meet the low frequency specifications is 25cm and the velocity on the line is $v = 0.65c$, determine the high cut-off frequency of the transformer.

    Determine the cut-off frequency if the transformer is compensated for the non-optimum characteristic impedance of the transmission line used. Calculate the capacitance values of the compensating capacitors required to compensate the transformer.

16. Determine the inductance and capacitance required to improve the high-frequency response of the transmission-line transformer in the previous problem if a transmission line with the optimum characteristic impedance is used.

17. Compare the high cut-off frequencies of the balanced and unbalanced 1:4 transmission-line transformers.

18. The input impedance of the balanced 9:1 transmission-line transformer shown in Fig. 5.35 is given by the equation [1]:

$$Z_{in} = 9R\ \frac{4 + 5\cos(\beta l) + j6r\sin(\beta l)}{9\cos(\beta l) + j[6/r]\sin(\beta l)} \tag{5.56}$$

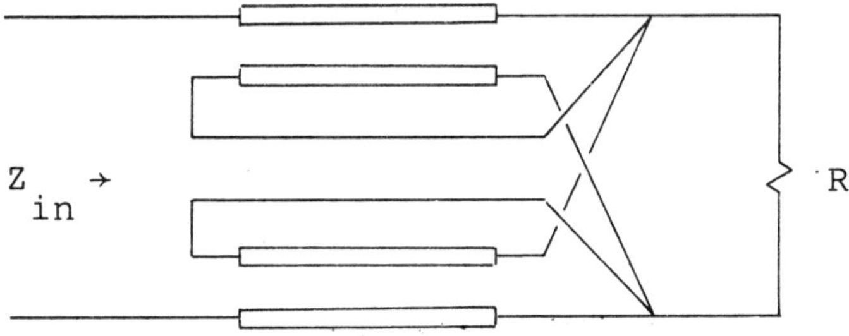

Figure 5.35   The Balanced 9:1 Transmission-Line Transformer

    Determine the cut-off frequency of the transformer if transmission lines with the optimum characteristic impedance are available. Compare the cut-off frequency with those of the 1:4 balanced and unbalanced transmission-line transformers.

19. A hybrid divider and combiner are used to obtain 100W from two 50W power amplifier modules. Two 30cm lines with $Z_0 = 35\Omega$ are used. If $R_L = R_s = 50\Omega$, determine the high cut-off frequencies of the hybrid transformers.

20. Determine the positions of the low frequency compensation capacitors in a balanced 1:4 transmission-line transformer. If $R_L = 50\Omega$, calculate the required capacitance of the compensating capacitors.

References and Additional Reading

1. Hilbers, A.H., "On the Design of HF Wideband Transformers (ECO 6907)," Philips C.A.B. Group, Eindhoven, 1969.

2. Ruthroff, C.L., "Some Broad-Band Transformers," *Proc. IRE,* August 1959.

3. Rotholtz, E., "Transmission-Line Transformers," *IEEE Trans. Microwave Theory and Tech.,* MTT-29, No. 4, April 1981.

4. Krauss, H.L., and C.W. Allen, "Designing Toroidal Transformers to Optimize Wideband Performance," *Electronics,* August 16, 1973.

5. Dutta Roy, S.C., "A Transmission-Line Transformer Having Frequency Independent Properties," *Int. J. Circuit Theory App.,* Vol. 8, January 1980, pp. 55-64.

6. Van Nierop, J.H., "Evolution of a 4:1 Impedance Transforming Balun," *IEEE Trans. Antennas Propag.,* Vol. AP-30, No. 4, July 1982.

7. Abrie, P.L.D., Impedance Matching Networks and Bandwidth Limitations of Class B Power Amplifiers in the HF and VHF Ranges," Master's Thesis, University of Pretoria, 1982.

CHAPTER 6
THE DESIGN OF WIDEBAND
IMPEDANCE-MATCHING NETWORKS

6.1 INTRODUCTION

Impedance-matching networks serve the dual purpose of matching a load to a source in the pass band and attenuating unwanted signals outside it.

When the load and source are purely resistive, LC networks can be designed relatively easily to fulfill the filter specifications in wideband matching networks. It is, however, difficult, if not impossible, to scale impedances over wide bandwidths by using only a limited number of inductors and capacitors. This transformation function can only be done with transformers when the bandwidth-transformation product becomes large.

It is, however, possible to transform resistances over large distances with LC networks, if the required bandwidth is relatively small.

When the load or source impedance is reactive, part of the impedance transformation function of the matching network is to remove this reactiveness. The extent to which this can be done is a function of the load impedance itself, as well as the transducer power gain *versus* frequency response required.

The limitations of a specific load impedance for a specific frequency response can be determined in at least three ways. Fano set up a set of integral equations which can be used to determine these constraints [13], while Youla formulated the constraints in terms of Laurent series expansions [21]. Carlin advanced an iterative procedure for this purpose [3].

Because of its relative simplicity, only the iterative technique developed by Carlin will be presented here.

With the limitations of a particular load (or source) known, a network that will provide the required power gain *versus* frequency response can be designed by using direct synthesis or iterative techniques. Both of these approaches will be discussed in this chapter.

Networks for matching a complex load to a complex source are often required. A theoretical approach to solving this class of problems was developed by Chen and Satyanarayana [15], and more recently an alternative and simplified theory was introduced by Carlin and Yarman [17]. Carlin and Yarman also developed iterative techniques for matching a complex load to a complex source [16, 17]. Because of its relative simplicity and the superior results obtainable by using it [18], only iterative techniques for matching a complex load to a complex source will be considered.

It is often possible to design matching networks for complex terminations by initially assuming the terminations to be purely resistive. The reactances are then absorbed parasitically into the network when the design is completed.

Because the effort required to design a network in this way is minimal, when it can be done, this approach will also be considered in this chapter.

Impedance-matching networks are often required to provide a power gain *versus* frequency response with a positive slope in the pass band. LC networks can be designed either iteratively or directly to fulfill this requirement relatively easily. There is, however, the disadvantage that the source will inevitably be mismatched at the lower frequencies in the pass band. Because this does not necessarily apply to RLC impedance-matching networks, the design of these networks will also be examined in this chapter.

6.2 DETERMINING AN IMPEDANCE FUNCTION FOR A SET OF IMPEDANCE VERSUS FREQUENCY COORDINATES

When impedance-matching networks are designed, impedance (or admittance) functions that will approximate a set of discrete impedance *versus* frequency coordinates are often required.

The set of coordinates might be the measured input impedance of a transistor or antenna, or it could be the output or input impedance (admittance) of a network to be designed.

It is sometimes possible to approximate the measured impedance of a device with simple RC, RL, or RLC equivalent circuits. This can usually be done when the resistive part of the impedance or admittance is approximately constant over the frequency range of interest.

The components of such an equivalent circuit can be determined by setting up an equation for the input impedance or admittance of the network chosen, and equating its real and imaginary parts to the measured values.

Although this technique has its use, more sophisticated techniques are often required.

A major problem in finding an impedance function that will fit a given set of coordinates is realizability. The function obtained must be positive-real.

A technique that usually gives good results is one based on the fact that the reactance (susceptance) of a minimum-impedance (admittance) function can be determined when the resistance (conductance) is known [9]. The equivalent circuit of a minimum-impedance function has a parallel capacitor or inductor as the last element, is shown in Fig. 6.1.

Because the reactance can be determined when the resistance is known, it follows that the impedance itself is known when its resistive part is known. In terms of equations, if

$$R(\omega) = \left[\sum_{2n} C_j/(\omega-\omega_j) + R_0/2\right] + \left[\sum_{2n} \overline{C}_j/(\omega-\overline{\omega}_j) + R_0/2\right] \tag{6.1}$$

then

$$Z(j\omega) = \sum_{2n} 2C_j/(\omega-\omega_j) + R_0 \tag{6.2}$$

The poles ω_1, ω_2, ..., ω_{2n} are the first and second quadrant poles of the resistance function (which is an even function), while C_1, C_2, ..., C_{2n} are the residues of these poles. The poles $\overline{\omega}_1$, $\overline{\omega}_2$, ..., $\overline{\omega}_{2n}$ and the residues $\overline{C}_1$, $\overline{C}_2$, ..., $\overline{C}_{2n}$ are the conjugates of the first and second quadrant poles and the residues, respectively.

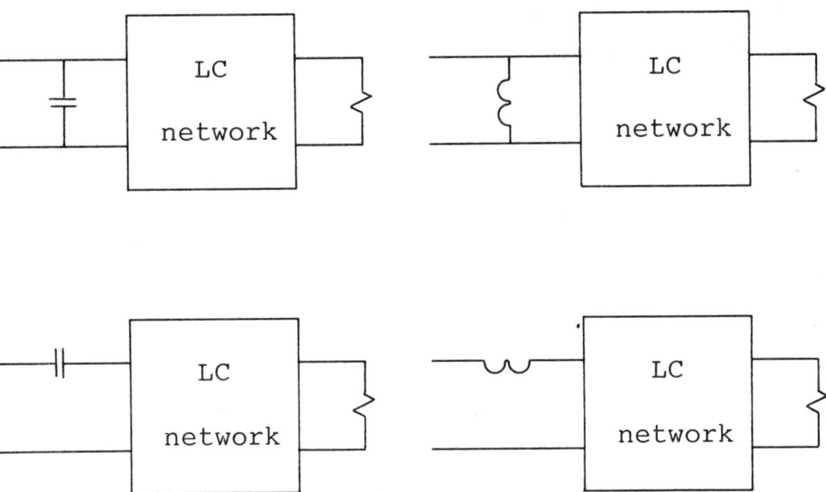

Figure 6.1   Equivalent Circuits for (a) Minimum-Impedance and (b) Minimum-Admittance Functions

If a rational function for the resistive part of the minimum impedance function can be found, the impedance function itself can be found by using (6.1) and (6.2).

If the resistance function is assumed to be of the form

$$R(\omega) = \omega^{2p}/(a_{2m}\omega^{2m} + a_{2m-2}\omega^{2m-2} + \ldots + a_0) \tag{6.3}$$

the unknown coefficients in the function can be found easily by using the equations relevant to fitting a polynomial to a given set of coordinates with least-square error.

Equation (6.3) can be changed to have the following form

$$T(\omega) = \omega^{2p} / R(\omega)$$
$$= a_{2m} (\omega^2)^m + a_{2m-2} (\omega^2)^{m-1} + \ldots + a_0$$
$$= b_m x^m + b_{m-1} x^{m-1} + \ldots + b_0 \tag{6.4}$$

where

$$x = \omega^2$$

Because $R(\omega)$ and the number of zeros at the origin (the designer must choose the number of zeros) are known, the function $T(\omega)$ is also known at discrete frequencies.

The coefficients $b_0, b_1, \ldots, b_m$ can now be determined by solving the following set of equations:

$$s_0 b_0 + s_1 b_1 + \ldots + s_m b_m = t_0$$
$$s_1 b_0 + s_2 b_1 + \ldots + s_{m+1} b_m = t_1$$
$$\cdot$$
$$\cdot$$
$$\cdot$$
$$s_{m+1} b_0 + s_{m+2} b_1 + \ldots + s_{2m} b_m = t_m \tag{6.5}$$

where

$$s_0 = h$$
$$s_1 = \Sigma \, x_i$$
$$s_{2m} = \Sigma \, x_i^{2m} \tag{6.6}$$

$$t_0 = \Sigma \, T(\omega_i)$$
$$t_1 = \Sigma \, T(\omega_i) \, x_i^2$$
$$t_m = \Sigma \, T(\omega_i) \, x_i^{2m} \tag{6.7}$$

with h the number coordinates $(x_i, \, T(\omega_i))$.

With these equations solved, the coefficients in (6.4) and, therefore, those in (6.3) are known.

The minimum-impedance function itself can now be determined by using (6.1) and (6.2). In order to use these equations, the poles of the resistance function must be determined.

The impedance function determined, as described above, will be positive-real if care is taken to ensure that the approximation function has no real zeros in the ω-plane.

It is possible that the input impedance, as given by the synthesized minimum impedance function, deviates slightly from the initial set of coordinates. This can usually be remedied by adding a pole at the origin of infinity to the impedance function. Alternatively, the minimum-admittance function corresponding to the set of impedance coordinates can be determined, or a computer optimization program can be used to improve the match between the two sets of impedances.

An approximation function for the resistive part of an impedance function can be found by using the program PLNM FORTRAN in Appendix A. With $R(\omega)$ known, the program ZVR FORTRAN in Appendix B can then be used to find the corresponding minimum-impedance function.

It happens occasionally that the polynomial determined by the program PLNM FORTRAN is not positive-real. There are two reasons for this. The first is that the number of coordinates specified in areas where $T(\omega)$ approaches zero is insufficient. This can be remedied easily by specifying more coordinates in these areas.

The second reason is that the increase in $T(\omega)$ is too slow at high frequencies. In such a case the polynomial will have a zero on the real axis of the ω-plane. This, in turn, implies a real pole in the resistance function $R(\omega)$, and therefore a pole on the $j\omega$-axis for the function $R(s)$, which is of course not allowable.

In order to overcome this problem, the facility to add an extra data point at a frequency 1.5 times that of the highest frequency specified was built into the program.

When this option is chosen, the initial value of the function $T(\omega)$ at the new frequency relative to that at the highest frequency in the data set, as well as the increment factor and the number of polynomials to be determined in this way, must be entered into the program.

Example 6.1

As an example of the application of the programs PLNM FORTRAN and ZVR FORTRAN, an equivalent circuit for the input impedance of the Motorola MFR406 power transistor will be determined.

The input impedance of the MFR406, as specified by the manufacturer, is tabulated in Table 6.1. It can be seen by inspection of the data in this table that the resistance approaches a constant at low frequencies. The approximation function for the resistance should therefore be of low-pass form, that is, the function must be of the form

$$R(\omega) = 1/[a_{2m}\ \omega^{2m} + a_{2m-2}\ \omega^{2m-2} + \ldots + a_0] \qquad (6.8)$$

In polynomial form, this becomes

$$T(\omega) = 1/R(\omega) = b_m\ (\omega^2)^m + b_{m-1}\ (\omega^2)^{m-1} + \ldots + b_0 \qquad (6.9)$$

The coefficients $b_m, \ldots, b_0$ will be determined by using the program PLNM FORTRAN (App. A). In order to do this, the following data must be entered into the program:

1)   The degree of the polynomial (m); the number of coordinates to be given; a one or a zero depending on whether problems were experienced with zeros on the real axis of the ω-plane in a previous attempt.

2)   The coordinates (f_i, $T(\omega_i)$), where f_i is one of the frequencies (Hz) of interest.

3)   If zeros on the real axis were a problem, an initial value for the function $T(\omega)$ at the new frequency relative to that at the last frequency entered in the data set, ($x_3 = T(\omega_a)/ T(\omega_n)$), as well as an increment factor (x_2, where $T(\omega_{a,i+1})$ $= x_2\ T(\omega_{a,i})$) and the number of times $T(\omega_a)$ should be increased with this factor (n_1) must be entered at this point in the sequence n_1, x_2, x_3. As an example, suitable entries would be 5, 1.5, and 1.0.

The input data for this particular problem is shown in Table 6.2.

Table 6.1

The Input Impedance of the MRF406 as a Function of the Frequency

Frequency (MHz)	Input Impedance (Ω)
2	$7.5 - j2.6$
5	$5.2 - j2.4$
10	$3.1 - j1.9$
15	$2.3 - j1.8$
20	$1.7 - j1.7$
25	$1.3 - j1.4$
30	$1.0 - j1.0$

The polynomial obtained from the program is

$$T(\omega) = 0.14956 + 0.41041 \cdot 10^{-4}\omega^2 - 0.10217 \cdot 10^{-8}\omega^4 + 0.1538 \cdot 10^{-13}\omega^6$$

Therefore, the resistance function is

$$R(\omega) = 1/[0.1496 + 0.4104 \cdot 10^{-4}\omega^2 - 0.1022 \cdot 10^{-8}\omega^4 + 0.1538 \cdot 10^{-13}\omega^6]$$

The poles of this function are

$$\omega = \pm\ 210.665 \pm j97.424$$
$$000.000 \pm j57.884$$

Table 6.2

The Input Data for the Program PLNM FORTRAN

```
3,7,0
2.0E0,0.1333E0
5.0E0,0.1923E0
10.0E0,0.3225E0
15.0E0,0.4248E0
20.0E0,0.5882E0
25.0E0,0.7692E0
30.0E0,1.0000E0
```

There are four poles in the complex ω-plane and two poles on the imaginary axis.

These poles as well as the numerator of the resistance function can be entered into the program ZVR FORTRAN in order to determine the minimum impedance function corresponding to the resistance function.

The input data must be specified in the following form:

1)   The degree of the numerator of the resistance (conductance) function.

2)   The coefficients of the numerator polynomial, starting with the lowest order coefficient:

b_0

b_1

et cetera

These coefficients must be scaled by a factor a_{2m} (see Eq. (6.3)) to ensure that the coefficient of the ω^{2m} term in the denominator of the resistance function is equal to one.

3)   The number of poles in the first quadrant (those on the imaginary axis are excluded).

4)   The poles in the first quadrant:

(r_1, x_1)

(r_2, x_2)

et cetera

5)   The number of poles on the positive side of the imaginary axis.

6)   The poles on the positive side of the imaginary axis:

$(0, y_1)$

$(0, y_2)$

et cetera

7)   The number of frequencies (an even number) at which the impedance (admittance), as given by the minimum impedance function, must be calculated.

8)   The frequencies (entered in pairs) at which the impedance (admittance) must be calculated.

The input data for the program is shown in Table 6.3.

Table 6.3

The Input Data for the Program ZVR FORTRAN

```
0
6.5020.10¹³
1
(210.665,97.424)
1
(0.00000,57.884)
8
2.0,5.0
7.5,13.0
15.0,20.0
25.0,30.0
```

The minimum-impedance function obtained from the program is

$$Z(j\omega) = \frac{-0.3948 \cdot 10^3\omega^2 + 0.9977 \cdot 10^5 j\omega + 0.2085 \cdot 10^8}{-j\omega^3 - 0.2527 \cdot 10^3\omega^2 + 0.6515 \cdot 10^5 j\omega + 0.3118 \cdot 10^7}$$

The impedance function is given as a function of s by the equation:

$$Z(s) = \frac{0.3948 \cdot 10^3 s^2 + 0.9977 \cdot 10^5 s + 0.2085 \cdot 10^8}{s^3 + 0.2527 \cdot 10^3 s^2 + 0.6515 \cdot 10^5 s + 0.3118 \cdot 10^7} \tag{6.10}$$

The impedance as given by this equation is compared to the measured impedances in Table 6.4.

Table 6.4

The Input Impedance of the MFR406 Compared to the Impedance as given by Eq. (6.10), as well as the Input Impedance of the Equivalent Circuit from Fig. 6.2

Frequency	Input impedance of transistor	Impedance as given by Eq. (6.10)	Input impedance of the equivalent circuit shown in Fig. 6.2
(MHz)	(Ω)	(Ω)	(Ω)
2.0	$7.5 - j2.6$	$6.4 - j1.3$	$6.4 - j1.2$
5.0	$5.2 - j2.4$	$5.3 - j2.6$	$5.3 - j2.4$
7.5	$3.9 - j2.1$	$4.2 - j3.1$	$4.2 - j2.8$
13.0	$2.6 - j1.8$	$2.6 - j3.0$	$2.6 - j2.5$
15.0	$2.3 - j1.8$	$2.3 - j2.9$	$2.3 - j2.3$
20.0	$1.7 - j1.7$	$1.7 - j2.6$	$1.7 - j1.8$
25.0	$1.3 - j1.4$	$1.3 - j2.4$	$1.3 - j1.3$
30.0	$1.0 - j1.0$	$1.0 - j2.3$	$1.0 - j1.0$

It can be seen by inspection of Table 6.4, that the impedance given by Eq. (6.10) correlates well with the input impedance of the transistor. The match can be improved, however, by adding a pole at infinity (series inductor) to the impedance function. The resulting network is shown in Fig. 6.2.

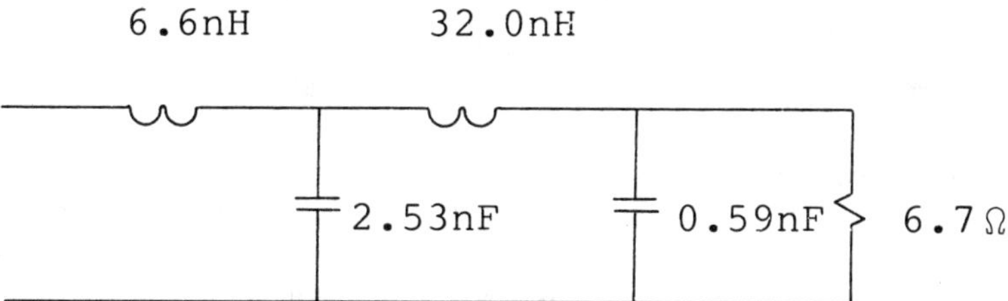

Figure 6.2   An Equivalent Circuit for the Input Impedance of the MRF 406 Transistor

If necessary, the correlation between the impedance of the transistor and that given by the new equation can be improved by entering the network into an optimization program. The alternative is to determine the minimum-admittance function corresponding to the measured data.

6.3 THE ANALYTICAL APPROACH TO IMPEDANCE MATCHING

Impedance-matching networks can be designed directly (analytically) or iteratively. The direct approach to the design of these networks will be discussed in this section.

The simplest form of the impedance-matching problem is where the load and source impedances are purely resistive and equal. In this case, the matching network has only a filtering function.

When an explicit function for the required transducer power gain *versus* frequency response is known, the network required to meet the filtering specifications can be designed easily with the well-known Darlington synthesis technique. Darlington synthesis will be discussed in sec. 6.3.1.

When a band-pass network is designed according to Darlington synthesis, the source or load resistance of the network synthesized often does not have the required value.

When the bandwidth is relatively narrow (less than two octaves) and the network contains L-sections consisting only of inductors or capacitors, it is sometimes possible to transform the source or the load to have the required value by using LC transformers. The design of these networks will be discussed in sec. 6.3.2.

The only solution possible when the required bandwidth transformation is large, is to use conventional or transmission-line transformers or, at microwave frequencies, tapered lines.

Although transformation over wide bandwidths can be accomplished by using transmission-line transformers, the easily obtainable transformation ratios are limited (1/4, 1/9, *et cetera*).

It is often possible to eliminate the need for resistance transformation by designing a network with a semi-low-pass or semi-high-pass transducer power gain *versus* frequency response (that is, matching networks without transmission zeros at the origin or infinity, respectively. This technique can also be used to provide matching between unequal load and source resistances when the transformation bandwidth is small enough.

It should be noted that the transformation distance-bandwidth product of LC impedance-matching networks are limited.

That this product should be limited, follows from the fact that high transformation Q-factors are required to transform resistances considerably. A high transformation Q, in turn, implies a network Q and, therefore, a narrow bandwidth.

The gain-bandwidth products of impedance-matching networks are also limited when the load or source impedance is reactive. The extent to which this reactiveness can be removed is limited because negative inductors and capacitors do not exist.

When only the load (or source) impedance is reactive, the gain-bandwidth constraints imposed by the load (or source) on a particular transducer power gain *versus* frequency response can be determined by using the integral constraints formulated by Fano, the Laurent series constraints of Youla, or the iterative approach of Carlin.

The constraints on several types of loads, usually for Chebyshev responses, are available in the literature in explicit form. Only the limitations imposed by a simple parallel RC (or series RL) load will be considered here. This will be done in sec. 6.3.3.

The constraints imposed by any other load can be determined by using Carlin's iterative approach which will be discussed in sec. 6.4.

The analytical design of networks for matching a complex load to a purely resistive source is discussed in sec. 6.3.4. Two analytical approaches to solving impedance-matching problems belonging to this class will be considered.

When the technique discussed in sec. 6.3.4.1 is used, the complexity of the load is immaterial. As long as the specified transducer power gain *versus* frequency response is realizable, any load can thereby be matched to a resistive source.

The parasitic absorption approach discussed in sec. 6.3.4.2 can only be used when the terminations can be approximated with simple equivalent circuits.

When the load is parasitically absorbed into an impedance matching network, it is initially assumed to be purely resistive. A network with a suitable topology is then designed and, if the gain-bandwidth constraints imposed by the reactive load on the transducer power gain *versus* frequency response chosen are taken into account, it will be possible to absorb the reactive part of the load into the designed network.

Although it is limited to simpler problems, this technique has the advantages that less effort is required in designing the network and it can also be used to solve simple problems where both the source and the load terminations are reactive.

The principle of parasitic absorption is illustrated in Fig. 6.3.

Although it is also possible to match a complex load to a complex source analytically [15,17], the relevant theory will not be considered here. The main reason for this is that much better results can be obtained with considerably less effort by using iterative techniques [18].

The additional theory necessary to design commensurate distributed networks will be covered in sec. 6.3.6.

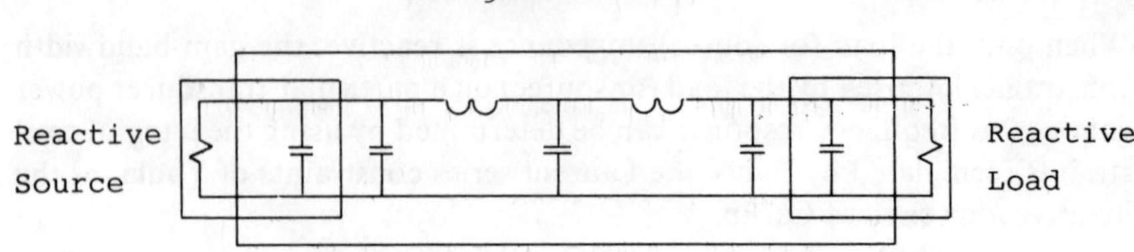

Figure 6.3   Illustration of the Principle of Parasitic Absorption

Richards' transformation, Kuroda and Norton's identities, unit elements, and their extraction will be considered. Under Richards' transformation all of the theory applicable to the design of lumped element networks also apply to commensurate distributed networks.

6.3.1   Darlington Synthesis of Impedance-Matching Networks

When a resistive load is matched to a resistive source, a network that will provide the required transducer power gain *versus* frequency response can be designed by following the procedure outlined here.

Bear in mind that the source (or load) resistance of the network designed by following this procedure will often not be equal to the specified value.

Transformers can be used to change the impedance levels in wideband designs. When the network designed contains band-pass L-sections, it is sometimes also possible to use LC transformers for this purpose.

If a network without transmission zeros at $\omega = 0$ or $\omega \to \infty$ is designed and the gain-bandwidth limitations are not a problem, the source resistance will always have the required value.

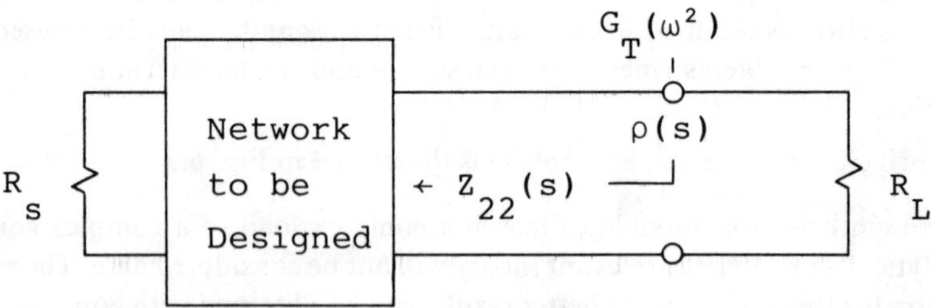

Figure 6.4   The Design of an Impedance Matching Network when the Terminations Are Purely Resistive

Specifications: The transducer power gain versus frequency response required and the values of the load and the source resistances.

Darlington Synthesis

1) Replace all ω^2 terms in the specified transducer power gain function $G_T(\omega^2)$, with $-s^2$ terms.

Determine the product $\rho(s)\,\rho(-s)$, where $\rho(s)$ is the reflection coefficient $s_{22}(s)$ corresponding to the specified transducer power gain function:

$$\rho(s)\,\rho(-s) = 1 - G_T(-s^2) \tag{6.11}$$

2) The next step is to determine $\rho(s)$:

Assign all the left-hand plane (LHP) poles of $\rho(s)\,\rho(-s)$ to $\rho(s)$. It is not necessary to assign only LHP zeros to $\rho(s)$. Any combination of zeros can be assigned to it, as long as they are assigned in conjugate pairs and the relationship between $\rho(s)$ and $\rho(-s)$ is kept in mind.

When a minimum phase network (that is, a network with minimum phase variation in the pass band) is required, all the LHP zeros must be assigned to $\rho(s)$.

When the parasitic absorption approach is followed, the right-hand plane zeros must be assigned to $\rho(s)$ if the source is reactive. When both the load and the source impedances are reactive, it is usually best to try all possible combinations.

The sign assigned to $\rho(s)$ is often important. When low-pass networks are designed, and the load and the effective source resistance as viewed from the load terminals at $\omega = 0$ are not equal, the sign of $\rho(s)$ is determined by its value at the origin. The sign must be such that

$$\rho(0) = \frac{Z_{22}(0) - R_L}{Z_{22}(0) + R_L} \tag{6.12}$$

When the two resistances are equal, a plus or a minus sign may be assigned to $\rho(s)$.

When the value of $\rho(s)$ is plus one (open-circuit) at infinity, the first element (as viewed from the load terminals) of the network will be a series inductor. When the sign is negative (short-circuit), the first element will be a parallel capacitor. The topology of the network is, therefore, a function of the sign of $\rho(s)$ at infinity.

When high-pass networks are designed and the resistances are unequal, the sign of $\rho(s)$ must be such that

$$\rho(s) = \frac{Z_{22}(\infty) - R_L}{Z_{22}(\infty) + R_L} \tag{6.13}$$

Where the load and source resistances are equal and a network with a series capacitor as first element is required, the sign of $\rho(s)$ must be such that

$$\rho(0) = +1$$

When

$$\rho(0) = -1$$

the first element will be a parallel inductor.

The relationship between the value of the reflection coefficient at zero or infinity and the topology of the network, is summarized in Table 6.5 for low-pass, high-pass, and band-pass networks.

The information in this table is useful when networks are designed to absorb the reactive part of the load impedance parasitically.

3)   Find the impedance function corresponding to the reflection coefficient $\rho(s)$ by using the equation:

$$\frac{Z_{22}(s)}{R_L} = \frac{1 + \rho(s)}{1 - \rho(s)} \tag{6.14}$$

4)   Synthesize the required network by using standard filter theory. If the topology is important, the transmission zeros and the poles at the origin and infinity must be extracted in the proper sequence.

5)   If the source resistance of the network does not have the required value, transformers or LC transformers can be used to change the impedance level as required.

Example 6.2

A third-order Butterworth network will be synthesized as an example of the application of this procedure. $R_L = 1\Omega = R_s$ and the required 3 dB cut-off frequency is 1 rad/s. A low-pass network with an inductor as the first element is required.

1)   The transducer power gain function for the third-order Butterworth characteristic is

$$G_T(\omega^2) = 1/[1 + \omega^6]$$

By substituting each ω^2 term with $-s^2$, this becomes

$$G_T(-s^2) = 1/[1 + (-s^2)^3] = 1/[1 - s^6]$$

The product $\rho(s)\,\rho(-s)$ can now be determined:

$$\rho(s)\,\rho(-s) = 1 - G_T(-s^2)$$

$$= \frac{s^6}{s^6 - 1}$$

$$= \frac{\pm s^3}{s^3 + 2s^2 + 2s + 1} \quad \frac{\pm s^3}{-s^3 + 2s^2 - 2s + 1}$$

2) All the LHP poles and half of the $j\omega$-axis zeros are assigned to $\rho(s)$.

Because $R_L = 1\,\Omega = R_s$, a positive or a negative sign can be assigned to $\rho(s)$, that is, if the topology was not important. Because an inductor is required as the first element in this example, the output impedance of the network will be high at high frequencies and, therefore, a positive sign must be assigned to $\rho(s)$:

$$\rho(s) = s^3/(s^3 + 2s^2 + 2s + 1)$$

3) The output impedance of the network to be designed is given by

$$\frac{Z_{22}(s)}{R_L} = \frac{1 + \rho(s)}{1 - \rho(s)} \tag{6.14}$$

$$= \frac{2s^3 + 2s^2 + 2s + 1}{2s^2 + 2s + 1}$$

4) The network can now be synthesized by continued fractionation of the impedance function:

$$\frac{Z_{22}(s)}{R_L} = s + \cfrac{1}{2s + \cfrac{1}{s + 1/1}}$$

Because the source resistance has the required value, no transformers or LC transformers are required in this particular case.

The designed network is shown in Fig. 6.5.

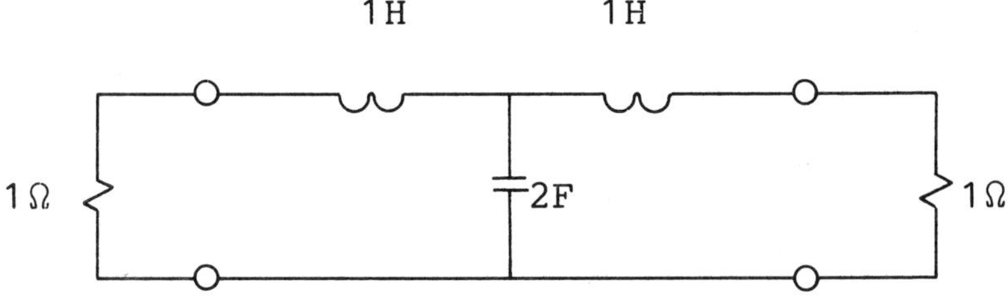

Figure 6.5   A Network with Third-Order Butterworth Response

Table 6.5

The Relationship between the Reactive Part of the Output Impedance of a Network and the Value of the Corresponding Reflection Coefficient ($\rho_{22}(s)$) at the Origin and Infinity

Type of Network	$\rho(s)$	$Z_{22}(s)$ at $\omega=0$	$Z_{22}(s)$ at $\omega\to\infty$	Example
Lowpass	$\rho(0) = \dfrac{R_s-R_L}{R_s+R_L}$ $\rho(\infty) = 1$	Resistive	Inductive	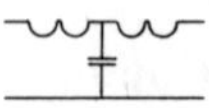
	$\rho(0) = \dfrac{R_s-R_L}{R_s+R_L}$ $\rho(\infty) = -1$	Resistive	Capacitive	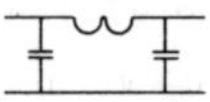
Highpass	$\rho(0) = 1$ $\rho(\infty) = \dfrac{R_s-R_L}{R_s+R_L}$	Capacitive	Resistive	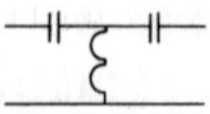
	$\rho(0) = -1$ $\rho(\infty) = \dfrac{R_s-R_L}{R_s+R_L}$	Inductive	Resistive	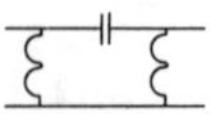
Bandpass	$\rho(0) = 1$ $\rho(\infty) = 1$	Capacitive	Inductive	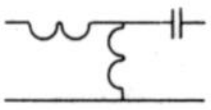
	$\rho(0) = -1$ $\rho(\infty) = 1$	Inductive	Inductive	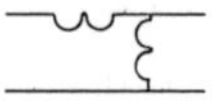
	$\rho(0) = 1$ $\rho(\infty) = -1$	Capacitive	Capacitive	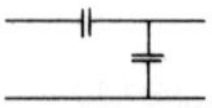
	$\rho(0) = -1$ $\rho(\infty) = -1$	Inductive	Capacitive	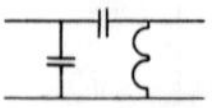

6.3.2   LC Transformers

The impedance level in a band-pass network containing an L-section consisting of capacitors or inductors only, can be changed by replacing the L-section with a suitable T- or Π-section.

In a similar way to the L-sections discussed in Ch. 3, the input impedance will be transformed downward when the element of the L-section to the right is a parallel element and it will be transformed upward when it is a series element.

The T- and Π-section equivalents for the band-pass L-sections are shown in Table 6.6.

The maximum transformation distances of these sections (n^2) are limited by the ratio of the series and the parallel reactances of the original section, as can be seen by inspection of (6.15) to (6.18).

The use of these sections will be illustrated with an example.

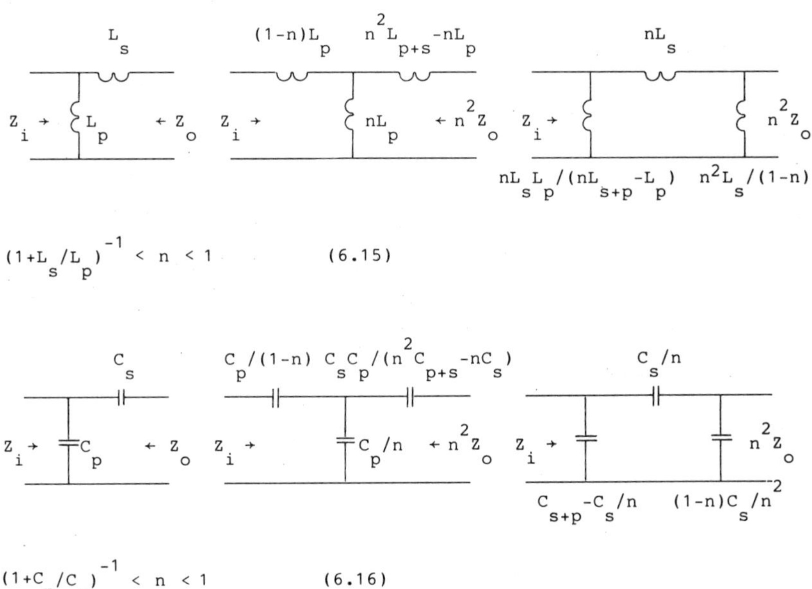

Figure 6.6a   LC Transformers Yielding a Downward Transformation

Example 6.3

A band-pass network with a second-order Chebyshev response is shown in Fig. 6.7. The ripple in the pass band is 0.5 dB, the center frequency f_0, and the bandwidth B.

The input impedance of the network can be transformed downward (the element to the right is a parallel element) by replacing either of the two band-pass L-sections with an equivalent T- or Π-section.

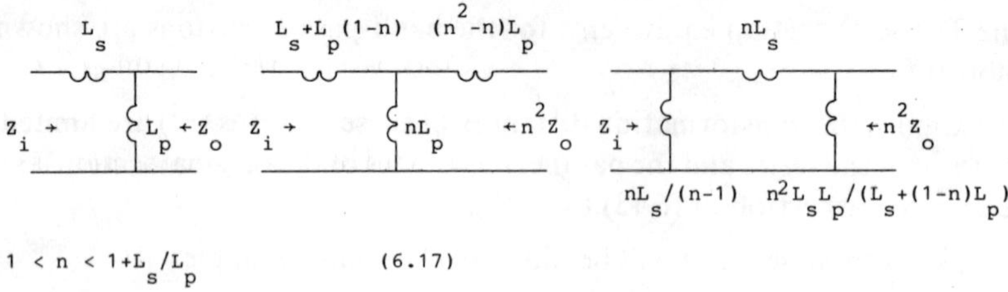

$$1 < n < 1 + L_s/L_p \qquad\qquad (6.17)$$

Figure 6.6a   LC Transformers Yielding an Downward Transformation

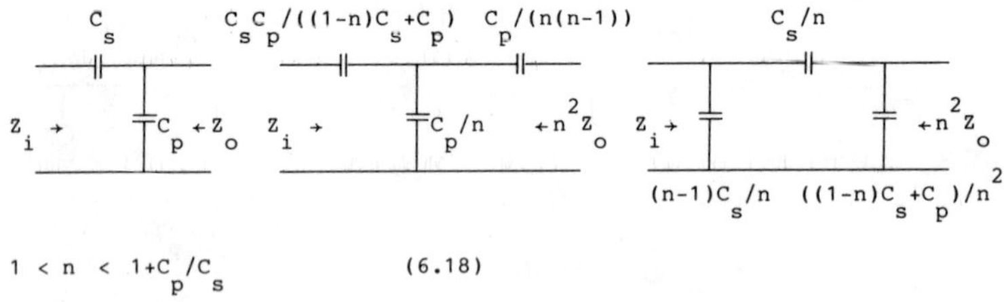

$$1 < n < 1 + C_p/C_s \qquad\qquad (6.18)$$

Figure 6.6b   LC Transformers Yielding an Upward Transformation

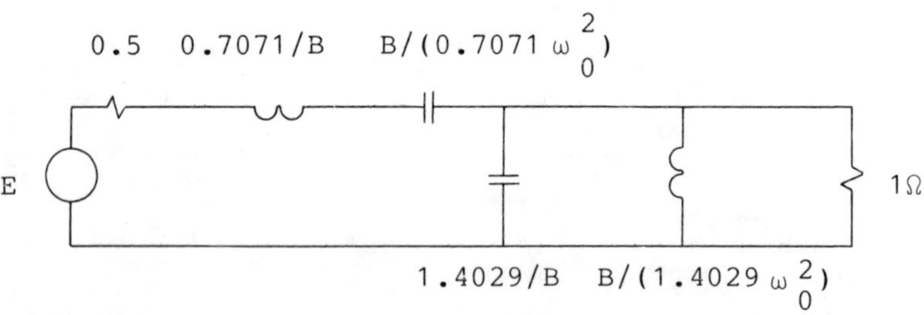

Figure 6.7   A Band-Pass Network with a Second-Order Chebyshev Response (Center Frequency ωw_0 rad/s; Bandwidth B rad/s)

The maximum transformation distance (n^2) possible by replacing either of the two band-pass L-sections in Fig. 6.7 with its equivalent T- or Π-section, can be determined by using (6.12) to (6.15):

$$n_{max}^2 = \left[\cfrac{1}{1 + \cfrac{1.4209}{B} \Big/ \cfrac{B}{0.7071\omega_0^2}}\right]^2$$

$$= \left[\cfrac{1}{1 + (\omega_0/B)^2}\right]^2 \qquad (6.19)$$

The transformation distance obtainable is, therefore, a function of the ratio of the center frequency and the bandwidth of the network. (This ratio can be defined as the Q-factor of the network).

The maximum transformation distance for different values of the relative bandwidth (f_H/f_L) is shown in Table 6.6. The transformation distance obtainable is large for small bandwidths and small for large bandwidths.

Table 6.6

The Maximum Transformation Distance for an LC Transformer in the Network shown in Fig. 6.7, as a Function of the Relative Bandwidth (f_H/f_L)

Relative Bandwidth	Transformation distance of the LC transformer
2	9.0
3	3.0
5	1.7
10	1.2

The transformed network for $\omega_0 = 1.732$ rad/s, $B = 3$ rad/s ($f_H/f_L = 3$), $R_L = 1\Omega$, and $R_s = 1/6\ \Omega$ is shown in Fig. 6.8. Note that the inductor to the left of the replaced capacitive L-section is scaled with a factor of 3. This must be done because of the change in impedance level caused by the LC transformer.

Because of the difference in the load and source impedances imposed by the ripple specification and the order of the network, the transformation distance of the network in Fig. 6.8, is twice that of the LC transformer.

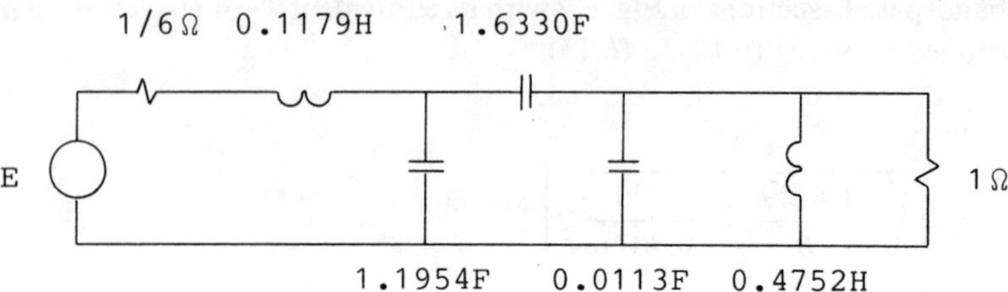

Figure 6.8:   A Network for Matching a 1Ω Load to a Source with 1/6Ω Internal Resistance [Chebyshev Response, 1/2 dB Ripple, $\omega_0 = 1.732\,rad/s$, B = 3 rad/s]

6.3.3   The Gain-Bandwidth Constraints Imposed by a Parallel RC and a Series RL Load

The constraints imposed by a parallel RC load on the gain-bandwidth product of a lossless network with reflection coefficient $\rho\,(s)$ can be expressed in the form [13]:

$$\int_0^{\infty} \ln \frac{1\omega}{|\rho\,(j\omega)|}\, d\omega \leq \frac{\pi}{RC} \tag{6.20}$$

Because the value of $1/|\rho(j\omega)|$ is high in the pass band and low outside it, this integral equation clearly illustrates the trade-off possible between the gain and the bandwidth of the matching network: the area under the curve of the inverse of the magnitude of the reflection coefficient as a function of the frequency is limited by the inverse of the time constant (RC) of the reactive load. Consequently, any increase in the bandwidth will bring about a decrease in the gain when the gain-bandwidth product exceeds the limit imposed by the load.

Assuming that the matching network has an ideal response ($G_{T,max}$ in the pass band and zero outside it), (6.20) can be manipulated to obtain the absolute maximum transducer power gain corresponding to a bandwidth B (Hz):

$$G_{T,max} \leq 1 - e^{-1/(RCB)} \tag{6.21}$$

Instead of using (6.20), a more convenient expression for the constraints imposed by a parallel RC load (or any other reactive load) can be derived by using the gain-bandwidth theory developed by Youla [21].

When the transducer power gain function is a low-pass Chebyshev function with ripple factor ϵ and cut-off frequency ω_c (rad/s), the maximum gain in the pass band (K_n) can be determined by using the equation [2]:

$$[1-K_n]^{1/2} > \epsilon \sinh \left\{ n \sinh^{-1} \left[\sinh \left(\frac{1}{n} \sin^{-1} \frac{1}{\epsilon} \right) - \frac{2\sin \dfrac{\pi}{2n}}{RC\omega_c} \right] \right\} \qquad (6.22)$$

where n is the order of the network.

If the product $RC\omega_c$ is replaced with $\omega_c L/R$, the restrictions on a low-pass Chebyshev matching network for a series RL load can also be determined by using (6.22).

Curves illustrating the relationship between the maximum realizable gain (K_n) as a function of the RCf_c product are given in Fig. 6.9 for a Chebyshev response with 0.5 dB ripple in the pass band [2].

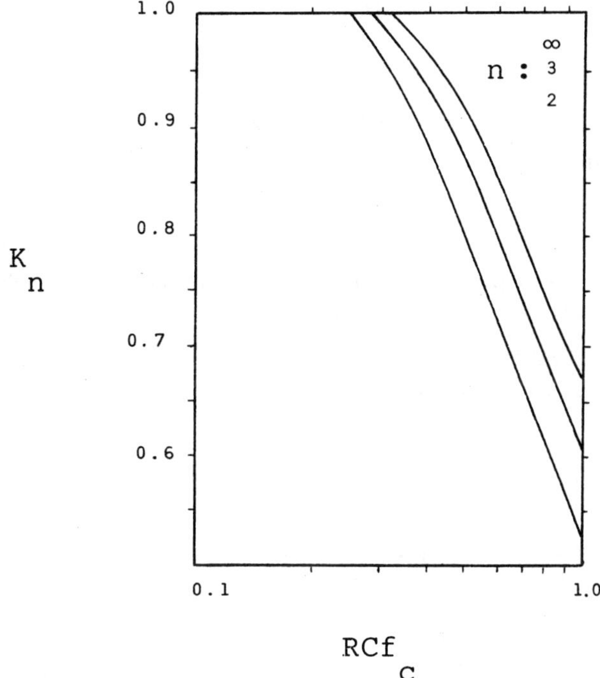

Figure 6.9:   The Maximum Realizable Power Gain (K_n) of a Parallel RC Load as a Function of the RCf_c Product of the Load (low-pass Chebyshev response with 0.5 dB ripple in the pass band) [after Chen]

It can be seen from these curves that the maximum power gain will be less than one when the RCf_c product is greater than approximately 0.3. When the RCf_c product increases above this value, the maximum realizable gain drops rapidly.

If more elements (n) are used in the impedance-matching network, the gain-bandwidth product increases. The improvement is, however, small when more than four elements are used.

6.3.4 The Direct Synthesis of Impedance-Matching Networks when the Load (or Source) Is Reactive

If the specified transducer power gain can be realized, a network for matching a reactive load to a resistive source can be designed by following the procedure outlined here [2].

Similar to Darlington synthesis, the source resistance obtained will often not be equal to that specified. The required transformation can usually be obtained by using transformers or LC transformers.

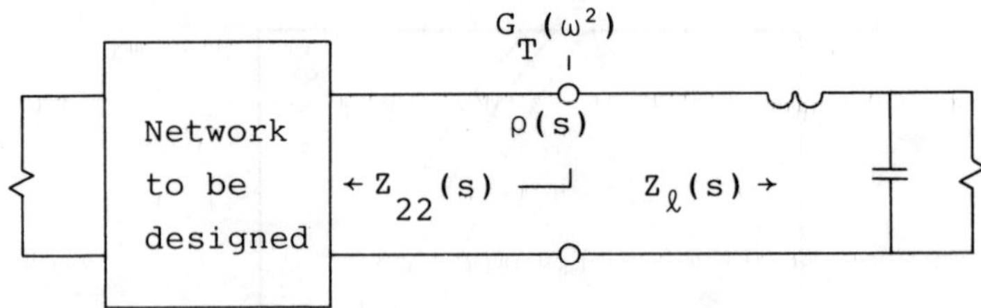

Figure 6.10   Illustration of the Design of Impedance-Matching Networks between a Resistive Source and a Reactive Load

Specifications: A realizable transducer power gain *versus* frequency function, the source resistance and the load impedance, $Z_l(s)$.

Design Procedure

1)   Replace all ω^2 terms in the specified transducer power gain *versus* frequency function ($G_T(\omega^2)$) with $-s^2$ terms.

Determine the product $\rho(s)\,\rho(-s)$ by using (6.11):

$$\rho(s)\,\rho(-s) = 1 - G_T(-s^2) \tag{6.11}$$

2)   The next step is to determine $\rho(s)$:

Assign all LHP poles to $\rho(s)$. If a minimum phase solution is required, assign all the LHP zeros to $\rho(s)$. If not, any combination of the zeros can be assigned to it, as long as they are assigned in conjugate pairs, and the relationship between $\rho(s)$ and $\rho(-s)$ is kept in mind.

When the load and source resistances are not equal and a low- or high-pass network is designed, the sign of $\rho(s)$ can be determined by determining the value of $\rho(s)$ as defined by equation (6.23) at $\omega = 0$ and when $\omega \rightarrow \infty$, respectively.

$$\rho(s) = \frac{Z_{22}(s) - Z_l(-s)}{Z_{22}(s) + Z_l(s)} \, A(s) \tag{6.23}$$

$$A(s) = \prod_i [s-s_i] \, / \, [s+s_i] \tag{6.24}$$

where $A(s)$ is an all-pass function with poles equal to the open LHP poles of the load impedance function $Z_l(s)$. $A(s)$ ensures that all the poles of the reflection coefficient $\rho(s)$, will lie in the LHP by cancelling the RHP poles caused by $Z_l(-s)$ in (6.23).

$Z_{22}(s)$ is the output impedance of the network to be designed.

When the load and source impedances are purely resistive and band-pass networks are designed, the sign of $\rho(s)$ is indeterminate and either sign can be used, depending on the topology required. This does not always apply when the load impedance is reactive.

Whether there are any constraints on the sign of $\rho(s)$ can be determined by considering (6.23) at $\omega = 0$ and when $\omega \rightarrow \infty$.

3)   Determine the output impedance of the impedance-matching network as seen from the load terminals. This can be done by using the following equations:

$$r_l(s) = 0.5 \, [Z_l(s) + Z_l(-s)] \tag{6.25}$$

where $r_l(s)$ is the even (resistive) part of the load impedance function $Z_l(s)$.

$$F(s) = 2r_l(s) \, A(s) \tag{6.26}$$

$$Z_{22}(s) = \frac{F(s)}{A(s) - \rho(s)} - Z_l(s) \tag{6.27}$$

4)   Synthesize the required network by using standard filter theory. When the topology is important, the elements of the network must be extracted in the proper sequence.

Example 6.4

As an example of the application of this procedure, a network will be designed to match a load consisting of 1Ω resistor in parallel with a 1.39F capacitor to a source with 0.5Ω internal resistance. The transducer power gain *versus* frequency function is to be a second-order low-pass Chebyshev function with 0.5 dB ripple in the pass band.

$$\omega_c = 1 \text{ rad/s}; \; K_n = 1$$

The specified transducer power gain function is

$$G_T(\omega^2) = \frac{K_n}{1 + \epsilon^2 \, C_n^2 \, (\omega/\omega_c)}$$

$$= \frac{1}{1 + 0.12202 \, (4\omega^4 - 4\omega^2 + 1)}$$

(The ripple factor and the polynomial $C_2\,(\omega)$ were obtained from standard filter tables).

Because

$$RCf_c = 0.221,$$

it is clear from Fig. 6.8 ($n=2$) that the gain function specified is realizable.

Step 1

$$G_T\,(-s^2) = \frac{1}{1 + 0.12202\,(4s^4 + 4s^2 + 1)}$$

$$\rho\,(s)\,\rho\,(-s) = 1 - G_T\,(-s^2) \tag{6.11a}$$

$$= \frac{s^4 + s^2 + 0.2500}{s^4 + s^2 + 2.2988}$$

Step 2

By assigning the LHP poles of the product $\rho\,(s)\,\rho\,(-s)$ to $\rho\,(s)$, its numerator is found to be

$$p\,(s) = s^2 + 1.4257s + 1.5126$$

By assigning the LHP zeros to $\rho\,(s)$, the denominator is found to be

$$q\,(s) = s^2 + 0.5000$$

The reflection coefficient $\rho\,(s)$ is, therefore,

$$\rho\,(s) = \pm \frac{s^2 + 0.5000}{s^2 + 1.4256s + 1.5162}$$

Since

$$A\,(s) = \prod_i [s - s_i] \,/\, [s + s_i] \tag{6.24a}$$

$$= \frac{s - 0.7914}{s + 0.7914}$$

and

$$\rho(0) = \frac{Z_{22}(0) - Z_l(0)}{Z_{22}(0) + Z_l(0)} A(s) \qquad (6.23a)$$

$$= \frac{0.5 - 1.0}{0.5 + 1.0} \frac{0 - 0.7194}{0 + 0.7194}$$

$$= \frac{0.5}{1.5}$$

it follows that the positive sign must be assigned to $\rho(s)$.

Step 3

$$Z_l(s) = 1/[1 + 1.39s]$$

$$r_l(s) = 0.5[Z_l(s) + Z_l(-s)] \qquad (6.25a)$$

$$= -0.5176/[s^2 - 0.5175]$$

$$F(s) = -1.0351/[s + 0.7194]^2$$

$$Z_{22}(s) = F(s)/[A(s) - \rho(s)] - Z_l(s) \qquad (6.27a)$$

$$= \frac{1.4297s}{0.0185s^2 + 0.0129s + 2.0360}$$

Step 4

$$Z_{22}(s) = \cfrac{1}{0.0129s + \cfrac{1}{0.7071s + \cfrac{1}{\dfrac{2.036}}}}$$

Step 5

The source resistance is equal to the specified value and, therefore, no transformation is required.

The designed network is shown in Fig. 6.11.

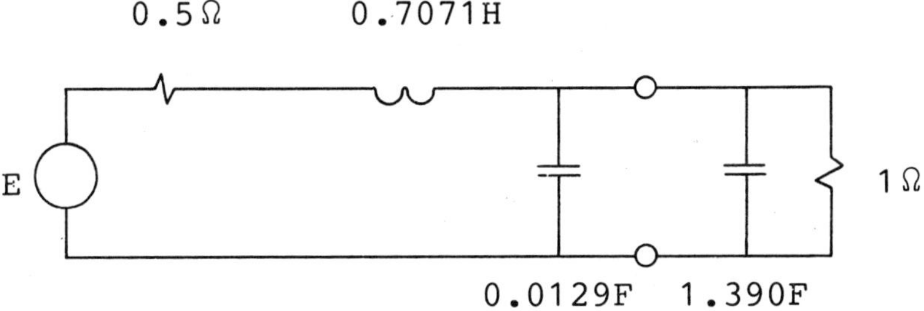

Figure 6.11   A Network for Matching a Capacitive Load to a Source with 0.5Ω Internal Resistance [0.5 dB ripple, second-order Chebyshev response]

6.3.5 Synthesis of Networks for Matching a Reactive Load to a Purely Resistive or a Reactive Source by Using the Principle of Parasitic Absorption

When the load or source impedances can be approximated with simple RC, RL, or RLC networks, impedance-matching networks for the reactive terminations can be designed by at first ignoring the reactiveness. If the gain-bandwidth constraints are taken into account and a network with a suitable topology is designed, it will be possible to absorb the reactive parts of the terminations into the designed network.

The topology of the network is a function of the order of the gain function chosen, its transmission zeros, and the sign of $\rho(s)$ as was explained in sec. 6.3.1.

When only the load or source impedance is reactive, and it can be approximated with a parallel RC or series RL network, the maximum gain in the pass band (K_n) can be determined for a Chebyshev transducer power gain *versus* frequency response with a specified ripple factor ϵ, by using (6.22).

Although this equation gives the optimum K_n corresponding to a specified ripple factor, it gives no indication as to which ripple factor will cause the lowest insertion loss in the pass band, in other words, the optimum ripple factor is not known.

The optimum ripple factors corresponding to some values of the load or source quality factors at the highest frequency in the pass band ($2\pi RCf_c$) were determined iteratively by substituting various values for the ripple factor into (6.22). The corresponding values for the highest and lowest gains in the pass band (K_n; $K_n/(1+\epsilon^2)$) are tabulated for different values of the Q-factor and the number of elements used in the network in Table 6.7.

When both the source and load impedances are reactive and can be approximated with parallel RC or series RL networks, the optimum values for the maximum gain in the pass band (K_n) and the ripple factor (ϵ) can be determined by using the following set of equations [6]:

$$X = [1/Q_n + 1/Q_1]\sin\frac{\pi}{2n} \tag{6.28}$$

$$Y = [1/Q_n - 1/Q_1]\sin\frac{\pi}{2n} \tag{6.29}$$

$$A = \sinh^{-1} X$$
$$= \ln[X + \sqrt{X^2 + 1}] \tag{6.30}$$
$$B = \ln[Y + \sqrt{Y^2 + 1}] \tag{6.31}$$
$$= 1/\sinh[nA] \tag{6.32}$$

$$C = 0.5 \, \frac{\sinh^2 [nB]}{\sinh^2 [nA]} + 0.5 \, \sqrt{\frac{\sinh^4 [nB]}{\sinh^4 [nA]} + \frac{4}{\sinh^2 [nA]}} \qquad (6.33)$$

$$K_n = 1/[C^2 \sinh^2 (nA)] \qquad (6.34)$$

$$G_T (\omega^2) = K_n/[1 + \epsilon^2 \, C_n^2 (\omega)] \qquad (6.35)$$

Table 6.7

The Values of the Highest and the Lowest Transducer Power Gain in the Pass Band (K_n; $K_n/(1+\epsilon^2)$) of the Optimum Low-Pass Chebyshev Function as a Function of the Load or Source Q-Factor at the Highest Frequency in the Pass Band and the Number of Elements Used

Q	K_n $K_n/(1+\epsilon^2)$			
	$n=2$	$n=3$	$n=4$	$n=5$
0.25	1.0000 0.9998	1.0000 1.0000	1.0000 1.0000	1.0000 1.0000
0.50	0.9997 0.9969	0.9999 0.9994	1.0000 0.9998	1.0000 1.0000
0.75	0.9997 0.9876	0.9989 0.9961	0.9992 0.9981	0.9994 0.9988
1.00	0.9929 0.9703	0.9951 0.9876	0.9962 0.9925	0.9968 0.9946
1.25	0.9814 0.9459	0.9875 0.9729	0.9894 0.9816·	0.9905 0.9856
1.50	0.9685 0.9165	0.9749 0.9527	0.9789 0.9655	0.9805 0.9715
1.75	0.9515 0.8839	0.9589 0.9284	0.9626 0.9451	0.9665 0.9533
2.00	0.9319 0.8499	0.9419 0.9016	0.9453 0.9216	0.9492 0.9319
2.25	0.9111 0.8157	0.9206 0.8731	0.9256 0.8963	0.9294 0.9083
2.50	0.8877 0.7821	0.9004 0.8442	0.9046 0.8699	0.9082 0.8834
2.75	0.8666 0.7496	0.8776 0.8152	0.8828 0.8431	0.8861 0.8580

Table 6.7 (cont'd)

3.00	0.8440	0.8548	0.8609	0.8638
	0.7184	0.7868	0.8165	0.8325
3.25	0.8223	0.8344	0.8391	0.8415
	0.6887	0.7592	0.7903	0.8073
3.50	0.7997	0.8126	0.8177	0.8195
	0.6606	0.7326	0.7648	0.7827
3.75	0.7800	0.7915	0.7968	0.8000
	0.6340	0.7071	0.7402	0.7587
4.00	0.7597	0.7695	0.7749	0.7791
	0.6090	0.6827	0.7165	0.7355
4.50	0.7206	0.7317	0.7352	0.7379
	0.5632	0.6373	0.6720	0.6918

Q_1 and Q_2 are respectively, the source and load Q-factors at the highest frequency in the pass band.

When the load or source impedance can be approximated with series or parallel RLC resonant circuits, the inductance or capacitance can be increased to cause resonance at the center frequency of the pass band (ω_0) and the band-pass problem can be transformed to an equivalent low-pass problem ($\omega_c = 1$ rad/s) by using the standard transformation formulas repeated in Fig. 6.12. The optimum Chebyshev gain function can then be determined as described above.

The low-pass Q-factors corresponding to the band-pass Q (at the center frequency) can be determined quickly by using the equation:

$$Q_L = Q_B / [\omega_0 / B] \tag{6.36}$$

This equation can be derived easily by using (6.39) and (6.40).

Example 6.5

The optimum values of K_n and ϵ will be determined for the load of *Example 6.4* ($1\Omega \parallel 1.39F$) for a two-element Chebyshev matching network.

Since

$$Q_L = \omega_c RC = 1.39$$

it follows by inspection of the first column of Table 6.7 that,

$$K_n \simeq 0.9814 - \frac{1.39-1.25}{1.50-1.25} \, [0.9814-0.9685]$$

$$= 0.9742$$

and

$$\frac{K_n}{1+\epsilon^2} \simeq 0.9459 - \frac{1.39-1.25}{1.50-1.25} \, [0.9459-0.9165]$$

$$= 0.9294$$

It follows by manipulation of the last equation that the optimum value of the ripple factor is

$$\epsilon = 0.2196$$

Therefore, the optimum two-element gain function

$$G_T(\omega^2) = \frac{0.97}{1 + 0.0482 C_2^2(\omega)}$$

The maximum value of the insertion loss is 0.32 dB.

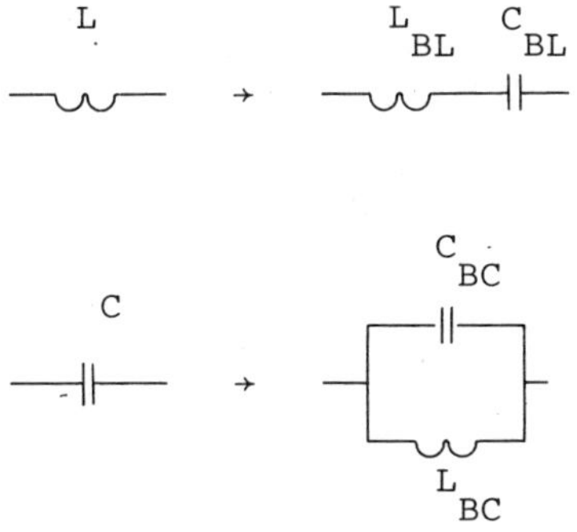

$$L_{BL} = L/B \tag{6.37}$$

$$C_{BL} = 1/[\omega_0^2 L_{BL}] \tag{6.38}$$

$$C_{BC} = C/B \tag{6.39}$$

$$L_{BC} = 1/[\omega_0^2 C_{BC}] \tag{6.40}$$

Figure 6.12   Formula for Transforming a Low-Pass Network [$\omega_c = 1$ rad/s] to a Band-Pass Network with Center Frequency ω_0 and Bandwidth B [rad/s]

6.3.6 The Analytical Approach to Designing Commensurate Distributed Impedance-Matching Networks

Using Richards' transformation [20], the analytical theory applicable to the design of lumped-element networks also applies to commensurate distributed networks (distributed networks in which the line lengths are all equal). Open-ended lines are transformed to lumped capacitors and short-circuited lines to lumped inductors with this transformation.

Unlike short-circuited and open-ended stubs, the series transmission lines used in distributed designs have no lumped equivalents using Richards' transformation. The influence of unit elements (series lines in commensurate designs) on the gain function and their extraction from an impedance function when a network is synthesized will be discussed in sec. 6.3.6.1, together with Richards' transformation.

The series short-circuited stubs, which are often part of a network designed by using Richards' transformation, are not realizable in planar form. When the designed network is to be realized in planar form, these unwanted stubs can be removed by using Kuroda's low-pass identities.

As in the case of lumped networks, impedance scaling is often required for a designed impedance-matching network. This impedance scaling function can be performed by using Kuroda's high-pass identities and Norton's band-pass identities.

Kuroda and Norton's identities will be discussed in sec. 6.3.6.2.

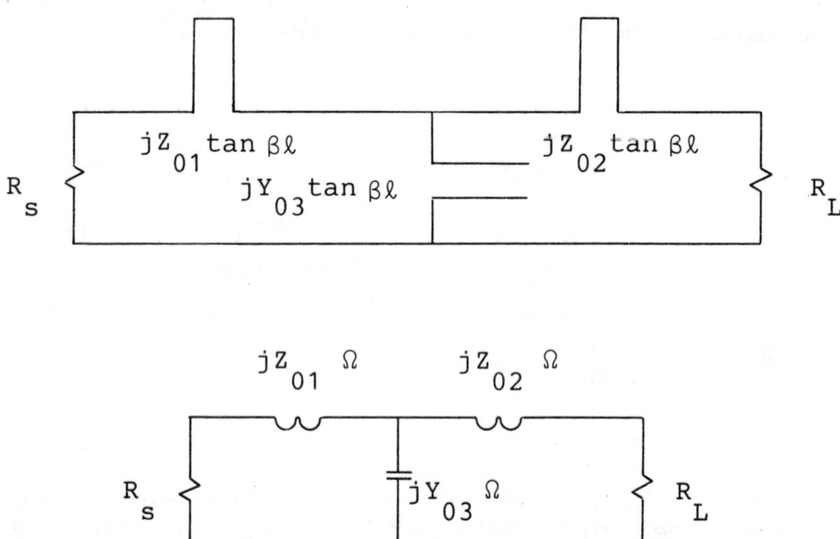

Figure 6.13   [b] *A Lumped-Element Equivalent for the Distributed Network in [a] under Richards' Transformation*

6.3.6.1 Richards' Transformation

Using Richards' transformation [20]:

$$S = j\Omega = j\tan[\beta l] = j\tan\left[\frac{\pi}{2}\frac{\omega}{\omega_0}\right] \tag{6.41}$$

open-ended and short-circuited stubs are mapped to capacitors and inductors in the *S*-plane. The inductance and capacitance of the lumped equivalents are respectively equal to the characteristic impedance and admittance of the short-circuited and open-ended lines in the distributed network. This is illustrated in Fig. 6.13.

The frequency responses of the two networks in Fig. 6.13 are compared in Fig. 6.14. Note that the response of the distributed network is periodic (βl *versus* tan βl characteristic), and that the gain at the even (including $\omega=0$) and uneven harmonics of ω_0 is equal to that of the lumped equivalent at $\omega=0$ and $\omega \rightarrow \infty$, respectively. The distributed response is simply a compressed, periodic version of its lumped equivalent.

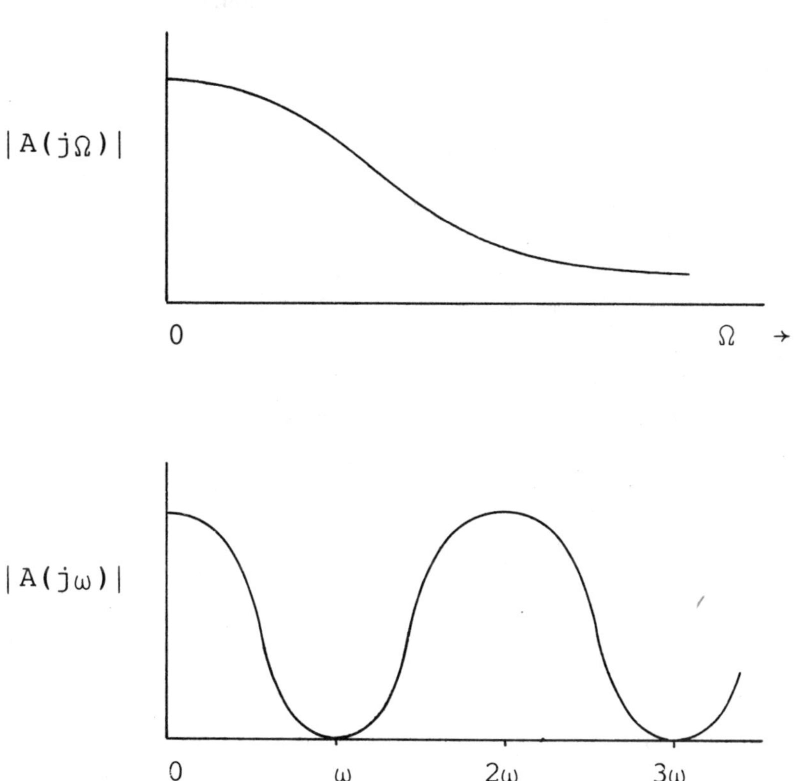

Figure 6.14   The Change in the Frequency Response of Low-Pass Network with Richards' Transformation

Series transmission lines are often used in distributed designs. The transmission matrix for a unit element (series transmission lines of equal length in a commensurate distributed network) is

$$
T = \begin{bmatrix} \cos\left[\dfrac{\pi}{2}\dfrac{\omega}{\omega_0}\right] & jZ_0\sin\left[\dfrac{\pi}{2}\dfrac{\omega}{\omega_0}\right] \\[2ex] jY_0\sin\left[\dfrac{\pi}{2}\dfrac{\omega}{\omega_0}\right] & \cos\left[\dfrac{\pi}{2}\dfrac{\omega}{\omega_0}\right] \end{bmatrix}
\tag{6.42}
$$

Using Richards' transformation, the transmission matrix becomes

$$
T = \frac{1}{\sqrt{1-S^2}}\begin{bmatrix} 1 & Z_0\,S \\ Y_0\,S & 1 \end{bmatrix}
\tag{6.43}
$$

The transducer power gain of a commensurate distributed cascaded network with N_u unit elements, N_h high-pass elements, and order N is given by an expression of the form

$$
s_{21}(S)\,s_{21}(-S) = K_0\,\frac{S^{2N_h}\left[1-S^2\right]^{N_u}}{G_N(S^2)}
\tag{6.44}
$$

where $G_N(S^2)$ is an Nth degree polynomial in S^2. Each unit element, therefore, contributes a factor $(1-S^2)$ to the numerator of the transducer power gain function.

With the gain function chosen, the input impedance of the corresponding network can be determined by Darlington synthesis, as described previously for lumped impedance-matching networks. With the input impedance corresponding to a specified gain response determined, the network can be synthesized. This can be done as before except for the extraction of unit elements. A unit element can be extracted from the impedance function when the even part of the input impedance at $S=1$ is equal to zero. The characteristic impedance of the element is given by

$$
Z_0 = Z_{IN}(S)\Big|_{S=1}
\tag{6.45}
$$

With the unit element extracted, the remaining input impedance can be determined by using the expression [20]:

$$
Z_{IN}' = Z_{IN}(1)\,\frac{S\,Z_{IN}(1)-Z_{IN}(S)}{S\,Z_{IN}(S)-Z_{IN}(1)}
\tag{6.46}
$$

This impedance function will always have a common factor S^2-1 in its denominator and numerator which can be cancelled.

With its lumped-element equivalent known, the design of the required distributed matching network is completed, that is, if impedance scaling is not required.

Example 6.6

The extraction procedure for a unit element will be illustrated by synthesizing the network with input impedance

$$Z_{IN}(S) = \frac{75\, S^2 + 125\, S}{1.5\, S^2 + 1.5\, S + 1.0}$$

$$= \frac{[75\, S^2] + 125\, S}{[1.5\, S^2 + 1.0] + 1.5\, S}$$

Because the numerator of the even part of $Z_{IN}(S)$ at $S=1$ is given by

$$NZu\ \Big|_{S=1} = [75\, S^2][1.5\, S^2 + 1.0] - [125\, S][1.5\, S]\ \Big|_{S=1} = 0$$

and

$$Z_{IN}(1) = 50$$

a unit element of 50Ω can be extracted.

The input impedance with the unit element removed can be determined by applying (6.46):

$$Z_{IN}' = Z_{IN}(1)\ \frac{S\, Z_{IN}(1) - Z_{IN}(S)}{S\, Z_{IN}(S) - Z_{IN}(1)} \tag{6.46}$$

$$= 50\ \frac{75\, S^3 - 75\, S^2}{75\, S^3 + 50\, S^2 - 75\, S - 50}$$

$$= 50\ \frac{[S^2 - 1]\, 75\, S}{[S^2 - 1][75\, S + 50]}$$

$$= \frac{1}{1/[75\, S] + 1/50}$$

The synthesized network is shown in Fig. 6.15.

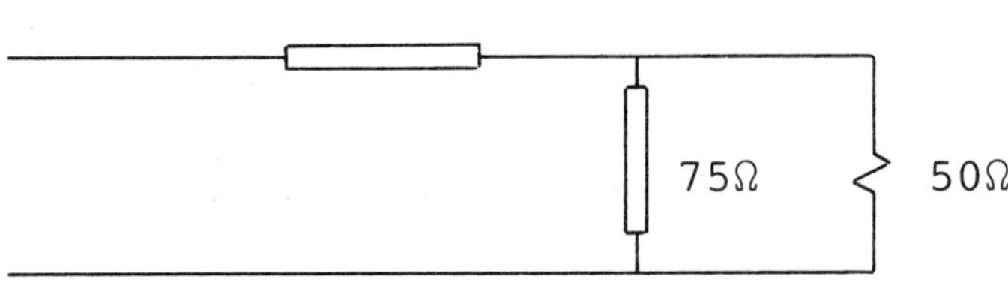

Figure 6.15   The Network Synthesized in Example 6.6

6.3.6.2 Kuroda and Norton's Identities

Kuroda's low-pass identities are shown in Fig. 6.16. These identities can sometimes be used to transform unrealistic impedances to more realistic levels, but they are more frequently used to remove unwanted series short-circuited stubs from planar designs.

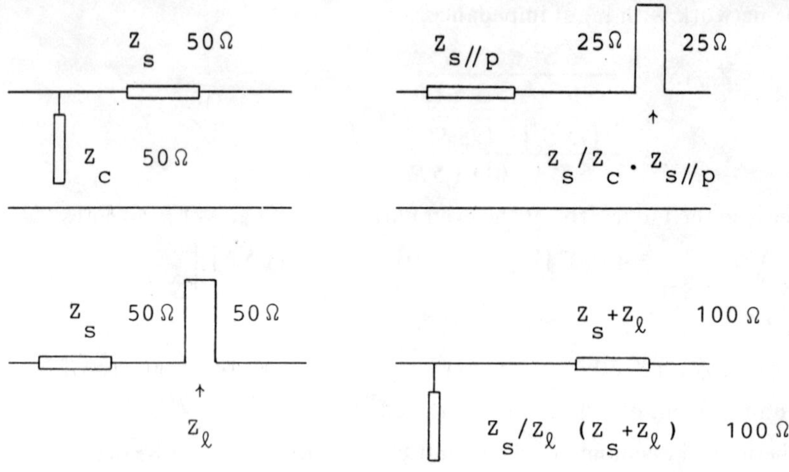

Figure 6.16   *Kuroda's Low-Pass Identities*

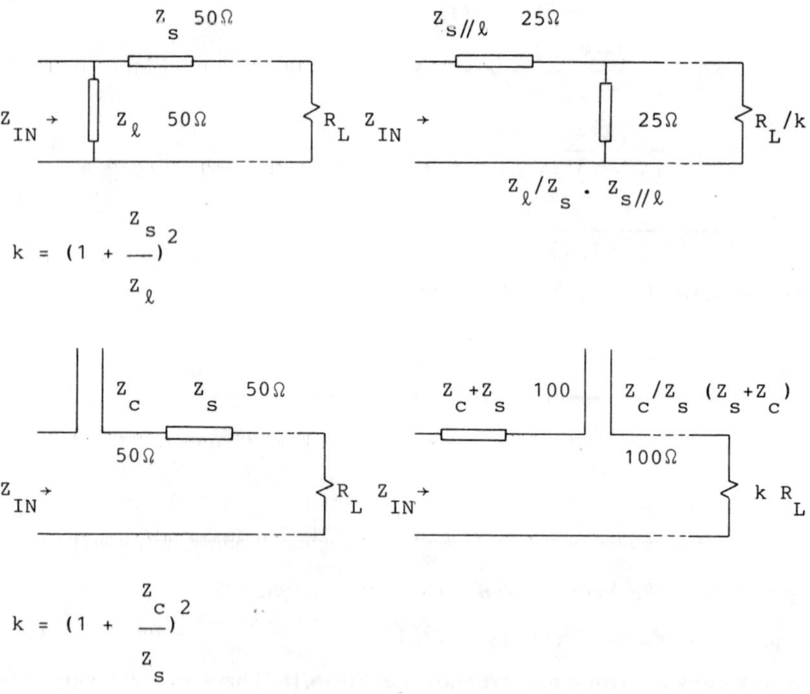

Figure 6.17   *Kuroda's High-Pass Identities*

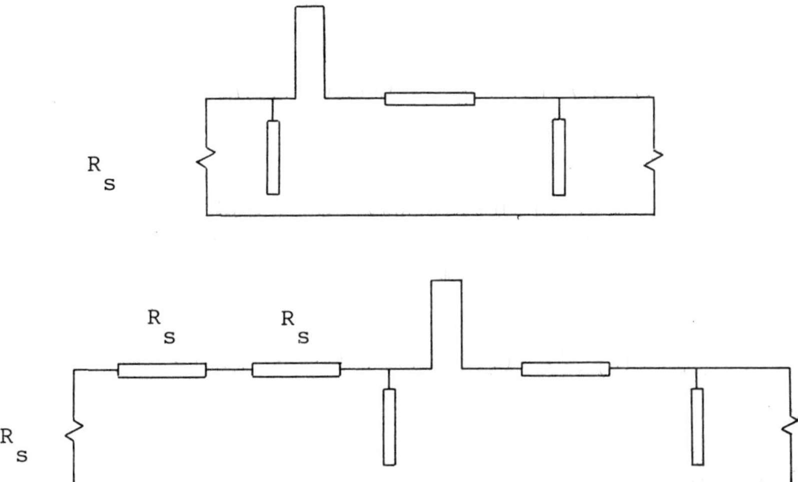

Figure 6.18   Adding Unit Elements to a Network with a Resistive Source

Kuroda's high-pass identities are shown in Fig. 6.17. These identities can be used to change the impedance level in a matching network, as illustrated. The impedance level to the right of the transformed components is scaled with a factor k, which is also defined in Fig. 6.17, while the input impedance (Z_{IN}) remains unchanged.

In applying Kuroda's identities it is useful to know that when the load or source impedance is purely resistive any number of unit elements with characteristic impedance equal to it can be added in series with it without changing the amplitude response. This is illustrated in Fig. 6.18.

The impedance level in a network can also be changed by using Norton's identities. Unlike Kuroda's identities in which unit elements are always involved, Norton's identities are applied to L-sections consisting only of open-ended or short-circuited stubs. These identities are shown in Fig. 6.19.

6.4   THE ITERATIVE DESIGN OF IMPEDANCE-MATCHING NETWORKS

Instead of following the analytical approach, impedance-matching networks can be designed iteratively. The "real-frequency" iterative techniques considered here have a major advantage over analytical and other techniques in that equivalent circuits for the load or source impedances, or an analytical expression for the transducer power gain *versus* frequency response, is not required. The networks synthesized by using iterative techniques are generally simpler in form with superior gain properties [16, 17, 18].

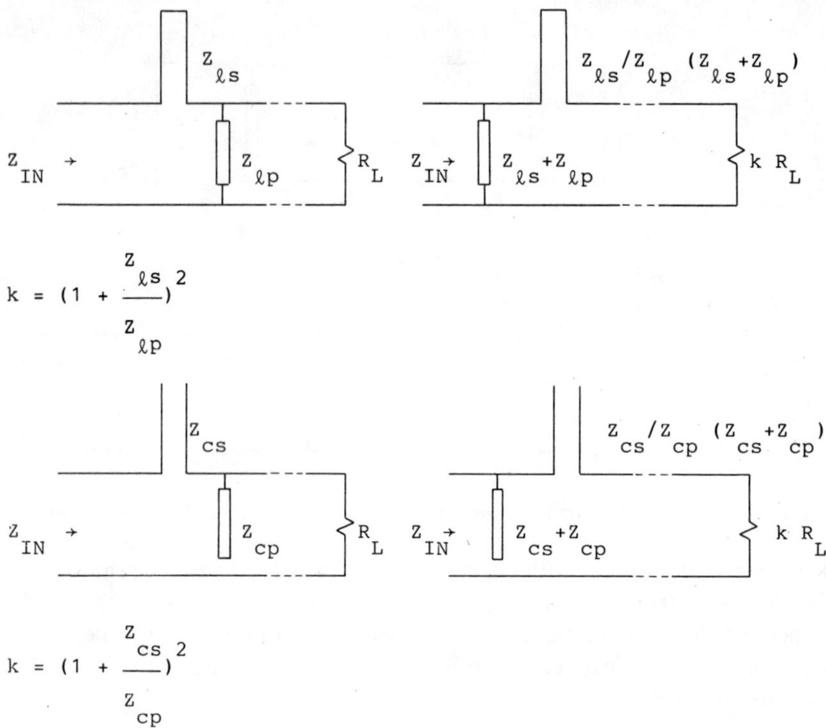

Figure 6.19   *Norton's Band-Pass Identities*

The first "real-frequency" technique was introduced by Carlin [3]. A complex
load can be matched to a purely resistive source by using this technique in
which a piece-wise linear approximation of the output resistance or conductance of the network to be designed is optimized by using a simple least-square
optimization routine.

The convergence properties of this technique are very good and it can generally by used to determine the gain-bandwidth constraints imposed by any complex load.

It has the disadvantage that the response outside the pass band is usually
unnecessarily constrained, and occasionally it will not give the best results
obtainable without considerable experimentation with the response outside
the pass band. When band-pass networks are designed, the reactance of the
network may not approximate the expected reactance well because of the

difficulty of detecting and approximating the narrow spikes which can occur in the resistance function of a band-pass network. In these cases, the actual response will be poorer than that expected from the line-segment results [19, p. 117-120].

Despite these disadvantages, the networks synthesized by using this technique are superior to those obtainable by direct application of analytical theory. This technique will be discussed in detail in sec. 6.4.1.

Apart from matching a complex load to a purely resistive source, the "real-frequency" technique introduced by Yarman and Carlin [16] can also be used to match a complex load to a complex source. In this technique, the numerator coefficients of the input reflection parameter (s_{11}) of the network terminated in a purely resistive load are optimized.

Compared to the line-segment technique where only one of the terminations is complex, the reflection coefficient technique has the advantage that it has no approximation step.

Initialization of the reflection coefficient procedure is not as simple as in the case of the line-segment technique where the unknown output impedance of the network to be synthesized is taken to be equal to the resistive part of the known reactive load. However, excellent results can be achieved if the results obtained by using the line-segment technique are employed for initialization.

Although the solution achieved may not necessarily be the best solution obtainable, it is as a rule much better than anything obtainable by direct application of analytical theory.

The reflection coefficient technique will be discussed in detail in sec. 6.4.2.

In another technique proposed by Yarman and Carlin to solve double-matching problems (that is, problems in which the load and the source are complex) [17], the output resistance of the matching network terminated at the input in a purely resistive load is optimized. Because this procedure has no significant advantage over the reflection coefficient technique, it will not be covered here and the interested reader is referred to [17].

The double-matching problem can also be solved very effectively by doing a grid-search and subsequent optimization of the transformation Q-factors of a network as defined in Ch. 3. This approach has the distinct advantage that it is not dependent on a good initial solution. If the grid-search is done thoroughly enough, the probability of finding the optimum solution to any impedance-matching problem is very high.

Other major advantages of the last approach is that many solutions instead of only one are obtained and that transformers are never required in the solutions synthesized. This technique will be considered in detail in sec. 6.4.3.

6.4.1 The Line Segment Approach for Matching a Complex Load to a Resistive Source

The gain-bandwidth constraints imposed by a reactive load (or source) can be determined iteratively by assuming that the output impedance (or admittance) of the network is a minimum-impedance (admittance) function. When this is done, the output reactance of the network is known when the resistance is known. The optimum resistance can then be determined by minimizing the mean-square deviation between the desired and the actual transducer power gain.

If the resistance is approximated with line segments, the problem is well-behaved and the approximation can be done with a simple least-square optimization routine.

A detailed description of the procedure [3,4]. and the mathematics involved, follows.

Line-Segment Algorithm

1)   Assume that the output impedance (admittance) of the optimum network is a minimum-impedance (admittance) function.

2)   Assume as a first approximation that the resistive part ($R(\omega)$) of the output impedance of the optimum network is equal to that of the measured load impedance (R_L), that is,

$$R(\omega_i) = R_L(\omega_i) \tag{6.47}$$

3)   Approximate the rational output resistance of the network with a piecewise linear function, as illustrated in Fig. 6.21.

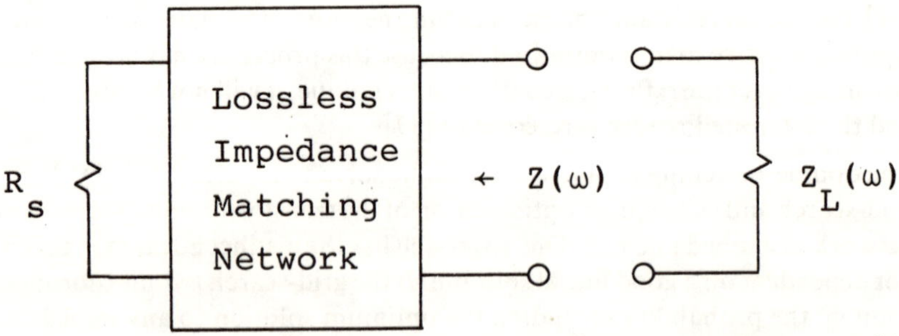

Figure 6.20   The Impedance-Matching Problem under Consideration.

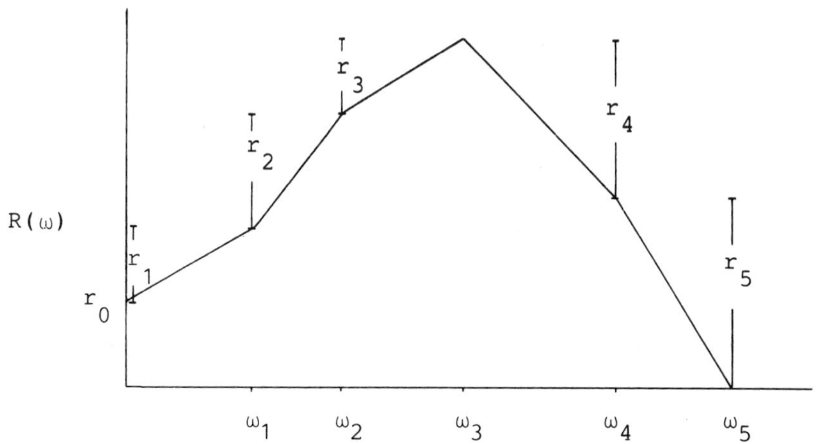

Figure 6.21 Approximation of the Output Resistance of the Matching Network with a Piece-Wise Linear Function

Enough increment frequencies (the frequencies at which the slope of the linear function changes) must be chosen to ensure a reasonable approximation of the unknown resistance. This can usually be done by choosing the frequencies to ensure a good approximation of the measured load resistance.

For the sake of simplicity, the resistance $R(\omega)$, is assumed to equal zero at frequencies greater than the last increment frequency (ω_n).

The linear resistance function can be considered the sum of the semi-infinite functions $a_1(\omega)$, $a_2(\omega)$, ..., $a_n(\omega)$ shown in Fig. 6.22, each with an appropriate weight factor r_k:

$$R(\omega) = r_0 + \sum_k r_k a_k(\omega) \tag{6.48}$$

where

$$a_k(\omega) = \begin{cases} 1 & \text{if } \omega_k < \omega \\[2ex] \dfrac{\omega - \omega_{k-1}}{\omega_k - \omega_{k-1}} & \text{if } \omega_{k-1} < \omega_k \\[2ex] 0 & \text{if } \omega < \omega_{k-1} \end{cases} \tag{6.49}$$

and

$$r_0 = Z(0) \tag{6.50}$$

When the optimum low-pass network is determined,

$$r_0 = R_s \tag{6.51}$$

where R_s is the source resistance as shown in Fig. 6.20.

In all other cases, r_0 is equal to zero.

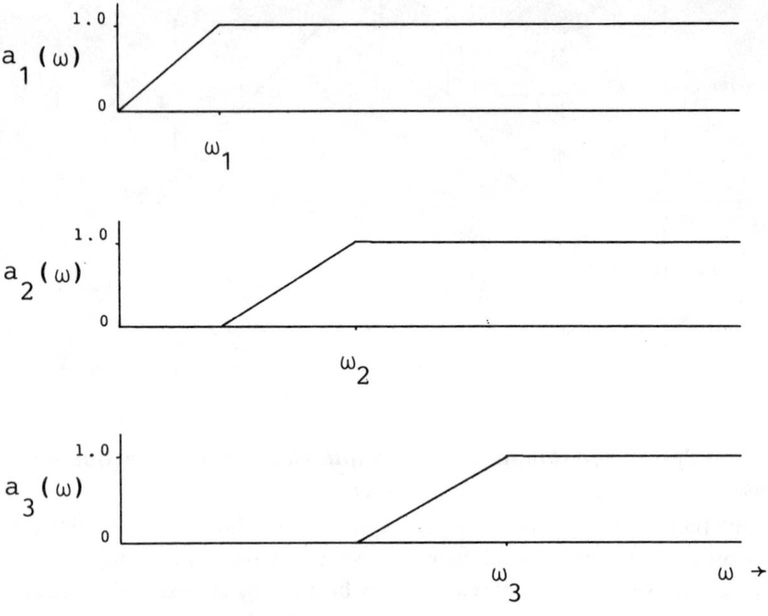

Figure 6.22   Illustration of the functions $a_1(\omega)$, $a_2(\omega)$, et cetera.

Since the resistance $R(\omega)$ equals zero when the frequency is greater than the last increment frequency (ω_n), the increment factor (weight factor) r_n is not independent of the other increment factors. The following equation applies:

$$r_n = -[r_0 + r_1 + r_2 + \ldots + r_{n-1}] \tag{6.52}$$

When this value for r_n is substituted into (6.48), it changes to

$$R(\omega) = [1 - a_n(\omega)]\, r_0 + \sum_k r_k\, [a_k(\omega) - a_n(\omega)] \tag{6.53}$$

In vector form this becomes

$$R(\omega) = [1 - a_n(\omega)]\, r_0 + \overline{a}^T(\omega)\, \overline{r}' \tag{6.54}$$

where

$$\overline{a}^T(\omega) = [a_1(\omega) - a_n(\omega),\ a_2(\omega) - a_n(\omega),\ \ldots,\ a_{n-1}(\omega) - a_n(\omega)] \tag{6.55}$$

and

$$\overline{r}' = \begin{bmatrix} r_1 \\ r_2 \\ \cdot \\ \cdot \\ r_{n-1} \end{bmatrix} \tag{6.56}$$

The resistance value at any particular frequency can be calculated by using (6.54).

Because the impedance function was assumed to be a minimum-impedance function, the reactance associated with the resistance $R(\omega)$ is known. It can be determined by using the equation:

$$X(\omega) = \overline{b}^{T}(\omega)\,\overline{r} \tag{6.57}$$

where

$$\overline{b}^{T}(\omega) = [b_1(\omega), b_2(\omega), \ldots, b_n(\omega)] \tag{6.58}$$

and

$$b_k(\omega) = \frac{1}{\pi[\omega_k - \omega_{k-1}]} \int_{\omega_{k-1}}^{\omega_k} \ln\left|\frac{y+\omega}{y-\omega}\right| dy \tag{6.59}$$

The value of the integral in the last equation is obviously high at frequencies close to ω and decreases as the frequency deviates from ω. By using this fact and inspecting (6.57), it follows that the reactance associated with the resistance at a particular frequency, will be high when the resistance changes rapidly at nearby frequencies.

The integral in (6.59) has a simple closed form evaluation [9], which is useful in determining its value.

If the dependence of r_n on the other increment factors is taken into account, then (6.57) becomes

$$X(\omega) = [0 - b_n(\omega)]\,r_0 + \overline{b}'^{T}(\omega)\,\overline{r}' \tag{6.50}$$

where

$$\overline{b}'(\omega) = \begin{bmatrix} b_1(\omega) - b_n(\omega) \\ b_2(\omega) - b_n(\omega) \\ \cdot \\ \cdot \\ b_{n-1}(\omega) - b_n(\omega) \end{bmatrix} \tag{6.61}$$

The resistance and reactance corresponding to a particular increment vector $\overline{r}'$, can be determined at any particular frequency by using (6.54) and (6.61).

4) Calculate the transducer power gain associated with the initial value $(\overline{r}_0)$ of the increment vector $\overline{r}\,'$ at the various frequencies of interest. This can be done by using the equation:

$$G_T(\omega) = 1 - |s_{11}|^2$$

$$= 1 - \left| \frac{Z_L(\omega) - Z^*(\omega)}{Z_L(\omega) + Z(\omega)} \right|^2$$

$$= 1 - \left| \frac{[R_L(\omega) + jX_L(\omega)] - [R(\omega) + jX(\omega)]^*}{[R_L(\omega) + jX_L(\omega)] + [R(\omega) + jX(\omega)]} \right|^2$$

$$= \frac{4R_L(\omega)\,R(\omega)}{[R_L(\omega) + R(\omega)]^2 + [X_L(\omega) + X(\omega)]^2} \tag{6.62}$$

where $Z_L(\omega)$ is the measured load impedance.

5) Determine the optimum value of the increment vector $\overline{r}\,'$ iteratively by minimizing the sum of the relative difference in the actual $(G_T(\omega))$ and the desired transducer power gain $(G_I(\omega))$ squared (E). This can be done by using a least-square optimizing routine.

The relevant equations are

$$E = \sum_j e^2(\overline{r}\,', \omega_j) \tag{6.63}$$

$$= \sum_j \left[\frac{G_T(\overline{r}\,', \omega_j)}{G_I(\omega_j)} - 1 \right]^2 \tag{6.64}$$

$$\overline{f}\,'(\omega) = \frac{\partial e(\overline{r}\,', \omega)}{\partial \overline{r}_0'} \tag{6.65}$$

$$= \frac{\partial e(R(\omega), X(\omega))}{\partial R(\omega)} \frac{\partial R(\omega)}{\partial \overline{r}_0'} + \frac{\partial e(R(\omega), X(\omega))}{\partial X(\omega)} \frac{\partial X(\omega)}{\partial \overline{r}_0'} \tag{6.66}$$

$$= \frac{\partial e(R(\omega), X(\omega))}{\partial R(\omega)} \overline{a}(\omega) + \frac{\partial e(R(\omega), X(\omega))}{\partial X(\omega)} \overline{b}(\omega) \tag{6.67}$$

where $\overline{r}_0'$ is the current initial value of the increment vector and $\overline{f}\,'(\omega)$ is the gradient vector associated with the error function $e(\overline{r}\,', \omega)$,

$$e(R(\omega), X(\omega)) = G_T(R(\omega), X(\omega)) \,/\, G_I(\omega) - 1 \tag{6.68}$$

with $G_T(R(\omega), X(\omega))$ as defined in (6.53) and

$$\sum_j \overline{f}\,'(\omega_j) \overline{f}\,'^T(\omega_j)\,\overline{\delta} = -\sum_j e(\overline{r}_0', \omega_j)\,\overline{f}\,'(\omega_j) \tag{6.69}$$

where $\overline{\delta}$ is defined by the equation:

$$\overline{r}\,' = \overline{r}_0' + \overline{\delta} \tag{6.70}$$

where $\overline{r}\,'$ is the new initial value of the increment vector.

With the optimum increment vector known, the gain-bandwidth constraints imposed by the load, as well as the output impedance (admittance) of the optimum network are known.

6)   The next step is to determine the optimum network.

Since the optimum increment vector is known, the output resistance of the network is known at any particular frequency. A rational approximation function and the corresponding minimum-impedance (admittance) function can be obtained by following the procedure outlined in sec. 6.2.

The order of the network and the number of zeros at the origin are variables in the approximation stage.

With the minimum-impedance and minimum-admittance functions known, the optimum network can be synthesized easily.

Whether a minimum-impedance or minimum-admittance function will be the best solution to a particular problem is usually not known at the outset. If good results are not obtained by using the one, the other can be tried.

When the load resistance is higher than the source resistance a minimum impedance solution often yields better results.

When the gain-bandwidth product in a particular problem is a limiting factor, it will be found that the results are dependent on the position of the last increment frequency. Some experimentation with this frequency is then necessary.

When band-pass networks are designed, both the first and the last increment frequencies have a significant influence on the results when the gain-bandwidth product is limited.

The program LSM FORTRAN in Appendix C can be used to determine the optimum increment vector and the gain-bandwidth limitations associated with any reactive load. The input data for the program consist of the following:

1. The number of frequencies at which the load impedance (or admittance) will be specified.

2. The load impedance and the required transducer power gain at each frequency (f, R, X, G_T).

3. The number of increment frequencies and the output resistance of the network to be designed at $\omega = 0$.

4. The initial values of the increments at each increment frequency (f_j, r_j).

5. The number of iterations to be done and a number specifying the detail of the printout required (1: detailed, 0: otherwise).

6. The amount (in Hz) by which the last increment frequency must be increased in search of the optimum value and the number of times it must be done.

The output data for the program consist of the elements of the optimized increment vector, the optimum value of the highest increment frequency, a comparison of the transducer power gain obtainable to that specified, and the output impedance (admittance) of the optimum network at the frequencies of interest.

Example 6.7

As an illustration of using the program LSM FORTRAN, the gain-bandwidth constraints imposed on a low-pass network by the input impedance of the transistor in *Example 6.1* will be determined. The source resistance is selected as 6.25Ω. The goal will be to achieve a flat response across the 2-30 MHz pass band. Only the results obtained for a minimum-admittance net-

Table 6.8

The Input Data of the Program LSM FORTRAN

```
13
2.0D6, 0.120D0, 0.042D0, 0.90D0
3.0D6, 0.133D0, 0.050D0, 0.90D0
5.0D6, 0.160D0, 0.072D0, 0.90D0
7.5D6, 0.200D0, 0.108D0, 0.90D0
10.0D6, 0.234D0, 0.144D0, 0.90D0
13.0D6, 0.262D0, 0.185D0, 0.90D0
15.0D0, 0.272D0, 0.210D0, 0.90D0
18.0D0, 0.292D0, 0.262D0, 0.90D0
20.0D6, 0.294D0, 0.294D0, 0.90D0
23.0D6, 0.333D0, 0.356D0, 0.90D0
25.0D6, 0.256D0, 0.384D0, 0.90D0
28.0D6, 0.415D0, 0.453D0, 0.90D0
30.0D6, 0.490D0, 0.500D0, 0.90D0
6, 0.160D0
2.5D6, -0.0335D0
7.0D6, +0.0655D0
11.0D6, +0.0513D0
19.0D6, +0.0497D0
26.0D6, +0.0827D0
40.0D6, -0.375D0
9,0
5.0D6,5
```

work need to be considered. (The input resistance is lower than 6.25Ω at the higher frequencies).

In order to determine the constraints, the gain was decreased progressively until the ripple in the pass band became very small. The last incremental frequency was adjusted in each case to minimize the ripple. The input data for a transducer power gain of 0.90 are shown in Table 6.8 and the results are summarized in Table 6.9.

If the criterion of minimum insertion loss across the pass band is used, the best results will be obtained if the gain is chosen to be approximately 0.90. The insertion loss will then be less than 0.6 dB.

Table 6.9

The Optimum Values of the Highest Increment Frequency and the Associated Normalized Least-Square Error and Maximum Deviation from the Prescribed Transducer Power Gain for the MRF406 as a Function of the Transducer Power Gain Specified

Transducer power gain specified	Optimum value of the highest increment frequency (MHz)	Normalized least-square error	Maximum deviation from the prescribed tranducer power gain (%)
1.00	36	0.1637	23
0.98	37	0.1062	15
0.96	37	0.0559	11
0.94	38	0.0254	8
0.92	39	0.0099	5
0.90	43	0.0030	3
0.88	58	0.0009	2
0.86	83	0.0004	1
0.84	120	0.0002	1

6.4.2 The Reflection Coefficient Approach to Solving Double-Matching Problems

When both the load and source impedances are reactive, impedance-matching networks can be designed iteratively by using the algorithm developed by Carlin and Yarman [16].

In following this approach, the lossless matching network is modelled as a two-port network. When all the transmission zeros are at the origin or infinity, the scattering parameters (S-parameters) of the network are given by the following equations:

$$s_{11}(s) = h(s)/g(s) \tag{6.73}$$

$$s_{12}(s) = s_{21}(s) = \pm s^k/g(s) \tag{6.74}$$

$$s_{22}(s) = -(-1)^k[h(-s)/g(s)] \tag{6.75}$$

where k is an integer specifying the order of the zero of transmission at the origin and h (s) and $g(s)$ are polynomials.

Because the transfer and reflection parameters are related by the equation:

$$s_{11}(s)\,s_{11}(-s) = 1 - s_{21}(s)\,s_{21}(-s) \tag{6.76}$$

it follows that

$$g(s)\,g(-s) = h(s)\,h(-s) + (-1)^k\,s^{2k} \tag{6.77}$$

It is clear from this equation, and the fact that the polynomial $g(s)$ must be positive-real, that $g(s)$ is a function of $h(s)$ and the order of the zero of transmission at the origin only. Because the S-parameters of the network are completely determined by $h(s)$, $g(s)$, and k, it follows that the network itself is defined when $h(s)$ and k are defined. The optimum network, therefore, can be determined by finding the optimum coefficients of the polynomial $h(s)$ for a given value of k.

In order to optimize the coefficients of $h(s)$, an expression for calculating the transducer power gain at each relevant frequency is required. In terms of S-parameters, the gain is given by

$$G_T(\omega) = \frac{[1-|S_G(\omega)|^2][1-|S_L|^2]\,|s_{21}(\omega)|^2}{\left|1-s_{11}(\omega)\,S_G(\omega)\right|^2\left|1- s_{22}(\omega) + \left[\dfrac{s_{21}^2(\omega)\,S_G(\omega)}{1-s_{11}(\omega)\,S_G(\omega)}\right]S_L(\omega)\right|^2} \tag{6.78}$$

$$= \frac{[1-|S_G(\omega)|^2][1-|S_L|^2]\,|\omega^{2k}|}{\left|g(j\omega)-(-1)^k S_G S_L g(-j\omega) - S_G h(j\omega) + (-1)^k S_L h(-j\omega)\right|^2} \tag{6.79}$$

where

$$S_G(\omega) = \frac{Z_s(\omega)-R_0}{Z_s(\omega)+R_0} \tag{6.80}$$

$$S_L(\omega) = \frac{Z_L(\omega)-R_0}{Z_L(\omega)+R_0} \tag{6.81}$$

and R_0 is the normalizing resistance of the S-parameters.

The coefficients of $h(s)$ can be optimized by using a linear least-square routine. Because the gain is not a simple function of $h(s)$, the problem is, however, more complex than before and the choice of the initial values can be critical.

Good results can be achieved if the results obtained by designing an impedance-matching network to match the more complex termination to a resistive source are used to determine the initial values.

The optimization can be done by using the program RCDM FORTRAN in Appendix D. The input data of the program consist of the following:

1. The number of frequencies at which the source and the load impedances are to be specified.

2. The measured source and load impedances and the required transducer power gain at each frequency (f_i, R_s, X_s, R_L, X_L, G_T).

3. The degree of the numerator polynomial $h(s)$ and the number of transmission zeros at the origin.

4. The initial values of the coefficients h_0, h_1, ..., h_n, in that sequence.

5. The number of iterations to be done and a numeral specifying the detail of the printout required (1=detailed; 2=otherwise).

When initial values are assigned to the coefficients of $h(s)$, it is important to realize that there are some constraints on the values of the coefficients.

In the case of low-pass networks, the value of the transmission parameter $s_{21}(s)$ must be equal to one when $\omega=0$, and the same input and output normalizing resistances (R_0) are used for the S-parameters.

Since

$$s_{21}(s) = \pm \frac{s^k}{g(s)} \tag{6.82}$$

$$= \frac{\pm s^k}{g_n s^n + g_{n-1} s^{n-1} + \ldots + g_0} \tag{6.83}$$

and $k = 0$ for a low-pass network, it is clear that g_0 must equal one to ensure that $s_{21}(0)$ will be equal to one.

This restriction on the value of g_0 results in two constraints on the polynomial $h(s)$.

The first is that in determining initial values for $h(s)$, the numerator and denominator of the input reflection coefficient $s_{11}(s)$ must be scaled to ensure that g_0 will equal one.

The second restriction follows from (6.77) by setting $s = 0$:

$$g_0^2 = h_0^2 + 1 \tag{6.84}$$

Since $g_0 = 1$, it implies that h_0 must be equal to zero.

Following the reasoning above, it can be shown easily for high-pass networks that h_n must be equal to zero, and that the numerator and denominator of the input reflection coefficient must be scaled to ensure that $g_n = 1$ when initial values for the coefficients are determined.

When a band-pass problem is initialized none of the coefficients need to equal zero, but scaling is still required. It is obvious from (6.74) that if $h(s)$ and $g(s)$ are scaled by a factor a, the gain $|s_{21}|^2$ will change by a factor a^2.

An appropriate scale factor can be determined by using (6.74) to calculate the gain at any frequency in the pass band without a scaling factor. The required scaling factor can then be taken as $\omega^k/|g(j\omega)|$.

The constraints on the numerator and denominator coefficients of $\rho(s)$ are summarized in Table 6.10.

Table 6.10

**The Constraints on the Numerator (h(s)) and the Denominator (g(s))
Coefficients of ρ (s)**

Network	$h(s)$	$g(s)$
Low-Pass	$h_0 = 0$	$g_0 = 1$
High-Pass	$h_n = 0$	$g_n = 1$
Band-Pass	scaling factor a for $h(s)$	—

Example 6.8

As an example of the application of the impedance-matching programs discussed, a high-pass network will be designed to mismatch the source impedance in Table 6.11 to a 50Ω load, as indicated in the table, by using the programs LSM FORTRAN, PLNM FORTRAN, and ZVR FORTRAN. The solution obtained will then be used to design a network to mismatch the same source to the complex load of the same table by using the program RCDM FORTRAN.

The Single-Matching Problem

Table 6.11

**The Source Impedance, Load Impedance, and Transducer Power Gain
Corresponding to the Problem Solved in Example 5.6**

Frequency (MHz)	$R_s + jX_s$ (Ω)	$R_L + jX_L$ (Ω)	G_T (Ω)
100	146.0–j114.0	79.1–j72.6	0.224
110	138.5–j112.5	73.6–j68.7	0.262
120	131.0–j111.0	68.0–j64.8	0.299
140	137.0–j103.0	63.2–j56.8	0.400
160	144.0–j88.0	59.6–j47.9	0.559
180	140.0–j88.0	57.5–j47.3	0.709
190	136.5–j92.0	55.0–j41.9	0.764
200	133.9–j96.0	53.5–j40.4	0.818

In order to design the required high-pass network with LSM FORTRAN, the
specifications in Table 6.11 must be changed to those of the equivalent low-
pass problem using the transformation $s \to 1/s$. The new set of specifications is
shown in Table 6.12. The frequencies in the table were obtained from the
transformed frequencies ($\omega' = 1/\omega \to f' = 1/(4\pi^2 f)$) by using a scaling factor of
$4\pi^2 10^9$. The impedances are the conjugates of those in Table 6.11.

Table 6.12

**The Source Impedance and Transducer Power Gain Corresponding to the
Equivalent Low-Pass Problem**

Frequency (Hz)	R_s (Ω)	X_s (Ω)	G_T
5.00	133.0	96.0	0.818
5.56	140.0	88.0	0.709
6.25	144.0	88.0	0.559
7.14	137.0	103.0	0.400
8.33	131.0	111.0	0.299
10.00	146.0	114.0	0.224

Because the source impedance of the transistor is higher than the load impedance (50Ω), a minimum-impedance matching network will be designed. (The
first element of the network, therefore, will be in parallel with the output
terminals of the transistor). The data entered into the program LSM FOR-
TRAN are shown in Table 6.13.

Table 6.13

The Input Data of the Program LSM FORTRAN

```
6
5.00D0, 133.0D0, 96.0D0, 0.818D0
5.56D0, 140.0D0, 88.0D0, 0.709D0
6.25D0, 144.0D0, 88.0D0, 0.559D0
7.14D0, 137.0D0, 103.0D0, 0.400D0
8.33D0, 131.0D0, 111.0D0, 0.299D0
10.0D0, 146.0D0, 114.0D0, 0.224D0
5,50.0D0
4.9D0, 80.0D0
5.7D0, 10.0D0
8.0D0, -10.0D0
9.8D0, 16.0D0
40.0D0, -146.0D0
8, 1
0, 1
```

The results of the program are shown in Table 6.14. The gain-bandwidth
limitations of the network were found to be insensitive to the value of the
highest increment frequency. (This was not the case when a minimum-admittance solution was attempted).

Table 6.14

The Results Obtained with the Program LSM FORTRAN

Frequency (Hz)	Input impedance of the network to be designed (Ω)	Transducer power gain obtainable
5.00	$64.7 - j49.9$	0.835
5.56	$40.7 - j56.4$	0.677
6.25	$30.1 - j46.3$	0.541
7.14	$22.8 - j42.4$	0.427
8.33	$15.1 - j34.6$	0.291
10.00	$12.6 - j27.6$	0.225

A rational approximation for the resistive part of the input impedance of the
network to be designed was obtained by using the program PLNM FOR-
TRAN. The data entered into the program are shown in Table 6.15.

Table 6.15

The Input Data of the Program PLNM FORTRAN

```
4, 7, 1
0.00D0, 0.020D0
5.00D0, 0.015D0
5.560D0, 0.025D0
6.250D0, 0.033D0
7.140D0, 0.044D0
8.330D0, 0.066D0
10.00D0, 0.079D0
10, 1.10, 1.00
```

The polynomial obtained from the program is:

$$T(\omega) = 0.19885 \cdot 10^{-1} - 0.32954 \cdot 10^{-4}\,\omega + 0.40267 \cdot 10^{-7}\,\omega^2 - 0.10282 \cdot 10^{-10}\,\omega^3 + 0.80155 \cdot 10^{-15}\,\omega^4$$

Therefore, the resistance function is

$$R(\omega) = 1/T(\omega) = 1/(0.80155 \cdot 10^{-15}\,\omega^8 - 0.10282 \cdot 10^{-10}\,\omega^6 + 0.40267 \cdot 10^{-7}\,\omega^4 - 0.32954 \cdot 10^{-4}\,\omega^2 + 0.19885 \cdot 10^{-1})$$

The poles of this function are

$$\omega = \pm\,78.31033 \pm j12.02878$$

$$\pm\,24.69292 \pm j13.55472$$

The minimum-impedance function corresponding to the resistance function was found by entering the data shown in Table 6.16 into the program ZVR FORTRAN.

Table 6.16

The Input Data of the Program ZVR FORTRAN

```
0
1.247858 · 10^15
2
(78.31033, 12.02878)
(24.69292, 13.55472)
0
6
5.0, 5.56
6.25, 7.14
8.33, 10.0
```

The minimum-impedance function obtained is

$$Z(j\omega) =$$

$$\frac{0.2505 \cdot 10^9 + j0.1290 \cdot 10^8\, \omega - 0.10186 \cdot 10^6\, \omega^2 - j0.1989 \cdot 10^4\, \omega^3}{0.4981 \cdot 10^7 + j0.1893 \cdot 10^6\, \omega - 0.7723 \cdot 10^4 \cdot \omega^2 - j0.5117 \cdot 10^2\, \omega^3 + 1.000\omega^4}$$

As a function of s, this becomes

$$Z(s) =$$

$$\frac{0.2505 \cdot 10^9 + 0.1290 \cdot 10^8\, s + 0.10186 \cdot 10^6\, s^2 + 0.1989 \cdot 10^4\, s^3}{0.4981 \cdot 10^7 + 0.1893 \cdot 10^6\, s + 0.7723 \cdot 10^4\, s^2 + 0.5117 \cdot 10^2\, s^3 + 1.000 s^4}$$

$$(6.85)$$

The network corresponding to this function is shown in Fig. 6.23. It was synthesized by continued fractionation of the impedance function.

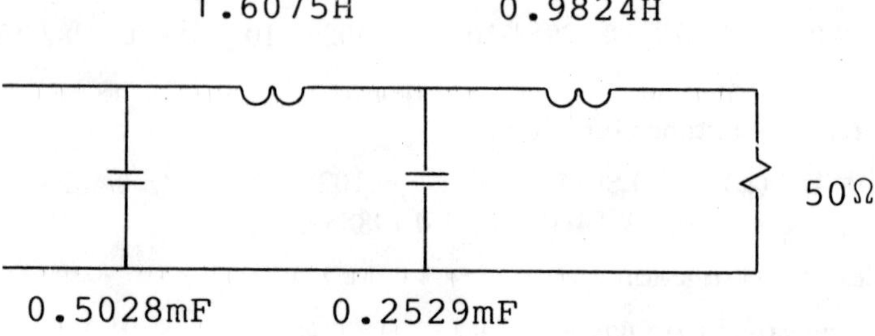

Figure 6.23   The Network Corresponding to Eq. (6.85)

The high-pass equivalent of this network under the transformation $s \rightarrow 1/s$ is shown in Fig. 6.24. Because a scaling factor of $4\pi^2\, 10^9$ was used to obtain the frequencies used in Table 6.12, this network must be frequency scaled with the same factor in order to obtain the matching network required. The frequency-scaled network is shown in Fig. 6.25.

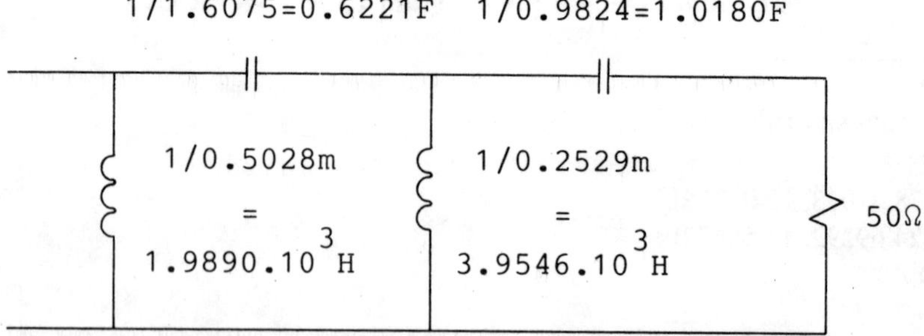

Figure 6.24   The Equivalent High-Pass Network for the Network in Fig. 6.23 under the Transformation $s \rightarrow 1/s$

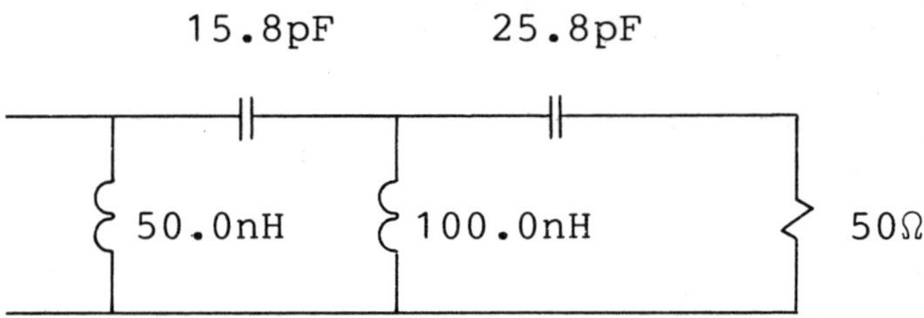

Figure 6.25   The Designed Matching Network

As a check on the results, the transducer power gain was calculated at the relevant frequencies. The results are compared to the specifications in Table 6.17.

The variation in gain is less than 0.24 dB over the frequency range of interest.

Table 6.17

Comparison of the Gain Obtained with the Network in Fig. 6.25 and the Specified Gain

Frequency (MHz)	Transducer Power Gain Specified	Transducer Power Gain Obtained
100	0.224	0.232
120	0.299	0.303
140	0.400	0.423
160	0.559	0.571
180	0.709	0.711
200	0.818	0.806

The Double-Matching Problem

Initial values for the design of a double-matching network can be obtained by calculating the reflection coefficient for the designed network shown in Fig. 6.25. The reflection coefficient is

$$s_{11}(s) = \frac{5.9\text{E-}13\,s^4+5.6\text{E-}8\,s^3-1.8\text{E-}4\,s^2-1.4\text{E-}2\,s-50.0}{2.1\text{E-}10\,s^4+3.6\text{E-}7\,s^3+3.1\text{E-}4\,s^2+1.2\text{E-}1\,s+50.0}$$

Because the values of the coefficients of $h(s)$ are limited to a range of approximately 1E20 (that is, h_0/h_n, $h_n/h_0 < 1\text{E}20$) by a subroutine in the program, it was necessary to scale the frequencies down to lower values. A scale factor of 1E6 was used in this case.

The required initial values can be obtained from this equation by scaling the numerator and denominator with a factor $1/(2.1E{-}10)$ in order to have $g_4 = 1.0$.

The initial values obtained are

   $h_0 = -2.4E11$

   $h_1 = -7.0E7$

   $h_2 = -8.9E5$

   $h_3 = 2.7E2$

   $h_4 = 0.0$

At this stage the double-matching problem and the initial values are defined, and RCDM FORTRAN can be used to determine the solution. The data entered into the program are shown in Table 6.18.

Table 6.18

The Data Entered into the Program RCDM FORTRAN

```
8
100.0, 146.0, -114.0, 79.1, -72.6, 0.224
110.0, 138.5, -112.5, 73.6, -68.7, 0.262
120.0, 131.0, -111.0, 68.1, -64.8, 0.299
140.0, 137.0, -103.0, 63.2, -56.8, 0.400
160.0, 144.0, -88.0, 59.6, -47.9, 0.559
180.0, 140.0, -88.0, 57.5, -47.3, 0.709
190.0, 136.5, -92.0, 55.5, -41.9, 0.764
200.0, 133.0, -96.0, 53.5, -40.4, 0.818
4,4
-2.4E11
-7.02E7
-8.87E5
+2.71E2
0.0
25,2
```

The matching network corresponding to the impedance function obtained is shown in Fig. 6.26. The maximum deviation from the specified gain response is 0.24 dB. The results obtained are compared to the specifications in Table 6.19.

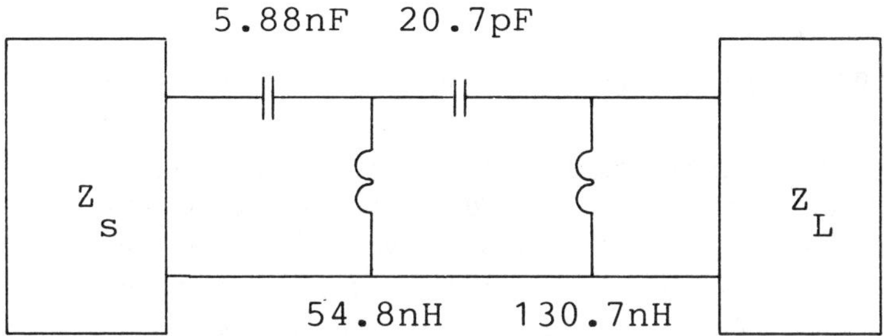

Figure 6.26   The Matching Network Designed with the Reflection Coeffi-cient Technique

Table 6.19

The Specifications and Gain Obtained with the Network in Fig. 6.26 Compared

Frequency (MHz)	Transducer Power Gain Specified	Transducer Power Gain Obtained
100	0.224	0.228
110	0.262	0.248
120	0.299	0.288
140	0.400	0.403
160	0.559	0.563
180	0.709	0.685
190	0.764	0.767
200	0.818	0.822

6.4.3   The Transformation Q Approach to the Design of Impedance-Matching Networks

The narrowband impedance-matching technique of Ch. 3 can be extended to increase the number of elements to an arbitrary number and to mismatch any complex load to any complex source by any specified amount at any single frequency. In order to do this, it will be shown in sec. 6.4.3.1 that the locus of input impedances — for which the source impedance of a network will be mismatched to the load by a specified amount — is a circle in the linear admittance or impedance plane and the parameters of these circles will be derived.

The necessary extentions to the single-frequency technique will be made in sec. 6.4.3.2.

The extended single-frequency matching techniques forms an excellent basis for the iterative design of wideband impedance-matching networks. The main reason for this is that the possible range of each transformation Q is very limited since high Q-factors inevitably lead to narrow bandwidths.

Because of this fact, it is feasible to do a grid-search on the transformation Q-factors in search of the optimum combination. In so doing the need for a good initial solution is eliminated. Furthermore, if the search is done thoroughly enough, the probability of finding the optimum solution will be very high.

With the grid-search completed, a number of the best sets of Q-factors can be optimized. This can be done with a least-square optimization routine, but better results in less time are obtainable by using a simple gradient optimization technique.

The mean-square error from the specified gain response can be used in the grid-search and during the optimization phase, but a better alternative is to use the maximum relative deviation from the optimum as the error criterion.

The topologies of the networks synthesized by following this approach can easily be limited to low-pass, high-pass, or band-pass form with no series capacitors.

The time required to solve a matching problem can be reduced significantly by constraining the gain at the frequency where the Q-factors are calculated to be higher than a specified minimum. The required running time is usually very short when networks with less than six elements are designed (typically 2.5 to 5.5 minutes on a popular microcomputer with a numeric co-processor).

Major advantages of the transformation Q technique over the techniques previously described are that many solutions instead of only one are obtained, that transformers are never required in the solutions, and that the probability of finding the optimum solution to a matching problem is very high when the search is done thoroughly enough.

The first advantage is important when the best solution obtained is not physically realizable or another topology is preferable.

6.4.3.1 Constraints on the Input Impedance of a Lossless Impedance-Matching Network if the Gain Is to Remain Constant at a Specific Frequency

The locus of input admittances for which the gain of a lossless impedance-matching network will remain constant can be derived by using the expression:

$$|S_s|^2 = \left| \frac{Z_{IN} - Z_s^*}{Z_{IN} + Z_s} \right|^2 \tag{6.86}$$

$$= \left| \frac{Y_{IN} - Y_s^*}{Y_{IN} + Y_s} \right|^2 \tag{6.87}$$

where S_s is the input reflection parameter with the actual source impedance Z_s as normalizing impedance, and $Z_{IN}(Y_{IN})$ is the input impedance (admittance) of the matching network.

By substituting

$$|S_s|^2 = 1 - G_T \tag{6.88}$$

$$Y_{IN} = G_{IN} + jB_{IN} \tag{6.89}$$

and

$$Y_s = G_s + jB_s \tag{6.90}$$

into (6.87), it follows after some manipulation that the locus of input transducer power gain (G_T) is a circle, and that the parameters (center and radius) of this circle are

$$G_0 + jB_0 = [2/G_T - 1] \, G_s - jB_s \tag{6.91}$$

and

$$R_{Y0} = 2 \, [1/G_T^2 - 1/G_T]^{1/2} \, G_s \tag{6.92}$$

The gain of a lossless network will remain constant for all values of the input impedance which fall on the circumference of this circle. This is illustrated in Fig. 6.27.

For all values of the input admittance which fall inside the constant gain circle, the gain will be higher than that on the circumference. The transducer power gain of a matching network, therefore, can be constrained to be higher than a specified minimum at any particular frequency by limiting its input admittance to the inside of the relevant constant-gain circle.

Because (6.86) and (6.87) are of exactly the same form, the locus of input impedances for which the gain of a lossless network will remain constant is also a circle, and the parameters of this circle can also be obtained from (6.91) and (6.92) with $G_s + jB_s$ replaced with $R_s + jX_s$, and $G_0 + jB_0$ with $R_0 + jX_0$. The resulting equations are

$$R_0 + jX_0 = [2/G_T - 1] \, R_s - jX_s \tag{6.93}$$

and

$$R_{Z0} = 2 \, [1/G_T^2 - 1/G_T]^{1/2} \, R_s \tag{6.94}$$

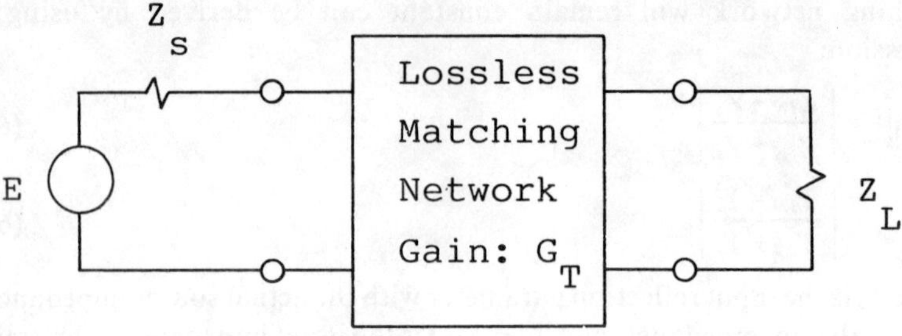

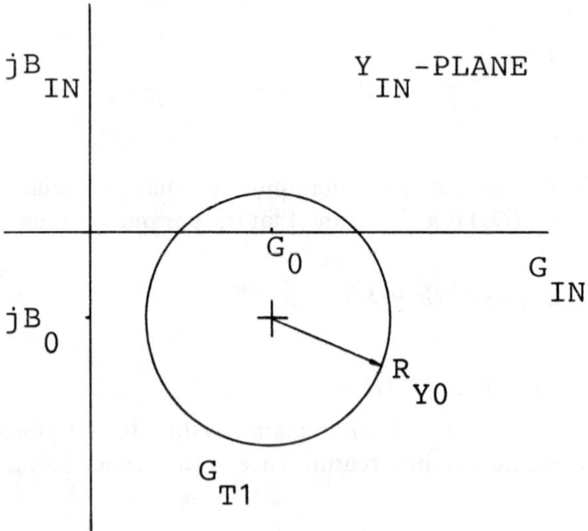

Figure 6.27   The Locus of Input Admittance for which the Gain of a Lossless Matching Network Will Remain Constant at a Specified Frequency

6.4.3.2   Extention of the Transformation Q Impedance-Matching Technique

In an impedance-matching network, the resistance level is changed by each element except the last which only serves to adjust the reactance or susceptance level. The change in the resistance of a four-element network with the first element a series element and no resonating sections is illustrated in Fig. 6.28. The resistance is transformed in each transformation step by a factor of the form

$$D_n(\omega) = 1 + Q_n^2(\omega) \tag{6.95}$$

where

$$Q_n(\omega) = \frac{X_n(\omega) + X_{rn}(\omega)}{R_{rn}(\omega)} \tag{6.96}$$

or

$$Q_n(\omega) = \frac{B_n(\omega) + B_{rn}(\omega)}{G_{rn}(\omega)} \tag{6.97}$$

depending on whether the transformation under consideration is series-to-parallel or parallel-to-series, respectively.

The factor $Q_n(\omega)$ is referred to as a transformation Q.

In (6.96) $X_n(\omega)$ is the reactance of the nth component, $X_{rn}(\omega)$ the effective reactance to the right of it, and $R_n(\omega)$ the effective resistance in series with it as illustrated in Fig. 6.29a. Similarly, $B_n(\omega)$ is the susceptance of the nth component, $B_{rn}(\omega)$ is the effective susceptance to the right of it, and $G_{rn}(\omega)$ the effective conductance in parallel with it. This is illustrated in Fig. 6.29b.

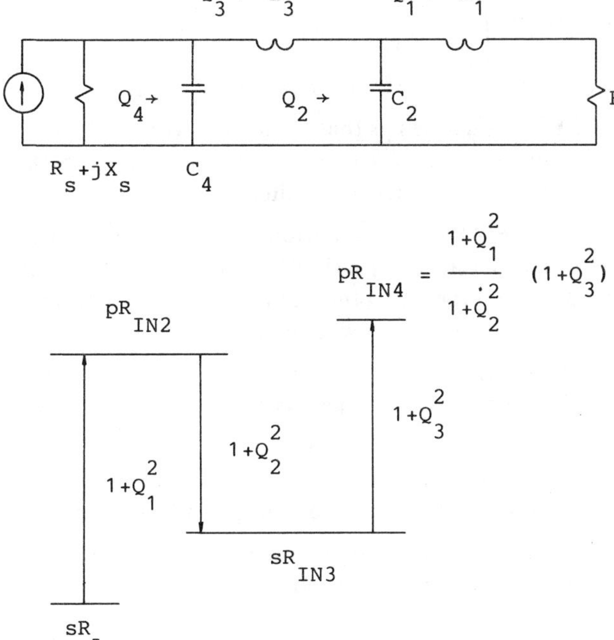

Figure 6.28   [a] Each Transformation Step at the Frequency of Interest and [b] a Schematic Illustration of the Change in Resistance Level of the Network

0.889 and the Q of the circuit is equal to 6.0

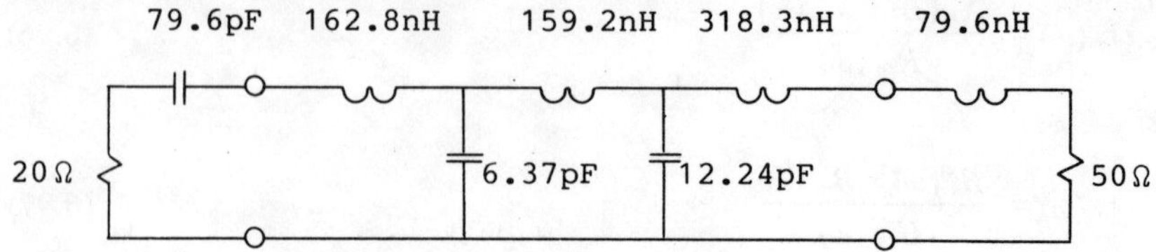

Figure 6.33   The Network Synthesized in Example 6.8

6.4.3.3   Optimization of the Transformation Q-factors of a Matching Network

The transformation Q-factors corresponding to an initial solution for a matching problem can be optimized by using a linear least-square optimization routine. Although good results can be obtained by doing this, better results are obtainable through following a different approach.

The first improvement is to use the maximum relative deviation:

$$MRD = MAX \left[\frac{G_T(\omega)}{G_{T\text{-}opt}(\omega)} - 1 \right]$$

as the error criterion instead of the mean-square error.

In (6.121) $MAX[X]$ is defined as that value of X with the highest magnitude with the sign retained. In contrast to the mean-square error, the MRD can therefore take on positive or negative values.

One advantage of the MRD error criterion is that the solution with the lowest insertion loss will be obtained when the ideal gain is set equal to unity. Another more obvious advantage is that the maximum deviation from the optimum rather than the average deviation will be minimized.

The second improvement is to optimize the error by using the steepest-decent method. The results obtained in doing this were superior to those corresponding to the least-square method.

The gradient vector required for optimizing the Q values can be determined by calculating the change in MRD corresponding to a small increment in each Q.

The new set of Q values ($\overline{Q}_N$) can be obtained from the previous set ($\overline{Q}_{N-1}$) and the current MRD by using the equation:

$$\overline{Q}_N = \overline{Q}_{N-1} + \alpha \, \frac{MRD}{|MRD|} \left| \frac{MRD}{|\partial MRD/\partial Q_1| + \ldots + |\partial MRD/\partial Q_N|} \right| \begin{bmatrix} \partial MRD/\partial Q_1 \\ \cdot \\ \cdot \\ \cdot \\ \partial MRD/\partial Q_N \end{bmatrix}$$

$$(6.122)$$

where the optimum value of α can be determined iteratively by using the following method [22, pp. 178-179]:

Start with a small value of α (α_1) and increase it during the next iterations (α_i) by using the expression:

$$\alpha_i = \alpha_1 \left[1 + l + l^2 + \ldots l^{i-1} \right] \quad i = 1,2,3, \ldots$$
$$l = 1.5 \tag{6.123}$$

until the MRD increases. This will result in the situation depicted in Fig. 6.34. A quadratic curve can now be fitted through the last three coordinates, and the value of α (α_M) for which the error will be a minimum can be estimated by using the expression:

$$\alpha_M = \frac{1}{2} \cdot$$

$$\frac{[\alpha_{n-1}^2 - \alpha_n^2]\, MRD_{n-2} + [\alpha_n^2 - \alpha_{n-2}^2]\, MRD_{n-1} + [\alpha_{n-2}^2 - \alpha_{n-1}^2]\, MRD_n}{[\alpha_{n-1} - \alpha_n]\, MRD_{n-2} + [\alpha_n - \alpha_{n-2}]\, MRD_{n-1} + [\alpha_{n-2} - \alpha_{n-1}]\, MRD_n} \tag{6.124}$$

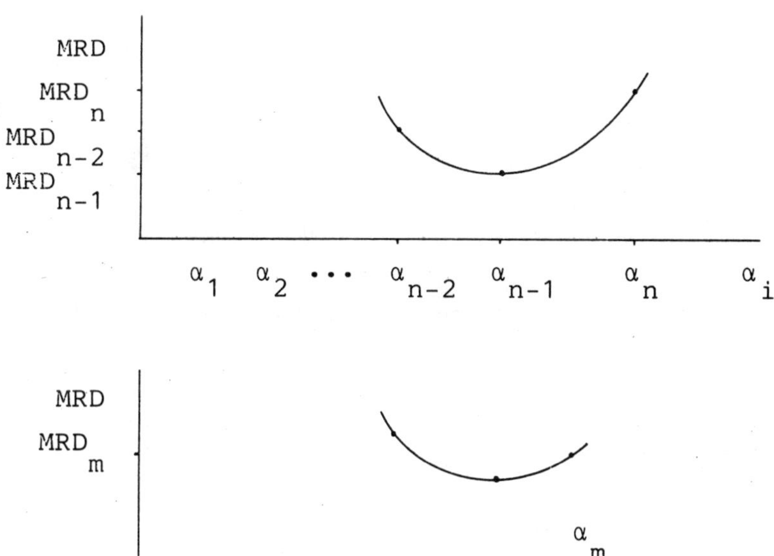

Figure 6.34   The Influence of the Scaling Factor α in Eq. (6.124) on the MRD

A series-to-parallel transformation will always transform the associated resistance upward, while downward transformations are effected with shunt elements.

It follows from (6.96) and (6.97) that the sign of a transformation Q is positive when the effective reactance in series with the effective resistance is inductive, or when the effective susceptance in parallel with the effective conductance is capacitive.

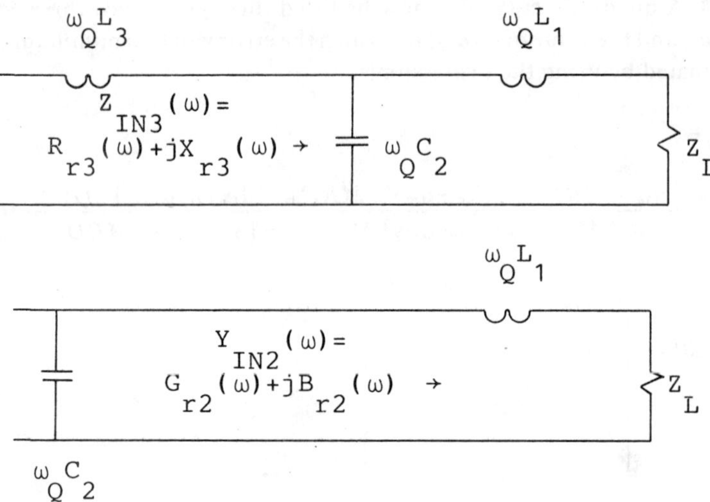

Figure 6.29   *Definition of the Symbols in Eqs. (6.96) and (6.97)*

When the first element of an N-element network is a series element, the input resistance is given after $(N\text{-}1)$ transformations by the expression:

$$R_{IN\text{-}N} = R_L \; \frac{1+Q_1^{\,2}}{1+Q_2^{\,2}} \; \frac{1+Q_3^{\,2}}{1+Q_4^{\,2}} \; \dots \; [1+Q_{N\text{-}1}^{\,2}] \tag{6.98}$$

or

$$R_{IN\text{-}N} = R_L \; \frac{1+Q_1^{\,2}}{1+Q_2^{\,2}} \; \dots \; \frac{1+Q_{N\text{-}2}^{\,2}}{1+Q_{N\text{-}1}^{\,2}} \tag{6.99}$$

depending on whether the last matching element is a parallel element or a series element, respectively.

When the first element is a parallel element, the input conductance is given by the expression:

$$G_{IN\text{-}N} = \frac{G_L}{1+Q_1^{\,2}} \; \frac{1+Q_2^{\,2}}{1+Q_3^{\,2}} \; \dots \; [1+Q_{N\text{-}1}^{\,2}] \tag{6.100}$$

or

$$G_{IN-N} = \frac{G_L}{1+Q_1^{\,2}} \; \frac{1+Q_2^{\,2}}{1+Q_3^{\,2}} \; \cdots \; \frac{1+Q_{N-2}^{\,2}}{1+Q_{N-1}^{\,2}} \qquad (6.101)$$

Because (6.98) and (6.100) as well as (6.99) and (6.101) are of the same form except that the resistance and conductance must be interchanged, it is only necessary to consider the design of matching networks with a series element as the first element. Exactly the same procedure can then be followed to design networks in which the first element is a parallel element, after replacing all impedance specifications with the equivalent admittances.

In low-pass and high-pass designs the number of Q-factors is equal to the number of elements in the network. In a band-pass network with N elements, but M resonating sections (series of parallel LC combinations), the number of Q-factors increases to $(N+M)$, since each transformation Q is associated with only one element. This is illustrated for a three-element network in Fig. 6.30.

When an element is absent, the associated Q is equal to the negative of the previous Q.

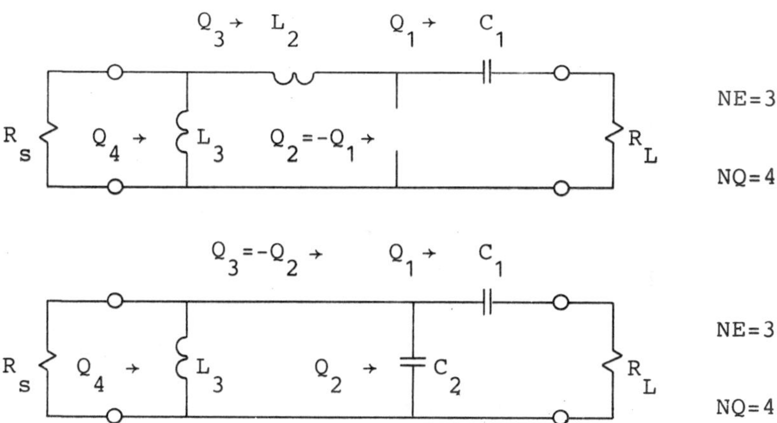

Figure 6.30   The Influence of Resonating Sections on the Number of Transformation Q-factors in a Three-Element Network

By using (6.98) through (6.101), it is very easy to calculate the input resistance of any impedance-matching network when the Q-factors are known.

In order to design a matching network to have a specified transducer power gain (G_T) at a particular frequency, it is only necessary to constrain the last two Q-factors to ensure that the input impedance (if the last element is a parallel element) will fall on the relevant gain circle as derived in the previous section.

When the last element is a series element, the next to last Q should be constrained to ensure that the input resistance (R_{IN}) will fall in the range

$$R_{IN,\ MIN} \leq R_{IN} \leq R_{IN,\ MAX} \tag{6.102}$$

where the bounds are defined in Fig. 6.31.

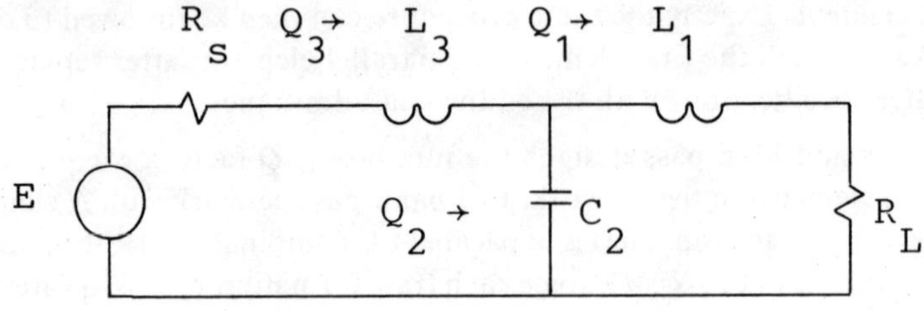

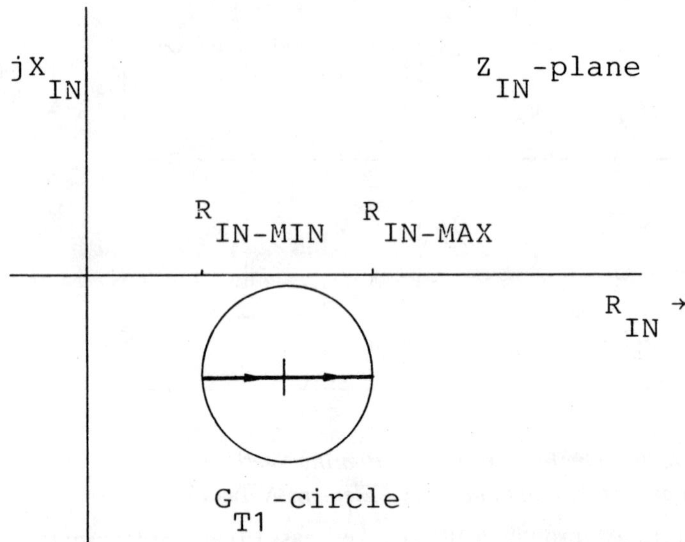

Figure 6.31   *The Constraints on the Input Resistance of a Matching Network with Last Element a Series Element when the Transducer Power Gain Should Be Higher than or Equal to a Specified Minimum at a Particular Frequency*

The bounds on the next to last Q follow easily from the values of the previous Q-factors by using (6.99) in conjunction with (6.102).

The resulting constraints are

$$Q_{N-1}^2 \geq \frac{R_L}{R_{IN-MAX}} \frac{1+Q_1^2}{1+Q_2^2} \dots [1+Q_{N-2}]^2 - 1 \tag{6.103}$$

and

$$Q_{N-1}^2 \leq \frac{R_L}{R_{IN-MIN}} \frac{1+Q_1^2}{1+Q_2^2} \dots [1+Q_{N-2}^2] - 1 \tag{6.104}$$

When the last element is a parallel element, the next to last Q should be constrained to ensure that the input conductance (G_{IN}) will be within the constraints imposed by the constant gain circle on the admittance plane.

The resulting constraints are

$$Q_{N-1}^2 \geq \frac{1}{R_L\, G_{IN-MAX}} \frac{1+Q_2^2}{1+Q_1^2} \dots [1+Q_{N-2}^2] - 1 \tag{6.105}$$

and

$$Q_{N-1}^2 \leq \frac{1}{R_L\, G_{IN-MIN}} \frac{1+Q_2^2}{1+Q_1^2} \dots [1+Q_{N-2}^2] - 1 \tag{6.106}$$

The constraints on the last transformation Q can be derived from Fig. 6.32. With the resistance (if the last element is a series element) or conductance (if the last element is a parallel element) in range, the reactance or susceptance must be such that the resulting impedance or admittance falls on the circumference of the gain circle.

When the last element is a series element, the reactance is constrained to

$$X_{IN} = X_0 \pm R_{Z0} \sin \left[\cos^{-1} \frac{R_{IN} - R_0}{R_{Z0}} \right] \tag{6.107}$$

The equivalent expression when the last element is a parallel element is

$$B_{IN} = B_0 \pm R_{Y0} \sin \left[\cos^{-1} \frac{G_{IN} - G_0}{R_{Y0}} \right] \tag{6.108}$$

Because R_{IN} or G_{IN} is known, it is a simple matter to calculate the Q corresponding to these reactances or susceptances.

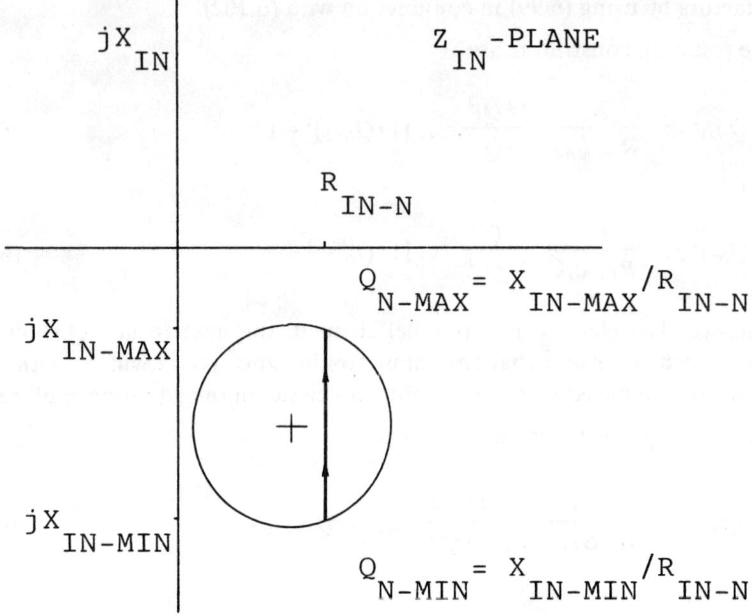

Figure 6.32   The Constraints on the Last Transformation Q of a Network if the Gain Is to be Higher than a Specified Minimum

From a single-frequency matching viewpoint, there are no constraints on the first *N*-2 transformation *Q*. If the response is also of interest over a narrow pass band, any change in the *Q*-factors should, as a first order approximation, be constrained to be smaller than twice the *Q* required for the circuit. The change in *Q* from one transformation to the next is given by

$$\Delta Q = [Q_n + Q_{n-1}]$$ (6.109)

The reason for the positive sign is that the sign of a *Q* of transformation changes under a series-to-parallel or parallel-to-series transformation. As an example of this, if the *Q* of a series combination is +3 (inductive), the *Q* of the parallel equivalent will be −3 (again inductive).

In summary, the following procedure can be followed to design an *N*-element matching network to match or mismatch a complex load by a specified amount to a complex source at a particular frequency and to have a specified (approximate) quality factor.

Specifications: load = Z_L; source = Z_s; gain at f = G_T; quality factor = Q.

Design Procedure

1)   If the first element of the network is to be a shunt element, change the source and load impedances to $1/Z_s$ and $1/Z_L$, respectively, and assume that the first element is now a series element.

2)   Choose any values for the first N-2 transformation Q-factors within the constraint that any change from one Q to the next must be smaller than $2Q$, that is,

$$Q_n + Q_{n-1} \leq 2Q \tag{6.110}$$

3)   If the last element is to be a parallel element, calculate the parameters of the constant gain circle on the admittance plane by using (6.91) and (6.92).

If the last element is a series element, use (6.93) and (6.94) to calculate the parameters of the constant gain circle on the impedance plane.

4)   Calculate the minimum and maximum allowable values for the input conductance or resistance by using the following equations, as applicable:

$$G_{IN,\ MAX} = G_0 + R_{Y0} \tag{6.111}$$

$$G_{IN,\ MIN} = G_0 - R_{Y0} \tag{6.112}$$

$$R_{IN,\ MAX} = R_0 + R_{Z0} \tag{6.113}$$

$$R_{IN,\ MIN} = R_0 - R_{Z0} \tag{6.114}$$

5)   Determine the constraints on the next to last transformation Q by using (6.103) and (6.104) when the last element is a series element. Otherwise use (6.105) and (6.106).

Choose a value for Q_{N-1} within these constraints and that imposed by (6.110).

6)   When the last element is a series element, calculates the two possible reactance values corresponding to the last transformation Q (X_{IN-MIN} and X_{IN-MAX}) by using (6.107). Otherwise, use (6.108) to calculate the allowable susceptance values (B_{IN-MIN} and B_{IN-MAX}).

Calculate R_{IN-N} or $G_{IN-N} = 1/R_{IN-N}$ by using (6.98) or (6.99), as applicable.

Calculate the two possible values for the last transformation Q by using the following equations:

$$Q_{N-MAX} = \frac{X_{IN-MAX}}{R_{IN-N}} \tag{6.115}$$

$$Q_{N-MIN} = \frac{X_{IN-MIN}}{R_{IN-N}} \tag{6.116}$$

or

$$Q_{N-MAX} = \frac{B_{IN-MAX}}{G_{IN-N}} \tag{6.117}$$

$$Q_{N-MIN} = \frac{B_{IN-MIN}}{G_{IN-N}} \tag{6.118}$$

Choose either of the two possible Q values within the constraint imposed by (6.110) on each transformation Q.

7)   Calculate the element values corresponding to the set of Q values. The reactance or susceptance of each component at the frequency where the Q values are calculated is given by an expression of the form

$$X_n = (Q_n + Q_{n-1})\ R_{rn} \tag{6.119}$$

or

$$B_n = (Q_n + Q_{n-1})\ G_{rn} \tag{6.120}$$

depending on whether it is a series or parallel element.

In these equations R_{rn} and G_{rn} are the effective series resistance and effective parallel conductance to the right of the component whose value is to be determined, respectively (refer to Fig. 6.29, if necessary).

8)   If the first element of the final network should be a shunt element, consider all inductors to be capacitors (i.e., 5 pH is 5 pF) and all capacitors to be inductors, and assign these values in sequence to the actual network.

As an illustration of this step, if element values of 3 pH (series element) 9 nF (shunt element), and 7 pF (series element) were obtained by following the procedure outlined above, the element values in the final circuit are 3 pF (shunt element), 9 nH (series element), and 7 pH (shunt element), respectively.

Example 6.8

As an example of the application of the procedure outlined above, consider the design of a five-element matching network with the first element a series element, no resonating sections, and the specifications:

1. $Z_L = 50+j50\Omega$
     $Z_s\ = 20-j20\Omega$
     $G_T = 0.89$
     $f_Q\ = 100\ \text{MHz}$
     $Q\ = 5$

2. $Q_1\ = 5\ (\Delta Q = 4)$
     $Q_2\ = 5\ (\Delta Q = 10)$
     $Q_3\ = -3\ (\Delta Q = 2)$

3. With no resonating sections and specifications as above, the last element of the network will be a series element, and therefore the constant gain circle on the input impedance plane is of interest. Application of (6.93) and (6.94) yields that the parameters of this circle are

$$R_0 + jX_0 = [2/0.89 - 1] \, 20 + j20 = 24.94 + j20.00\Omega$$

and

$$R_{Z0} = 2 \sqrt{1/0.89^2 - 1/0.89} \, 20 = 14.91$$

4. $R_{IN, \, MAX} = 24.94 + 14.91 = 39.85\Omega$

$\quad\quad R_{IN, \, MIN} = 24.94 - 14.91 = 10.03\Omega$

5. Application of (6.103) and (6.104) yields that

$$|Q_4| \geq \left[\frac{50}{39.85} \, \frac{1+5^2}{1+5^2} \, (1+3^2) - 1 \right]^{1/2} = 3.398$$

$$|Q_4| \leq \left[\frac{50}{10.03} \, \frac{1+5^2}{1+5^2} \, (1+3^2) - 1 \right]^{1/2} = 6.989$$

$$Q_{n-1} = 5 \, (\Delta Q = 2)$$

6. The two allowable values for the input reactance are found by using (6.107):

$$X_{IN} = 20 \pm 14.91 \sin \left[\cos^{-1} \frac{19.23 - 24.94}{14.91} \right]$$

$$= 6.23\Omega; \; 33.77\Omega$$

In the equation above, R_{IN} was found by applying (6.98):

$$R_{IN} = 19.23\Omega \; (10.03 \leq 19.23 \leq 39.85)$$

The last transformation Q is simply

$$Q_n = 6.23/19.23 = 0.32 \, (\Delta Q = 5.32)$$

7. $X_1 = Q_1 \, R_1 - X_L = 5(50) - 50 = 200\Omega \; (318.3 \, \text{nH})$

$$Y_2 = [Q_2 + Q_1] \, G_{L2} = \frac{10 \, (1)}{50 \, (1 + 5^2)} = 7.69 \text{ mS} \; (12.24 \text{ pF})$$

$$X_3 = [Q_3 + Q_2] \, R_{L3} = 2 \, (50) \, \frac{1 + 5^2}{1 + 5^2} = 100\Omega \; (159.2 \text{ nH})$$

$$Y_4 = [Q_4 + Q_3] \, G_{L4} = 2 \, \frac{1}{50} \, \frac{1 + 5^2}{1 + 5^2} \, \frac{1}{1 + 3^2} = 4.0 \text{ mS} \; (6.37 \text{ pF})$$

$$X_5 = [Q_5 + Q_4] \, R_{L5} = 5.32 \, (50) \, \frac{1 + 5^2}{1 + 5^2} \, \frac{1 + 3^2}{1 + 5^2} = 102.3\Omega \; (162.8 \text{ nH})$$

The designed network is shown in Fig. 6.33. The gain at 100 MHz is equal to

The actual value of the MRD at α_M can now be calculated. Depending on which of the resulting group of four errors is now the smallest, one of the four coordinates can be eliminated and the procedure can be repeated on the remaining three points.

By continuing this way, the optimum value of α can be determined.

Excellent results were obtained by optimizing the Q values of transformation as outlined above.

6.4.3.4   An Algorithm for the Design of Impedance-Matching Networks by Using the Transformation Q-factors of the Network

A procedure for designing a network to match a complex load to a complex source with a specified gain at a specified frequency was outlined in sec. 6.4.3.2. By taking the transducer power gain to be the minimum expected gain at the frequency where the Q values are evaluated (usually the highest frequency in the pass band, or that frequency where the gain required is a maximum), this narrowband technique forms the basis of an excellent approach to solving wideband impedance-matching problems.

It was shown that the first N-2 Q values in the single-frequency design can take arbitrary values and the constraints imposed on the last two Q values were derived. Since the range of possible transformation Q values is limited in a wideband design, it is feasible to do a grid-search on these Q values in order to find solutions which yield good results over the whole pass band. In this way, the dependence on a good initial solution is eliminated.

When the search is completed, a number of the best results obtained can be optimized as described in sec. 6.4.3.3. If the search was done thoroughly enough, the optimum solution to any matching problem will be obtained. A further advantage is that the local minima corresponding to other initial solutions will also be obtained, and consequently a large choice between networks with different element values and topologies exist.

An idea of the required range of Q values can be obtained from the desired Q of the network when applicable, the maximum Q of the load and source impedances, and the analytically derived constraints on simple reactive loads as summarized in Table 6.7.

As a rule, a minimum value of -4.2 and a maximum value of 4.2 yield excellent results. When some of the Q values of solutions obtained exceed these values and the optimum solution is required, the bounds must be extended. This will seldom be necessary when a wideband network is designed. Increment values in the range from 0.5 to 0.8 can be used.

ALGORITHM

1)   Decide on the frequency at which the Q values are to be evaluated (f_Q) and the number of elements.

Estimate the range of possible Q values of transformation and specify the incremental value to be used.

Estimate the minimum gain expected at f_Q.

Specify the number of sets of transformation Q values to be optimized.

2)   Generate an allowable set of Q values by using the theory outlined in sec. 6.4.3.2.

3)   Synthesize the equivalent network and calculate the gain error (MRD). Compare the results with the previous results obtained and store the solution if it is better than the M best solutions previously stored.

4)   Optimize the best results obtained in the search as described in sec. 6.4.3.3.

Example 6.9

As an example of the results obtainable with the transformation Q technique, consider the double-matching problem of *Example 6.7.*

With the gain set equal to the specified value of 0.818 at 200 MHz during the grid-search plus minimum, incremental, and maximum values of –4.2, 0.6, and 4.2, respectively, for the transformation Q values, the maximum deviation from the specified response was found to be 0.11 dB for the best four-element solution obtained. The corresponding network is shown in Fig. 6.35a. The Q values corresponding to this solution are –0.1545, 3.7736, 3.2741, and –0.1597, respectively.

The second best solution obtained is the low-pass network shown in Fig. 6.35b. The maximum deviation from the specified gain response is 0.13 dB, and the Q values are –0.6386, 2.1814, 3.1547, and –1.1868, respectively.

The best three-element solution obtained under the constraint that the topology must be of high-pass form is shown in Fig. 6.35c. The maximum deviation from the specified gain is 0.18 dB and the Q values are –0.8531, –0.1160, and –2.3795, respectively. This solution is basically the same as that obtained with the reflection parameter technique.

6.5   THE DESIGN OF RLC IMPEDANCE MATCHING NETWORKS

RLC impedance-matching networks are often used to compensate for the decrease in the gain of transistors at high frequencies. They have an advantage over lossless networks in that this can sometimes be done without mismatching the load to the source.

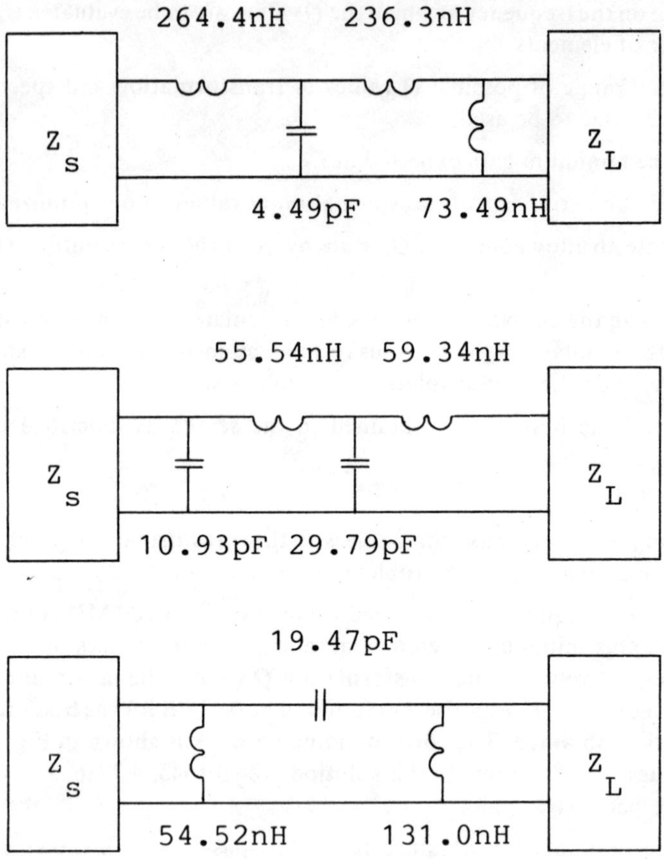

Figure 6.35   Some of the Solutions Obtained with the Transformation Q Technique for the Matching Problem of Example 6.7

These networks are usually designed by using a computer optimization program on a circuit with a suitable topology, after initial values have been assigned to its components.

The resistors in an RLC impedance-matching network have two functions: first, they provide the required attenuation at the lowest frequency in the pass band and, second, they match the load impedance to that of the source. A minimum of one series and one parallel resistor are required in order to do this.

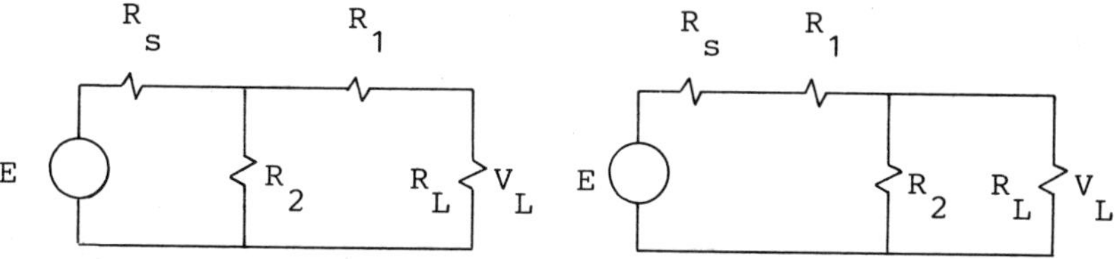

Figure 6.36   Impedance Matching with Resistors

When only one series and one parallel resistor are used, initial values can be assigned to them by using the following set of equations:

$$A = E/V_L = \frac{4R_s}{G_T R_L} \tag{6.125}$$

$$G_1^2 \left[1 + \frac{G_{IN}}{G_s} - A \right] + G_1 \left[2G_L \left(1 + \frac{G_{IN}}{G_s} \right) - A/2 \right] + G_L^2 \left[1 + \frac{G_{IN}}{G_s} \right] = 0 \tag{6.126}$$

$$G_2 = G_{IN} - \frac{G_1 G_L}{G_1 + G_L} \tag{6.127}$$

where G_T is the required transducer power gain at the lowest frequency in the pass band and G_{IN} is the required input admittance of the matching network at the lowest frequency. (If a perfect match is required $G_{IN} = G_s$).

Equations (6.125) to (6.127) apply to Fig. 6.36a. The equations relevant to Fig. 6.36b can be obtained by replacing G_s with G_L and G_L with G_s in these equations.

In order to minimize the insertion loss at the higher frequencies in the pass band, the resistors in an RLC network should be used in parallel with capacitors or in series with inductors, depending on whether they are used in a series or a parallel branch, respectively.

Apart from reducing the insertion loss, the capacitors and inductors used in the network also serve to match the load to the source at the higher frequencies.

The network shown in Fig. 6.37 is a typical example of an RLC network. Note that the elements which are not combined with resistors are used in low-pass positions.

Initial values can be assigned to the lossless elements of the network chosen, by considering the different elements to be part of independent L-, T-, and Π-sections.

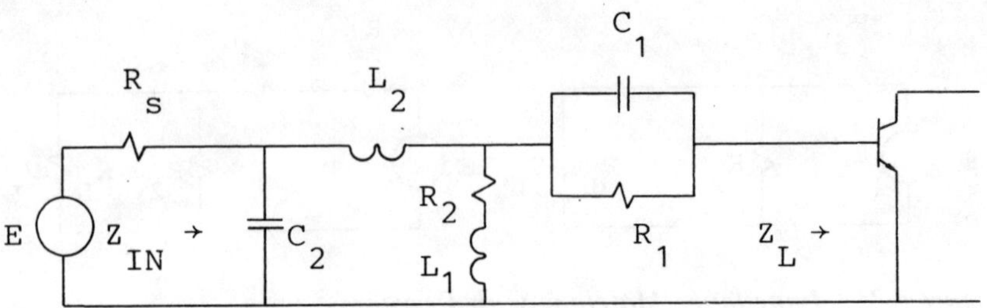

Figure 6.37   An Example of an RLC Impedance-Matching Network

As an example of this, C_2 and L_2 in the network shown in Fig. 6.37 forms a low-pass L-section that should be designed to match the load to the source at the highest frequency in the pass band. L_1 and C_1 should be designed to ensure that the insertion loss at the highest frequency will be as low as possible.

An alternative way of assigning initial values to the lossless components of an RLC network, is to follow the iterative approaches outlined earlier for designing a lossless band-pass network that will match the source to the load at the intermediate and higher frequencies in the pass band.

With initial values assigned to the lossless components and the resistors, an optimization program [10] can be used to optimize the network.

Example 6.11

The use of Eqs. (6.125) to (6.127) will be illustrated by applying them to the following problem:

$$R_L = 7.50\,\Omega$$
$$R_s = 6.25\,\Omega$$
$$G_T = 0.19$$
$$R_{IN} = 6.25\,\Omega$$

$$A = \frac{4R_s}{R_L\,G_T} = \frac{4\,(6.25)}{0.19\,(7.5)} = 4.19$$

$$G_1[2.00-4.19] + G_1\left[2\,\frac{2-4.19/2}{7.5}\right] + 1\,\frac{2}{7.5^2} = 0$$

$$G_1 = 0.1218;\ -0.1334\ S$$

$$G_2 = \frac{1}{6.25} - \frac{0.1218}{7.5\,[0.1218 + 1/7.5]}$$

$$= 0.096\ S$$

The initial values of the resistors are, therefore,

$$R_1 = 8.2\,\Omega \text{ and } R_2 = 10.4\,\Omega.$$

Problems

1)   The input impedance of a GaAs FET is tabulated in Table 6.20. Find an equivalent circuit for the input admittance.

Find the equivalent circuit by choosing a suitable topology, setting up an equation for its input admittance, and solving for the unknown components by equating the admittance to the values given in *Example 6.8*. If the components are determined by using different combinations of the measured admittances and the average of these values are taken, a good approximation can usually be obtained. This can only be done if there are more equations than independent variables.

Table 6.20

The Input Impedance of a GaAs FET at Different Frequencies

Frequency (GHz)	Input Impedance (Ω)
2	$85.7-j17.0$
3	$132.6-j35.2$
4	$97.6-j73.2$
5	$62.8-j65.8$
6	$43.9-j39.8$

2)   Find an equivalent circuit for the input admittance of the transistor in the previous problem by using the programs in Appendix A and Appendix B.

3)   A low-pass network is to be used to match the input impedance of the transistor in *Problem 6.1* to a 50Ω source. Determine the minimum value of the insertion loss.

4)   Design a network for matching the input impedance of the transistor in *Problem 6.1* to a 50Ω source. Do this with a low-pass as well as a band-pass network.

5)   How many different networks can be synthesized to have an input admittance of

$$Y_{IN} = \frac{2.125\,s^2 + 0.248\,s + 1.000}{1.992\,s^3 + 0.379\,s^2 + 1.531}\ ?$$

6)   Determine the gain-bandwidth limitations of the power transistor with input impedance as given in Table 6.21. Design a network for matching the transistor to a 12.5Ω source.

7)   Design a fourth-order 0.5 dB ripple Chebyshev interstage matching network with a positive slope of 6 dB/octave by following the Darlington approach. $R_s = 211\Omega$; $R_L = 50\Omega$.

The required slope can be obtained by adding a s^2 term to the numerator of the flat Chebyshev transducer power-gain function. When this is done the constant in the gain function must be adjusted to ensure that the maximum gain will be equal to one. This is best done iteratively.

Table 6.21

The Input Impedance of a Power Transistor at Different Frequencies in the Pass Band

Frequency (MHz)	Input Impedance (Ω)
100	$16.0-j9.0$
150	$15.5-j3.0$
200	$14.0-j0.0$
250	$11.0+j4.0$
300	$9.0+j9.0$
350	$7.5+j11.5$
400	$6.0+j15.0$

8)   A 50Ω source must be matched to the impedance shown in Fig. 6.38a over the frequency range 115-150 MHz. Design an impedance-matching network for this purpose. The insertion loss in the pass band must be as low as possible.

9)   Use Table 6.7 to determine the minimum insertion loss possible with the load shown in Fig. 6.38b, when a lossless three-element Chebyshev impedance-matching network is used. The cut-off frequency is to be 100 MHz.

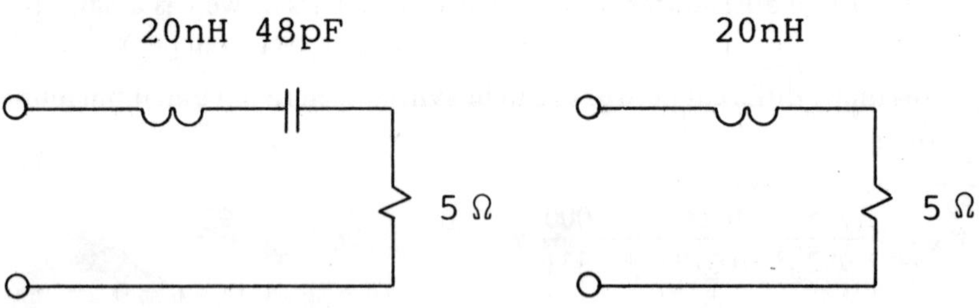

Figure 6.38   The Loads Relevant to Problems 6.8 and 6.9

10)   Find the optimum values of the maximum gain and the pass-band ripple if the load and source shown in Fig. 6.39 are to be matched with a four-element low-pass Chebyshev network. Assume $\omega_c = 1$ rad/s.

Figure 6.39   The Load and Source Relevant to Problem 6.10

11)   Design a low-pass network to match the load shown in Fig. 6.38a to a 15Ω source over the frequency range 115-150 MHz.

12)   Use Richards' transformation to design a commensurate distributed network to match the load shown in Fig. 6.40 to a 25Ω source over the pass band 1150-1500 MHz. Remove any series inductive stubs in the design by using Kuroda's low-pass identities.

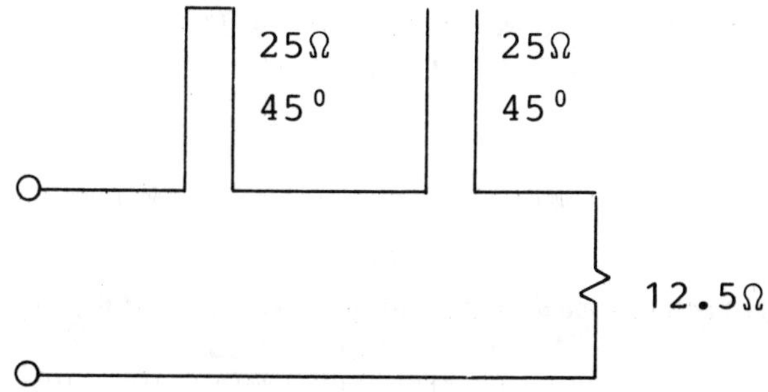

Figure 6.40   The Load to be Matched in Problem 6.12 [line lengths specified at the center frequency]

13)   Design a low-pass network to match the reactive load and source shown in Fig. 6.41 to each other over the frequency range 4 to 10 rad/s.

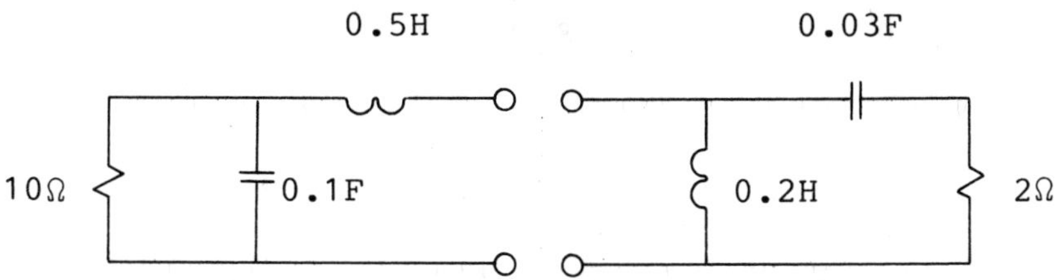

Figure 6.41   The Impedance to be Matched in Problem 6.13

14)   Design a lossless four-element network to mismatch a load of 50Ω to a 6.25Ω source with $G_T = 0.79$ at 600 MHz.

15)   Determine the poles and zeros of the RLC network shown in Fig. 6.42. Plot the response for different values of L and C.

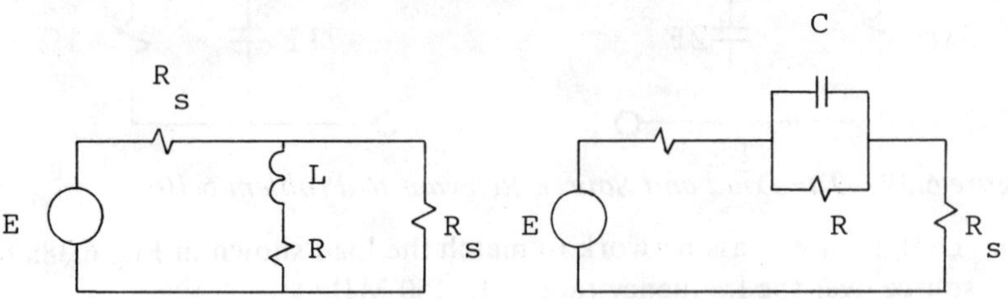

Figure 6.42   Simple RLC Impedance Matching Networks

16)   The input impedance and operating power gain of an HF power transistor are given in Table 6.22 at several frequencies. Design an RLC network for matching the transistor to a 2.78Ω source.

Table 6.22

The Input Impedance and Operating Power Gain of the Transistor Relevant to Problem 6.16

Frequency (MHz)	Input Impedance (Ω)	Power Gain (dB)
1.6	$4.8 - j1.2$	27.8
2.5	$4.3 - j1.6$	27.6
4.0	$3.8 - j1.9$	27.4
5.0	$3.1 - j1.9$	26.9
7.5	$2.4 - j1.8$	26.1
10.0	$1.9 - j1.4$	24.8
15.0	$1.6 - j1.1$	23.1
20.0	$1.4 - j0.7$	20.8
24.0	$1.3 - j0.5$	19.5
28.0	$1.3 - j0.4$	18.3

References and Additional Reading

1. Van Valkenburg, M.E., *Introduction to Modern Network Synthesis,* New York: John Wiley and Sons, 1960.

2. Chen, Wai-Kai, *The Theory and Design of Broadband Matching Networks,* London: Pergamon Press, 1976.

3. Carlin, H.J., "A New Approach to Gain-Bandwidth Problems," *IEEE Trans. Circuits Syst.* Vol. CAS-24, April 1977.

4. Carlin, H.J., and J.J. Komiak, "A New Method of Broadband Equalization Applied to Microwave Amplifiers," *IEEE Trans. Microwave Tech.,* Vol. MTT-27, No. 2, February 1979.

5. Ku, W.H., and W.C. Peterson, "Optimum Gain-Bandwidth Limitations of Transistor Amplifiers as Reactively Constrained Two-Port Networks," *IEEE Trans. Circuits Syst.* Vol. CAS-22, June 1975, pp. 523-533.

6. Levy, R., "Explicit Formulas for Chebyshev Impedance-Matching Networks, Filters and Interstages," *Proc. IEEE,* Vol. 111, No. 6, June 1964.

7. Mellor, D.J., and J.G., Linvill "Synthesis of Interstage Networks of Prescribed Gain versus Frequency Slopes," *IEEE Trans. Microwave Theory Tech.,* Vol. MTT-23, No. 12, December 1975.

8. Pitzalis, O., and R.A. Gilson, "Tables of Impedance Matching Networks which Approximate Prescribed Attenuation versus Frequency Slopes," *IEEE Trans. Microwave Theory Tech.,* Vol. MTT-19, No. 14, April 1971.

9. Bode H.W., *Network Analysis and Feedback Amplifier Design,* New York: Van Nostrand, 1945, pp. 205-207, 319.

10. Control Data Corporation, Cybernet Services, AMPSYN, CADSYN, RFOPT, 1978.

11. Matthaei, G.L., "Tables of Chebyshev Impedance Transforming Networks of Low-pass Filter Form," *Proc. IEEE,* August 1964.

12. Abrie, P.L.D., Impedance Matching Networks and Bandwidth Limitations of Class B Power Amplifiers in the HF and VHF Ranges, Master's Thesis, University of Pretoria, November 1982.

13. Fano, R.M., "Theoretical Limitations on the Broadband Matching of Arbitrary Impedances," *Journal of the Franklin Inst.,* Vol. 249, January/February 1950, pp. 57-83, 139-154.

14. Schoeffler, J.D., "Impedance Transformation Using Lossless Networks," *IRE Trans. Circuit Theory,* CT-8, June 1961, pp. 131-137.

15. Chen, Wai-Kai, and C. Satyanarayana, "General Theory of Broadband Matching," *IEE Proc. G (GB),* Vol. 129, No. 3, June 1982, pp. 96-102.

16. Yarman, B.S., and H.J. Carlin, "A Simplified 'Real-Frequency' Technique Applied to Broad-Band Multistage Microwave Amplifiers," *IEEE Trans. Microwave Theory Tech.,* Vol. MTT-30, No. 12, December 1982, pp. 2216-2222.

17. Carlin, H.J., and B.S. Yarman, "The Double Matching Problem: Analytic and Real Frequency Solutions," *IEEE Trans. Circuits Syst.,* Vol. CAS-30, No. 1, January 1983.

18. Carlin, H.J., and P. Amstutz, "On Optimum Broad-Band Matching," *IEEE Trans. Circuits Syst.,* Vol. CAS-28, No. 5, May 1981.

19. Yarman, B.S., Broad-Band Matching a Complex Generator to a Complex Load, Doctoral Dissertation, Cornell University, 1982.

20. Richards, P.I., "Resistor-Transmission-Line Circuits," *Proc. IRE,* February 1948, pp. 217-219.

21. Youla, D.C., "A New Thoery of Broad-Band Matching," *IEEE Trans. Circuits Syst.,* Vol. CT-11, March 1964, pp. 30-50.

22. Ha, T.T., The Design of Microwave Solid State Amplifiers, John Wiley and Sons, 1981.

23. Liu, L.C.T., and Walter H. Ku, "Computer-Aided Synthesis of Lumped Lossy Matching Networks for Monolithic Microwave Integrated Circuits (MMIC's)," *IEEE Trans. Microwave Theory Tech.* Vol. MTT-32, No. 3, March 1984.

24. Fletcher, R., and M.J.D. Powell, "A Rapidly Convergent Descent Method for Minimization," *Computer J.,* 6, 1963, pp. 163-168.

CHAPTER 7

MICROWAVE LUMPED ELEMENTS, DISTRIBUTED EQUIVALENTS, AND THE PARASITICS ASSOCIATED WITH MICROSTRIP TRANSMISSION LINES

7.1 INTRODUCTION

Impedance-matching networks can be realized in lumped form as long as the dimensions of the components used are small compared to a quarter-wavelength at the highest frequency of interest. When the dimensions are on the order of one-twelfth of a wavelength, the resulting phase shift across a component can cause a significant deviation from the expected response. Furthermore, if the associated incremental characteristic impedance is too low for inductors and too high for capacitors, the response will be degraded even more by the resulting parasitic capacitance or inductance. Because it is often a problem when high impedance circuits are designed, the bounds imposed by the phase shift across and the finite incremental characteristic impedance of practical inductors will be examined in this chapter. In order to do this the transforming properties of a series transmission line will be examined first. Different types of microwave inductors, capacitors, and resistors, and the design of microwave inductors will also be considered.

When the components cannot be realized with negligible phase shift and parasitics, matching networks must be realized in distributed form. Fabrication of distributed networks on microstrip, thin-film substrates or as MICs is simple, and where lumped spiral inductors or interdigital capacitors are the alternative, less design time is required. Excellent equations for the characteristic impedance and effective dielectric constant of microstrip lines have been developed by the many workers in the field [3] and were reviewed in Ch. 2. At higher frequencies it becomes necessary to incorporate the effect of distributed discontinuities into designs in order to obtain good results. The magnitude of these effects at the lower microwave frequencies [4] will be considered here along with a compensation technique for reducing them [5].

Lumped-element designs are often transformed into distributed designs by replacing inductors and capacitors with short- and open-circuited stubs and short sections of series transmission line. The range of series and shunt reactances which can be transformed with negligible error will be examined here. It will also be shown that significantly better results can be obtained by replacing low-pass T- and Π-sections instead of replacing only series inductors with sections of series transmission lines.

7.2 LUMPED MICROWAVE RESISTORS

Thin-film techniques are often used to manufacture lumped resistors at microwave frequencies. By keeping the dimensions of a resistor small, the associated capacitance and inductance can be minimized. The capacitance can be reduced further by depositing the thin-film on a low dielectric-constant substrate.

A thin-film resistor can be characterized as a lossy transmission-line. The relevant equations are:

$$r = R_s / W_{eff} \tag{7.1}$$

$$C = \sqrt{\epsilon_{eff}} / (3.0\text{E}8 \; Z_{0\text{-}LC}) \tag{7.2}$$

$$L = Z_{0\text{-}LC} \sqrt{\epsilon_{eff}} / 3.0\text{E}8 \tag{7.3}$$

$$\theta_1 = 0.5 \left[90.0 + \tan^{-1} (\omega L / r) \right] \tag{7.4}$$

$$\zeta = \sqrt{\omega C \sqrt{r^2 + (\omega L)^2}} \left[\cos \theta_1 + j \sin \theta_1 \right] \tag{7.5}$$

$$\theta_2 = -0.5 \tan^{-1} \left[r / (\omega L) \right] \tag{7.6}$$

$$Z_0 = \sqrt{ \frac{\sqrt{r^2 + (\omega L)^2}}{\omega C} } \left[\cos \theta_2 + j \sin \theta_2 \right] \tag{7.7}$$

$$\begin{bmatrix} V_I \\ I_I \end{bmatrix} = \begin{bmatrix} \cosh (\zeta l) & Z_0 \sinh (\zeta l) \\ \sinh (\zeta l) / Z_0 & \cosh (\zeta l) \end{bmatrix} \begin{bmatrix} V_L \\ I_L \end{bmatrix} \tag{7.8}$$

where $Z_{0\text{-}LC}$ is the characteristic impedance of a lossless line with identical dimensions, W is the width (meter), l is the length (m) of the resistor, and R_s the resistance per square of the thin-film. The angles θ_1 and θ_2 must be specified in degrees. The influence of the skin-effect on the resistance can be incorporated into the resistance per square, R_s.

Equation (7.8) is the transmission matrix equation for the series resistor, with V_I and I_I the input voltage and current, and V_L and I_L the load voltage and current, respectively.

Thin-films with resistances of 50 to 1000 ohms per square are available.

7.3 EVALUATION OF THE LIMITATIONS OF A SERIES TRANS-MISSION LINE USED AS A LUMPED ELEMENT

Series transmission lines are often used to replace series inductors when a lumped matching network is transformed into a distributed equivalent. On the other hand, all lumped inductors of finite dimensions have some capacitance to ground and as such can be considered transmission lines of high characteristic impedance, although this impedance would not be uniform for bonding wire inductors and square or spiral inductors.

In order to get an idea of the range of inductances that can be replaced with series transmission lines with negligible error, as well as bounds on the inductance of lumped inductors, it is necessary to consider the transformation properties of a series transmission line.

Assuming a load impedance of $Z_L = R_L + jQ \cdot R_L$, the input resistance and reactance of a series transmission line is given by

$$R_{IN} = R_L \left[1 + \tan^2 \theta\right] / Z \tag{7.9}$$

and

$$X_{IN} = j \left[QR_L (1 - \tan^2 \theta) + Z_0 \tan \theta - R_L^2 \tan \theta (1 + Q^2) / Z_0\right] / Z \tag{7.10}$$

where

$$Z = \left[1 - QR_L \tan \theta / Z_0\right]^2 + \left[R_L \tan \theta / Z_0\right]^2 \tag{7.11}$$

and

$$\theta = \beta l \tag{7.12}$$

In order to exhibit truly lumped behavior, the line length and characteristic impedance, respectively, must be short enough and high enough for the input impedance to be approximately

$$Z_{IN} = R_L + jQR_L + jZ_0 \tan \theta \tag{7.13}$$

For (7.13) to apply, the following inequalities must be satisfied:

$R_{IN} \simeq R_L$:

$\tan^2 \theta \ll 1$

$Z_0 / R_L \gg 2Q \tan \theta$

$$(Z_0 / R)_L^2 \gg \tan^2 \theta \tag{7.14}$$

$X_{IN} \simeq jQ R_L + jZ_0 \tan \theta$:

$\tan^2 \theta \ll 1$

$Z_0 / R_L \gg 2Q \tan \theta$

$(Z_0 / R_L)^2 \gg \tan^2 \theta$

$$(Z_0 / R_L)^2 \gg 1 + Q^2 \tag{7.15}$$

It follows from (7.9) and (7.11) that even if the characteristic impedance of the line was equal to infinity, the resistance would still be transformed to

$$R_{IN} = R_L \left[1 + \tan^2 \theta\right] \tag{7.16}$$

that is, the influence of the phase shift does not become negligible with increasing characteristic impedance.

With Z_0 approaching infinity, the input reactance of the line is given by

$$Z_{IN} = jQR_L[1 - \tan^2\theta] + jZ_0\tan\theta \qquad (7.17)$$

These values for the input resistance of the line can be used to provide upper bounds on the line length for which the input impedance will be approximately equal to that given by (7.13). One way to do this is to evaluate the reflection parameter of the circuit in Fig. 7.1 for an infinite value of the characteristic impedance of the line. The input reflection parameter for this circuit is then given by

$$s_{11} = \frac{R_L(1+\tan^2\theta) - R_L + j[X_L(1-\tan^2\theta) - X_L]}{R_L(1+\tan^2\theta) + R_L + j[X_L(1-\tan^2\theta) - X_L]}$$

$$= \frac{1 - jQ}{[2/\tan^2\theta + 1] - jQ} \qquad (7.18)$$

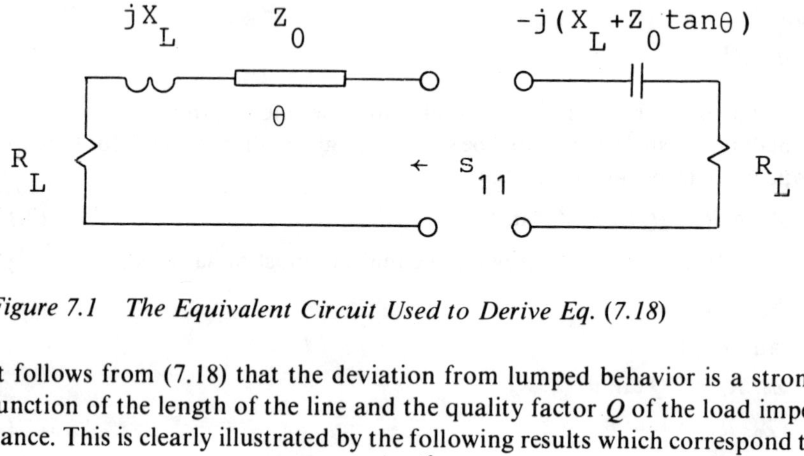

Figure 7.1   *The Equivalent Circuit Used to Derive Eq. (7.18)*

It follows from (7.18) that the deviation from lumped behavior is a strong function of the length of the line and the quality factor Q of the load impedance. This is clearly illustrated by the following results which correspond to an insertion loss of 0.25 dB ($G_T=1 - |s_{11}|^2$):

$Q = 0: \theta = 38°$
$Q = 1: \theta = 33°$
$Q = 2: \theta = 26°$
$Q = 3: \theta = 22°$
$Q = 4: \theta = 19°$

Because of the finite characteristic impedance of any physical line, the exact deviation will always be greater than that predicted by (7.18). The exact reflection parameter for any particular case can be calculated by substituting R_{IN} and X_{IN} as given by (7.9) to (7.12) into the equation:

$$s_{11} = \frac{[R_{IN} - R_L] + j[X_{IN} - X_L - Z_0\tan\theta]}{[R_{IN} + R_L] + j[X_{IN} - X_L - Z_0\tan\theta]} \qquad (7.19)$$

As an illustration of the combined influence of a reactive load and finite values for the characteristic impedance, the line lengths corresponding to an insertion loss of approximately 0.25 dB in the circuit shown in Fig. 7.1 are tabulated in Table 7.1 as a function of the ratio Z_0/R_L and the line length.

It is clear from the results in Table 7.1 that the range of characteristic impedances and line lengths over which a series transmission line can be considered as a lumped inductor, and over which the distributed nature of a series inductor can be ignored, is very limited, especially when the load Q is high.

7.4 LUMPED MICROWAVE INDUCTORS

Lumped microwave inductors can be fabricated in different forms. For low inductance values, strip inductors or bonding wire is frequently used, while larger inductance values are obtainable with spiral or solenoidal inductors.

Table 7.1

The Line Lengths Corresponding to an Insertion Loss of Approximately 0.25 dB in Fig. 7.1 as a Function of the Characteristic Impedance of the Series Line and the Q of the Load

Z_0/R_L	Line Length (degrees)					
	$Q{=}0$	$Q{=}1$	$Q{=}2$	$Q{=}3$	$Q{=}4$	$Q{=}5$
1.0	26	13	5	2.7	1.6	1.1
2.0	35	20	10	5	3	2.1
3.0	37	24	13	7	4	3
4.0	37	26	15	9	6	4
5.0	38	27	17	11	7	5
7.5	38	29	20	14	9	7
10.0	38	30	21	15	11	9
15.0	38	31	23	17	13	10
20.0	38	31	24	18	15	12

The inductance of an isolated (without ground plane), flat, ribbon inductor (or strip-inductor) is given approximately by [15]:

$$L \text{ (nH/mm)} = 0.2 \{\ln [l/(w+t)] + 1.193 + 0.224 (w+t)/l\} \tag{7.20}$$

where w is the width of the ribbon, t its thickness, and l its length.

An approximate expression for the Q of a ribbon inductor is [8]

$$Q = 2.15\text{E}3 \, \frac{L \text{ (nH)}}{K} \, \frac{\omega}{l} \left(\frac{\rho \text{ (Cu)}}{\rho}\right)^{1/2} \left(\frac{f \text{ (GHz)}}{2}\right)^{1/2} \tag{7.21}$$

where ρ is the resistivity of the material used and K is a correction factor for the current crowding occuring at the corners of the strip [15]. K is given approximately by the following expression:

$$K = 1.3565 - 0.2319 \ln [w/t] + 0.2386 [\ln (w/t)]^2 -$$
$$0.0536 [\ln (w/t)]^3 + 0.0043 [\ln (w/t)]^4 \tag{7.22}$$

It follows from (7.21) that the width of a ribbon inductor should be as wide as possible in order to obtain a high Q.

Equation (7.20) can also be used to calculate the inductance of a single-turn circular loop in those cases where the width of the strip is much smaller than the diameter. Alternatively, the following expression [17] can be used:

$$L \text{ (nH/mm)} = 0.2 [\ln (l/(w+t)) - 1.76] \tag{7.23}$$

For (7.23) to apply, the inequality

$$l \gg 2 (w + t) \tag{7.24}$$

must be satisfied.

The inductance associated with bonding wire of diameter d and length l can be calculated by using the equation:

$$L \text{ (nH/mm)} = 0.20 [\ln (l/d) + 0.386] \tag{7.25}$$

An approximate expression for the Q of a round-wire inductor is [8]:

$$Q = 3.38\text{E}3 \, L \text{ (nH)} \, \frac{d}{l} \left(\frac{\rho \text{(Cu)}}{\rho} \right)^{1/2} \left(\frac{f \text{(GHz)}}{2} \right)^{1/2} \tag{7.26}$$

Bonding-wire inductors have the advantage over strip-inductors that higher Q-factors can be expected because of the larger surface area. Furthermore, touch-up tuning is possible with bonding-wire inductors, while the inductance is fixed for strip-inductors.

For square spirals, the inductance (in the absence of any ground plane) is given approximately by [16]:

$$L \text{ (nH)} = 0.85 \sqrt{A} \, N^{5/3} \tag{7.27}$$

where A is the area in mm^2 and N the number of turns.

The associated line length (mm) is approximately

$$l_e = N [8r_i + d (4N - 3)] \tag{7.28}$$

The parameters in this equation are defined in Fig. 7.2.

Square spirals are often used as RF chokes in MICs.

The inductance of a circular spiral inductor can be calculated by using the equation:

$$L \text{ (nH)} = 3.930 \, a^2 \, N^2 / [0.8a + 1.1c] \tag{7.29}$$

where

$$a \text{ (mm)} = (d_0 + d_i)/4.0 \tag{7.30}$$

and

$$c \text{ (mm)} = (d_o - d_i)/2.0 \tag{7.31}$$

where d_i and d_o are respectively the inner and outer diameter of the spiral, s the spacing between two adjacent conductors, and N is the number of turns.

For minimum losses, the outer diameter of a spiral inductor should be approximately five times the inner diameter [6]. Under this constraint, the Q is given approximately ($\pm 20\%$) by [8]:

$$Q = \frac{1.3\text{E}2\, w}{K'} \sqrt{\frac{L}{d_0}} \left(\frac{\rho(\text{Cu})}{\rho} \right)^{1/2} \left(\frac{f(\text{GHz})}{2} \right)^{1/2} \tag{7.32}$$

where K' [15] is given approximately by

$$K' = 1.0090 + 0.8584 e^{-(s+w)/w} - 0.6376 e^{\{-(s+w)/w\}2}$$
$$+ 1.8431 e^{\{-(s+w)/w\}3} \tag{7.33}$$

with s the spacing between two adjacent conductors and w the width of the conductor.

In order for (7.32) to apply, d_o should be greater than $1.2d_i$, N greater than one, and the thickness (t) greater than five skin depths [8].

Typical values for the conducting strip width of a spiral inductor are 50-250 μm. For close to optimum results, a width-to-spacing ratio of unity is recommended [8].

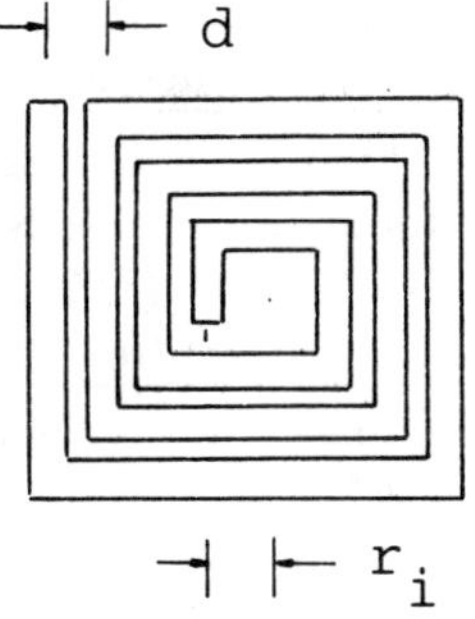

Figure 7.2   A Square Spiral Inductor

At microwave frequencies, solenoidal inductors are often used as RF chokes in hybrid circuits. When the size is not prohibitively small, they can also be used as inductors.

The inductance of a solenoidal coil is

$$L \text{ (nH)} = 10.0 \, r^2 \, N^2 / [2.29l + 2.54r] \tag{7.34}$$

where r is the radius (mm), l is the length (mm), and N is the number of turns of the coil.

In order to remain approximately lumped, an inductor must be electrically short. Reasonable results can be expected with shunt inductors when the associated electrical length is shorter than 30° (the deviation from the expected linear increase in reactance will then be less than 10%). In the case of a series inductor, the restrictions are more severe because the resistance in series with the inductor will be transformed due to the transmission-line effect. In order to provide an idea of the bounds on realizable series inductances, the inductance values for each of the inductors discussed above, corresponding to an electrical length of 38°, $Q=0$ (7.18), and $\epsilon_r=1$ at different frequencies, were calculated and are tabulated in Table 7.2. Because the inductive and capacitive coupling were ignored, the boards on the inductance of square spiral and solenoidal coil inductors are only approximate.

Table 7.2

Upper Bounds on the Series Inductance Realizable ($\epsilon_r=1$; $\theta=38°$) with Different Inductors as a Function of Frequency

Frequency	Inductance (nH)			
(GHz)	Bonding Wire (d=25μm)	Strip Inductor (w=50μm)	Square Spiral (r_i=20μm(25μm)) (d_i=10μm(50μm))	Solenoidal Coil (c=25μm)
1	48.0	48.0	109.0 (65.0)	144.0
2	22.0	22.0	41.0 (25.0)	50.0
4	9.7	9.9	15.0 (9.1)	17.0
6	6.1	6.2	8.2 (5.0)	9.4
8	4.3	4.4	5.3 (3.2)	6.1
10	3.3	3.4	3.8 (2.3)	4.3
12	2.7	2.7	2.9 (1.7)	3.1

Table 7.3

The Inductance of Different Inductors as a Function of the Length of the Conductor

Length (mm)	Inductance (nH)			
	Strip Inductor $w=50\mu$m	Bonding Wire $d=25\mu$m	Square Spiral $r_i=25\mu$m$(20\mu$m$)$ $d_i=50\mu$m$(10\mu$m$)$	Solenoidal Coil $c=25\mu$m
1.0	0.8	0.8	0.3 (0.6)	0.7
1.5	1.4	1.3	0.7 (1.2)	1.3
2.0	2.0	1.9	1.1 (1.9)	2.1
2.5	2.6	2.5	1.6 (2.6)	2.9
3.0	3.2	3.1	2.1 (3.5)	3.9
4.0	4.5	4.4	3.3 (5.4)	6.1
5.0	5.8	5.7	4.6 (7.6)	8.6
7.5	9.3	9.1	8.4 (14.0)	16.0
10.0	13.0	13.0	13.0 (21.0)	25.0
15.0	21.0	20.0	23.0 (38.0)	46.0
20.0	29.0	28.0	34.0 (57.0)	71.0
25.0	37.0	36.0	47.0 (78.0)	100.0
30.0	46.0	45.0	60.0 (101.0)	132.0

The inductance values in Table 7.2 are optimistic in the sense that the Q of the load was assumed to be zero, the relative dielectric constant was assumed to be unity, and the influence of the finite incremental characteristic impedance associated with the lumped inductors was ignored. The influence of the effective relative dielectric constant is to increase the electrical length of the inductor by a factor $\epsilon_r^{1/2}$, and the Q and Z_0 influences are tabulated in Table 7.1. An idea of the lowering in the inductance bounds caused by these factors can be obtained by using Table 7.3, in which the inductance of the different inductors is tabulated as a function of the approximate conductor length, in conjunction with Table 7.1.

The inductance of the solenoidal coil in Tables 7.2 and 7.3 was calculated by using (7.34) in conjunction with the following set of equations:

$$r_{opt} = 0.3788 \sqrt{l_e \cdot c} \tag{7.35}$$

$$l_{opt} = 0.4202 \sqrt{l_e \cdot c} + c \tag{7.36}$$

$$N = 0.4202 \sqrt{l_e / c} \tag{7.37}$$

where c is the wire thickness (mm), r_{opt} the optimum radius, l_e the conductor length, and l_{opt} the optimum coil length.

The wire thickness of the solenoidal coil should be chosen to optimize the Q (refer to Ch. 2, sec. 2.3.6).

Equations (7.35) to (7.37) were derived by setting the derivative of the inductance, as given by (7.34), equal to zero in order to find the highest inductance corresponding to a specified conductor length.

Example 7.1

The matching network in Fig. 7.3 was designed to match the output impedance of a GaAs FET to a 50Ω load over the pass band 2-6 GHz. As an example of the application of the material derived in the previous sections, the feasibility of realizing the inductors in the network in lumped form will be investigated.

Inspection of Table 7.2 yields that the maximum realizable inductance ($\epsilon_r=1$; $Z_0 \to \infty$; $Q_L = 0$; $|s_{21}| = -0.25$ dB) at 6 GHz (solenoidal coils excluded) is approximately 8.2 nH, which is higher than the inductance in Fig. 7.3.

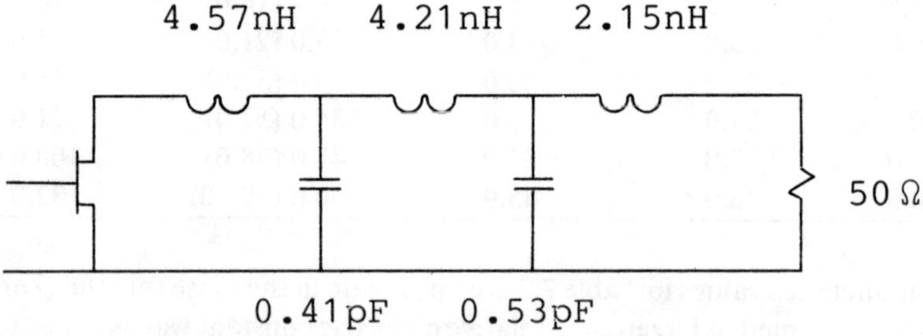

Figure 7.3   The Matching Network Considered in Example 7.1

It follows from Table 7.3 that a conductor of approximately 4mm length will be required to realize the 4.58nH inductor. Assuming the effective relative dielectric constant to be 2.17 (strip-inductor), it follows that the required electrical length is approximately

$$\theta = 120\text{E-}11\ l\ \sqrt{\epsilon_r}\,f = 120\text{E-}11\ 4\ \sqrt{2.17}\ 6\text{E}9 = 42°$$

Inspection of Table 7.1 yields that even with an infinite value for the characteristic impedance, the 4.58nH inductor cannot be realized without significantly degrading the match. The 4.21nH inductor presents an even bigger problem because it is located at a higher Q point (2.01 compared to 1.37).

The electrical length of the 2.15nH inductor is approximately 22.8° and the load Q at that point is equal to zero. It follows by inspection of Table 7.1 that this inductor can be realized in lumped form even with an incremental characteristic impedance as low as 100Ω. Application of (7.19) yields an approximate value of –0.07 dB for the error in gain with Z_0 taken as 100Ω.

7.5 LUMPED MICROWAVE CAPACITORS

Lumped microwave chip capacitors can be used up to very high frequencies. The self-resonant frequencies for some capacitance values as specified by one manufactuer [13] are tabulated in Table 7.4. The dimensions of these capacitors are as small as 0.154mm $\times$ 0.508mm and 2.032m $\times$ 2.540mm for capacitance values between 0.1 and 5.6pF and 3.0 and 62pF, respectively. The thicknesses vary between 0.076 and 0.254mm. The approximate series inductance is 0.05nH. It should be noted that the power which can be dissipated in capacitors with such small dimensions is limited.

Instead of using discrete capacitors, capacitors can be integrated into a microstrip, thin-film, or MIC design. These capacitors can be small plate capacitors, gap capacitors, or interdigital capacitors. Gap capacitors [10] are only used at the higher microwave frequencies.

Table 7.4

The Self-Resonant Frequencies for Some High Quality Microwave Chip Capacitors

Capacitance (pF)	Self-Resonant Frequency (GHz)
.1	50
1	28
10	9
100	3
1000	1

Interdigital capacitors with capacitors ranging from 0.1-15pF can be realized on MICs and thin-film. The approximate capacitance of an interdigital capacitor is given by the equation:

$$C(F) = [(\epsilon_r + 1)/W]\, l^2\, [(N\text{-}3)\, A_1 + A_2] \tag{7.28}$$

where N is the number of fingers, A_1 and A_2 are weighting factors associated with the inside and outside fingers, respectively, and l is the length of overlap as illustrated in Fig. 7.4. When the substrate is thick enough, these constants are 8.85826E-3pF/mm and 9.92125E-3pF/mm, respectively. For maximum capacitance the line widths and spacings should be equal [14]. Spacing of 10-25μm between the fingers is typical [8].

The parasitics associated with interdigital capacitors can be ignored as long as the capacitance-frequency product is smaller than 2.0E-3 [14].

Interdigital capacitors are analyzed in detail in [14].

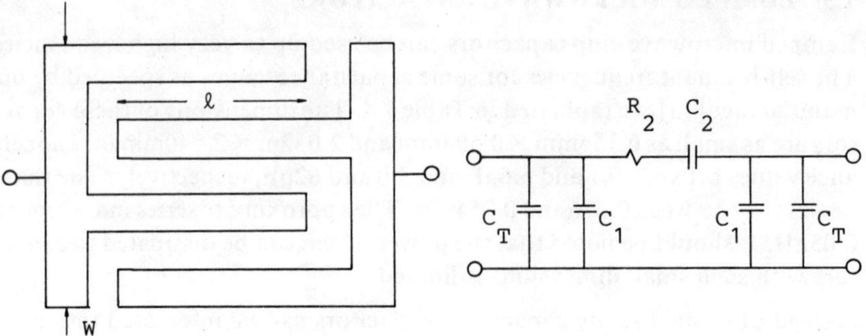

Figure 7.4 [a] *An Example of the Layout of an Interdigital Capacitor;* [b] *a Low-Frequency Equivalent Circuit for a Series Interdigital Capacitor*

7.6 DISTRIBUTED EQUIVALENTS FOR SHUNT INDUCTORS AND CAPACITORS

If the required inductance is low enough, a shunt inductor can be replaced with good approximation by a shorted transmission line. Similarly, a shunt capacitor can be replaced with an unterminated stub if the required capacitance is small enough. The accuracy with which these replacements can be made is dependent on the linearity of the tan function. To give an indication of the frequency range over which this function can be considered linear, the value of $(\tan\theta - \theta)/\theta$ is summarized for several values of θ (radians) in Table 7.5. If a 10% deviation is acceptable, the maximum electrical length for an equivalent line is 30°. The maximum deviation across the pass band can be reduced to less than 5% with the same line length by averaging the deviation across the pass band.

The equations relevant to an exact replacement at a frequency f_H are

$$Z_{0L} \tan (\beta l) = X_{HL} \tag{7.39}$$

and

$$Z_{0C}/\tan (\beta l) = X_{HC} \tag{7.40}$$

where X_{HL} and X_{HC} are the reactances to be replaced at frequency f_H, and Z_{0L} (shorted-circuited stub) and Z_{0C} (open-ended stub) are the corresponding characteristic impedances.

To give an idea of the range of reactances which can be replaced in this way, the minimum capacitive and maximum inductive reactance corresponding to a perfect match at low frequencies, and a 10% and 20% deviation at the highest frequency in the pass band are tabulated in Table 7.6 for $\epsilon_r = 2.17$ and $\epsilon_r = 10.3$. In deriving this table, the minimum and maximum width-to-height ratios

Table 7.5

$[\tan \theta{-}\theta]/\theta$ (in radians) as a Function of the Angle θ (in degrees)

θ	$\dfrac{\tan \theta - \theta}{\theta}$	θ	$\dfrac{\tan \theta - \theta}{\theta}$
(°)	(%)	(°)	(%)
5.0	0.3	35.0	14.6
7.5	0.6	37.5	17.2
10.0	1.0	40.0	20.2
12.5	1.6	42.5	23.5
15.0	2.3	45.0	27.3
17.5	3.2	47.5	31.6
20.0	4.3	50.0	36.0
22.5	5.5	52.5	42.2
25.0	6.9	55.0	48.8
27.5	8.5	57.5	56.4
30.0	10.3	60.0	65.4
32.5	12.3		

were taken as 0.3 and 10.0, respectively. The minimum width is determined by the amount of (unpredictable) under-etching and the acceptable resistive losses. The maximum ratio is determined by the electrical width of the stub.

In calculating the minimum capacitive reactance entered into Table 7.6, the capacitor was replaced with two parallel stubs (cross-junction).

As an example of the improvement possible by averaging the deviation across the pass band, the reactance corresponding to a 2-6 GHz pass band and maximum deviations of $\pm 4.4\%$ ($\theta=30°$) and $\pm 8.3\%$ ($\theta=40°$) are also given in Table 7.6. The equations used to calculate these reactances were

$$Z_{0L} = 1.808 \, X_{HL} \tag{7.41}$$

$$Z_{0C} = X_{HC}/1.808 \tag{7.42}$$

and

$$Z_{0L} = 1.209 \, X_{HL} \tag{7.43}$$

$$Z_{0C} = X_{HC}/1.209 \tag{7.44}$$

respectively.

Example 7.2

Consider the matching network in Fig. 7.5 (pass band 2-4 GHz). Assuming that the inductors are realizable in lumped form with negligible error, equivalent open-ended stubs will be determined for the capacitors ($\epsilon_r = 2.17$)

Table 7.6

Approximate Values for the Minimum Capacitive and Maximum Inductive Shunt Reactances which Can Be Replaced with Parallel Stubs

ϵ_r		2.17	10.3
Z_{0-MIN} Z_{0-MAX}	(Ω)	21/2 141	10/2 72
X_{HC-MIN} X_{HL-MAX}	(+10%)	34/2 85	16/2 43
X_{HC-MIN} X_{HL-MAX}	($\pm$4.4%; 2-6 GHz)	38/2 78	18/2 40
X_{HC-MIN} X_{HL-MAX}	(+20%)	24/2 118	12/2 60
X_{HC-MIN} X_{HL-MAX}	($\pm$8.3%; 2-6 GHz)	27/2 109	13/2 56

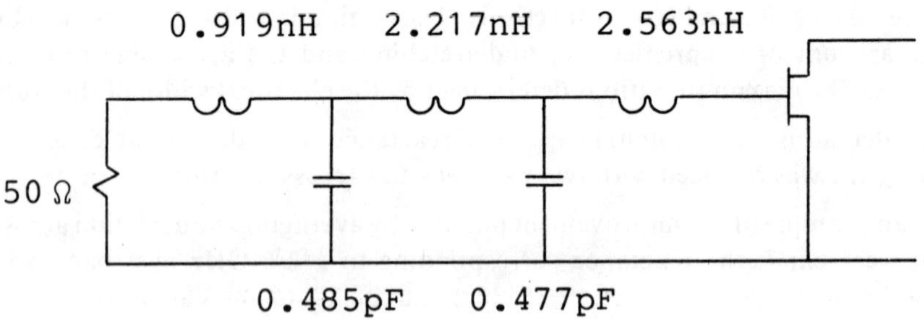

Figure 7.5   The Matching Network Considered in Example 7.2

It follows from Table 7.6 that the lowest practical characteristic impedance on a substrate with ϵ_r = 2.17, is approximately 25Ω. With Z_0 chosen to be 25Ω, the line lengths required for an exact equivalent at 4 GHz are

$$\beta l = \tan^{-1} \frac{Z_{0C}}{X_{HC}} = \tan^{-1}[25/82.04] = 16.9°$$

and

$$\beta l = 16.7°$$

respectively.

The expected errors at 2 GHz are

$$[25/\tan \theta_L - 1/(\omega_L C)]/[1/\omega_L C] = 2.6\%$$

and 2.1%, respectively.

Example 7.3

Because a significant reduction in the deviation in reactance is possible in wideband designs by averaging it across the pass band, an equation for the optimum characteristic impedance (admittance) as a function of the inductance (capacitance) to be replaced and the line length will be derived here.

When an inductor is replaced with a short-circuited stub, the error in reactance is given by

$$\Delta X = \frac{Z_0 \tan \theta - \omega L}{\omega L}$$

$$= \frac{\tan \theta - \omega L/Z_0}{\omega L/Z_0} \tag{7.45}$$

Under the equality

$$Z_0 = \frac{b}{\theta_{max}} \omega_{max} L \tag{7.46}$$

the error corresponding to (7.45) will be the same as that for

$$\Delta X' = \frac{\tan \theta - \theta/b}{\theta/b} \tag{7.47}$$

The optimum value for b can be determined by setting the error at θ_{max} in the passband equal to the negative of the error at $\theta_{min} = \theta_{max}/u$, where u is the relative bandwidth. The result is

$$b = 2 / \left[\frac{\tan \theta_{max}}{\theta_{max}} + \frac{\tan (\theta_{max}/u)}{\theta_{max}/u} \right] \tag{7.48}$$

The optimum value for the characteristic impedance can be obtained as a function of the phase shift at the highest frequency in the passband (θ_{max}) and the reactance to be replaced by substituting the result of (7.48) into (7.46). These impedances are tabulated in Table 7.7 together with the corresponding errors in reactance. The error in reactance is small when the bandwidth is relatively narrow and the electrical line length at the highest frequency in the pass band short.

The characteristic impedance required is clearly a weak funciton of the relative bandwidth, and a strong function of the stub length and reactance required at the highest frequency in the pass band.

Table 7.7

The Optimum Normalized Characteristic Impedance (Admittance) and the Corresponding Error in Reactance (Susceptance) for a Short-Circuited (Open-Ended) Stub as a Function of the Line Length at the Highest Frequency in the Pass Band and the Relative Bandwidth (u = f$_H$/f$_L$).

θ_{max}	$Z_{0-opt}/[\omega_H L]$; Reactance Error (%) $Y_{0-opt}/[\omega_H C]$; Susceptance Error (%)							
(°)	u=1.5		u=2.0		u=3.0		u=4.0	
10.0	5.687	±0.3	5.693	±0.4	5.697	±0.5	5.698	±0.5
11.0	5.162	±0.3	5.169	±0.5	5.173	±0.6	5.174	±0.6
12.0	4.724	±0.4	4.731	±0.6	4.736	±0.7	4.737	±0.7
13.0	4.353	±0.5	4.360	±0.7	4.365	±0.8	4.367	±0.8
14.0	4.033	±0.6	4.041	±0.8	4.407	±0.8	4.049	±0.9
15.0	3.756	±0.6	3.765	±0.9	3.771	±1.0	3.773	±1.1
16.0	3.513	±0.7	3.522	±1.0	3.529	±1.2	3.531	±1.2
17.0	3.298	±0.8	3.308	±1.1	3.315	±1.3	3.317	±1.4
18.0	3.107	±0.9	3.117	±1.3	3.124	±1.5	3.126	±1.6
19.0	2.935	±1.1	2.945	±1.4	2.953	±1.7	2.956	±1.8
20.0	2.780	±1.2	2.791	±1.6	2.799	±1.9	2.801	±2.0
21.0	2.639	±1.3	2.651	±1.7	2.659	±2.1	2.662	±2.2
22.0	2.511	±1.4	2.523	±1.9	2.531	±2.3	2.534	±2.4
23.0	2.393	±1.6	2.406	±2.1	2.415	±2.5	2.410	±2.6
24.0	2.285	±1.7	2.298	±2.3	2.310	±2.7	2.310	±2.9
25.0	2.185	±1.9	2.199	±2.5	2.208	±3.0	2.211	±3.1
26.0	2.092	±2.0	2.106	±2.7	2.116	±3.2	2.120	±3.4
27.0	2.006	±2.2	2.021	±3.0	2.031	±3.5	2.035	±3.7
28.0	1.926	±2.4	1.941	±3.2	1.952	±3.8	1.955	±4.0
29.0	1.851	±2.6	1.866	±3.5	1.877	±4.1	1.881	±4.3
30.0	1.780	±2.8	1.797	±3.7	1.808	±4.4	1.812	±4.6
31.0	1.714	±3.0	1.731	±4.0	1.742	±4.7	1.746	±4.9
32.0	1.652	±3.2	1.669	±4.3	1.681	±5.0	1.685	±5.2
33.0	1.593	±3.5	1.611	±4.6	1.623	±5.4	1.627	±5.6
34.0	1.537	±3.7	1.555	±4.9	1.568	±5.7	1.572	±6.0
35.0	1.485	±3.9	1.503	±5.2	1.516	±6.1	1.520	±6.4
36.0	1.434	±4.2	1.466	±5.6	1.471	±6.5	1.471	±6.8
37.0	1.387	±4.5	1.406	±5.9	1.419	±6.9	1.423	±7.3
38.0	1.341	±4.8	1.361	±6.3	1.374	±7.4	1.379	±7.7
39.0	1.298	±5.1	1.318	±6.7	1.331	±7.8	1.336	±8.2
40.0	1.256	±5.4	1.276	±7.1	1.290	±8.3	1.295	±8.7

41.0	1.216	±5.7	1.237	±7.5	1.251	±8.7	1.256	±9.2
42.0	1.178	±6.1	1.199	±8.0	1.213	±9.2	1.219	±9.7
43.0	1.141	±6.4	1.163	±8.4	1.177	±9.8	1.182	±10.2
44.0	1.106	±6.8	1.128	±8.9	1.142	±10.3	1.147	±10.8
45.0	1.072	±7.2	1.094	±9.4	1.109	±10.9	1.113	±11.4
46.0	1.039	±7.6	1.061	±9.9	1.076	±11.5	1.081	±12.0
47.0	1.007	±8.0	1.030	±10.4	1.045	±12.1	1.050	±12.6
48.0	0.977	±8.5	0.999	±11.0	1.015	±12.7	1.020	±13.2
49.0	0.947	±8.9	0.970	±11.6	0.985	±13.4	0.991	±14.0
50.0	0.918	±9.4	0.941	±12.2	0.957	±14.0	0.962	±14.7
51.0	0.890	±9.9	0.914	±12.8	0.929	±14.8	0.935	±15.4
52.0	0.863	±10.5	0.887	±13.5	0.902	±16.2	0.908	±16.2
53.0	0.837	±11.0	0.861	±14.2	0.876	±16.3	0.881	±17.0
54.0	0.811	±11.6	0.835	±14.9	0.851	±17.1	0.856	±17.8
55.0	0.786	±12.2	0.810	±15.7	0.826	±17.9	0.831	±18.7
56.0	0.761	±12.9	0.786	±16.5	0.801	±18.8	0.806	±19.6
57.0	0.738	±13.6	0.762	±17.3	0.777	±19.7	0.783	±20.5
58.0	0.714	±14.3	0.738	±18.2	0.754	±20.6	0.759	±21.5
59.0	0.691	±15.0	0.715	±19.1	0.731	±21.6	0.736	±22.5
60.0	0.669	±15.8	0.693	±20.0	0.708	±22.7	0.713	±23.5

7.7 A TRANSMISSION LINE EQUIVALENT FOR A SYMMETRIC LOW-PASS T- OR Π-SECTION

Series inductors in lumped designs are often replaced with high characteristic-impedance transmission lines. It was shown in sec. 7.3 that the range of inductances which can be replaced in this way is limited. Where an inductor forms part of a low-pass Π-section, significantly better results can be obtained by replacing the inductance and some of the capacitance with a series line. Similarly, shunt capacitors that are part of a low-pass T-section can also be replaced with series lines. These two possibilities are illustrated in Fig. 7.6.

An exact transmission line equivalent for any symmetric low-pass T- or Π-section can be obtained at any particular frequency by equating the transmission matrix of the section to be replaced to that of a transmission line.

The transmission-matrix of the T-section shown in Fig. 7.7a is

$$\begin{bmatrix} 1-\omega^2 LC & j\omega L (2-\omega^2 LC)/(1-\omega^2 LC) \\ j\omega C & 1-\omega^2 LC \end{bmatrix} \qquad (7.49)$$

By equating this to

$$\begin{bmatrix} \cos(\beta l) & jZ_0 \sin (\beta l) \\ j\sin(\beta l)/ Z_0 & \cos(\beta l) \end{bmatrix} \qquad (7.50)$$

it follows that a transmission line with the following parameters will be exactly equivalent to the T-section at the frequency ω (rad/s)

$$L' = \frac{L}{1 - \omega^2 LC} \, [2 - \omega^2 LC] \tag{7.51}$$

$$C' = \frac{C}{1 - \omega^2 LC} \tag{7.52}$$

$$Z_0 = \sqrt{\frac{L'}{C'}} \tag{7.53}$$

$$\beta l = \tan^{-1} (\omega \sqrt{L'C'}) \tag{7.54}$$

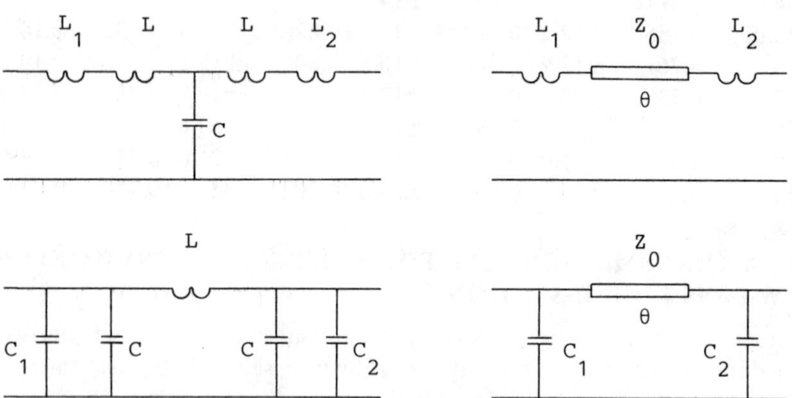

Figure 7.6   *The Partial Replacement of [a] a Low-Pass T-Section and [b] a Low-Pass Π-Section with a Series Line*

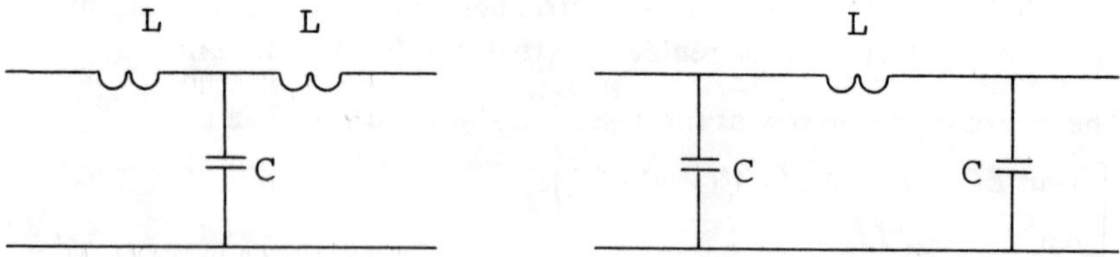

Figure 7.7   *[a] A Symmetrical Low-Pass T-Section; [b] Symmetrical Low-Pass Π-Section*

Excellent results can be expected when a T-section is replaced with a transmission line, and the difference between the characteristic impedances and line lengths required for exact equivalences at the low and high ends of the pass band is negligible. Alternatively, the capacitance and inductance corresponding to a chosen line section at the lowest and highest frequency in the pass band can be compared. The necessary equations for this purpose are

$$\omega L = Z_0 \; \frac{\sin(\beta l)}{1 + \cos(\beta l)} \tag{7.55}$$

$$\omega C = Y_0 \sin(\beta l) \tag{7.56}$$

where Y_0 is the inverse of Z_0.

The equations corresponding to the Π-section of Fig. 7.7b are

$$L' = \frac{L}{1 - \omega^2 LC} \tag{7.57}$$

$$C' = \frac{C}{1 - \omega^2 LC} \, [2 - \omega^2 LC] \tag{7.58}$$

$$Z_0 = \sqrt{\frac{L'}{C'}} \tag{7.59}$$

$$\beta l = \tan^{-1}(\omega \sqrt{L'C'}) \tag{7.60}$$

The inverse relationships are

$$\omega L = Z_0 \sin(\beta l) \tag{7.61}$$

and

$$\omega C = Y_0 \; \frac{\sin(\beta l)}{1 + \cos(\beta l)} \tag{7.62}$$

It follows easily from the equations given above that the length of the equivalent line for a T- or Π-section is only a function of the normalized reactance $\omega L/Z_0$ or the normalized susceptance $\omega C/Y_0$, respectively. The following equations can be used to calculate the required normalized susceptance $\omega C/Y_0$ and the line length corresponding to a specified normalized value for the reactance of the inductance in a Π-section:

$$\frac{\omega C}{Y_0} = \frac{Z_0}{\omega L} \left[1 - \sqrt{1 - \left(\frac{\omega L}{Z_0} \right)} \right]^2 \tag{7.63}$$

and

$$\beta l = \tan^{-1} \frac{\omega L/Z_0}{\sqrt{1 - (\omega L/Z_0)^2}} \tag{7.64}$$

With ωC, ωL and Y_0, Z_0 interchanged, the same set of equations applies to a T-section.

As a design aid and to give an idea of the characteristic impedances required to achieve very good results over different bandwidths, the normalized reactance and susceptance corresponding to different line lengths are tabulated in Table 7.8 together with the deviations in reactance or susceptance relative to the predicted linearly value based on the reactance or susceptance corresponding to a line length of $10°$ for a Π-section. With the necessary changes, Table 7.8 also applies to T-sections.

As an example of its application, the characteristic impedance required to replace a 50Ω inductor at a specific frequency with a line which is $45.0°$ long is 70.7Ω and the required capacitive susceptance is 5.86mS. If the relative bandwidth is $45.0/10.0=4.5$ and the frequency of exact equivalence is the highest frequency in the pass band, the error in inductance will be smaller than 9.5%, while the error in capacitance will be smaller than 5.2%. These errors can, of course, be decreased to approximately half of these values by averaging them across the pass band.

Example 7.4

As an example of the application of the Π-section transformation, a series transmission line equivalent for a 2nH inductor over the pass band 2-8 GHz will be determined.

With $Z_0=150\Omega$, application of (7.63) and (7.64) yields that the required capacitance and the line length corresponding to an exact equivalent at 8 GHz are

$\quad C = 0.051\text{pF}$

and

$\quad \beta l = 42.08°$

The Π-section equivalent for this line at 2 GHz ($\beta l=42.08/3=10.52°$) can be found by applying (7.61) and (7.62). The results are

$\quad L = 2.18nH$

and

$\quad C = 0.049\text{pF}$

which are very close to the original values (+9.0% and −7.3% respectively). Even better results can be obtained by minimizing the error across the pass band. This can be done by lowering the frequency of the exact transformation iteratively. By selecting this frequency as 5.8 GHz, the line length becomes $29.07°$ (at 5.8 GHz) and the difference in inductance 3.9% at 2 GHz and −3.9% at 8 GHz, while the difference in capacitance reduces to −1.9% and 2.0%, respectively.

Table 7.8

The Normalized Reactance (Susceptance) and Susceptance (Reactance) of the Components of the Π-Section (T-Section) Equivalent of a Series Transmission Line as a Function of the Line Length at the Frequency of Interest and the Percentage Deviation from the Expected Reactance and Susceptance Based on the Value Corresponding to a Line Length of 10°

βl	$\omega L/Z_0$	$\omega C/Y_0$
	$(\omega C/Y_0)$	$(\omega L/Z_0)$
(°)	— (%)	— (%)
10.0	0.1736 (0.0)	0.0875 (0.0)
12.5	0.2164 (−0.3)	0.1095 (0.1)
15.0	0.2588 (−0.6)	0.1317 (0.3)
17.5	0.3007 (−1.0)	0.1539 (0.5)
20.0	0.3420 (−1.5)	0.1763 (0.7)
22.5	0.3827 (−2.0)	0.1989 (1.0)
25.0	0.4226 (−2.6)	0.2217 (1.3)
27.5	0.4617 (−3.3)	0.2447 (1.7)
30.0	0.5000 (−4.0)	0.2679 (2.1)
32.5	0.5373 (−4.8)	0.2915 (2.5)
35.0	0.5736 (−5.6)	0.3153 (3.0)
37.5	0.6088 (−6.5)	0.3395 (3.5)
40.0	0.6428 (−7.4)	0.3640 (4.0)
42.5	0.6756 (−8.4)	0.3889 (4.6)
45.0	0.7071 (−9.5)	0.4142 (5.2)
47.5	0.7373 (−10.6)	0.4400 (5.9)
50.0	0.7660 (−11.8)	0.4663 (6.6)
52.5	0.7934 (−12.9)	0.4931 (7.3)
55.0	0.8192 (−14.2)	0.5206 (8.2)
57.5	0.8434 (−15.5)	0.5486 (9.0)
60.0	0.8660 (−16.9)	0.5774 (10.0)

Example 7.5

Consider the matching network shown in Fig. 7.8. A distributed equivalent over the pass band 2-6 GHz will be determined for it by replacing the two series inductors and some of the shunt capacitance with two series transmission lines (Z_0=150Ω) and the remaining capacitance with open-ended stubs. The relative dielectric constant of the material is taken as 2.17.

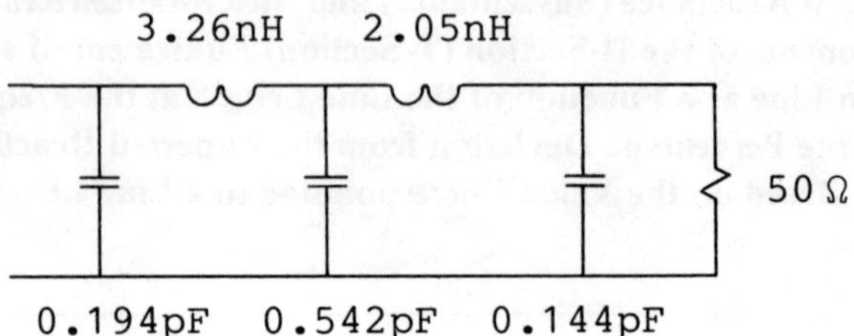

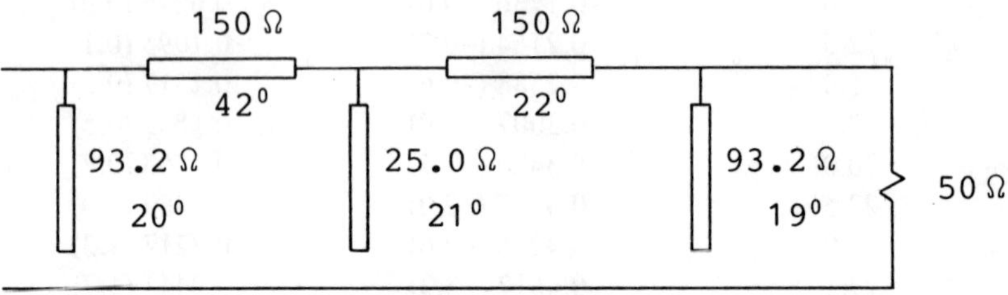

Figure 7.8   [a] The Matching Network Considered in Example 7.5 and [b] a Distributed Equivalent [electrical length specified at 6 GHz]

By applying (7.57) through (7.64) and changing the frequency of exact transformation iteratively, the optimum transformation frequency for both inductors is found to be approximately 7.4 GHz. The required line lengths and capacitance are 42°, 0.03pF (3.26nH) and 22.2°, 0.047pF (2.05nH), respectively. The maximum errors in inductance over the pass band are ±7.8% and ±2.0%, respectively.

After substracting the capacitances required for the series lines, the new values for the shunt capacitances are found to be 0.102pF (previously 0.194pF), 0.402pF (0.542pF), and 0.097pF (0.144pF), respectively.

The reactances of the first and last capacitors are very high and the error resulting from transforming them to equivalent stubs will be very small. It follows by inspection of Table 7.7 that the error in susceptance will be less than 1.9% if X_{HC}/Z_0 is equal to 2.799, that is Z_0=93.2Ω. With this value for the characteristic impedance, the required line lengths are appoximately 20° for the 0.107pF capacitor and 19° for the 0.097pF capacitor.

For minimum error, the 0.402pF capacitor should be replaced with a low characteristic impedance line. A 25Ω line will be used in this case. The corresponding X_{HC}/Z_0 ratio is then 2.647. Inspection of Table 7.7 yields that the error will be approximately equal to 1.9%. The required line length is approximately 21°.

The transformed circuit is shown in Fig. 7.8b. The output voltage standing-wave ratio (VSWR) of the two-stage amplifier in which this network was used, decreased from 1.72 to 1.65 with the transformation.

7.8 PARASITIC EFFECTS OF MICROSTRIP DISCONTINUITIES AT THE LOWER MICROWAVE FREQUENCIES

Microstrip discontinuities like open-ended stubs, changes in line width, and T-junctions add undesirable inductance and capacitance to designed circuits. The magnitude of these parasitics at the lower microwave frequencies (below X-band) [9] together with a compensation technique to reduce their effect [10] will be considered here.

Open-Ended Stubs

The effect of the parasitic fringing capacitance of a open-ended stub is similar to extending the length of the line slightly.

The equivalent additional line length is given empirically by [10]. The expression for the phase shift (in degrees) is

$$\Delta\theta_{oc} = 4.944\text{E-7hf} \sqrt{\epsilon_{r,\;eff}} \;\; \frac{\epsilon_{r,eff} + 0.300}{\epsilon_{r,eff} - 0.258} \;\; \frac{W/h + 0.264}{W/h + 0.800} \tag{7.65}$$

Table 7.9

The Electrical Line Length Associated with Different Dielectric Constants and Width-to-Height Ratios at 10 GHz with h=0.635mm

ϵ_r	Z_0	W/h	θ
—	(Ω)	(–)	(°)
	25	7.20	5.6
	50	2.80	5.0
2.5	75	1.35	4.4
	100	0.70	3.7
	125	0.38	3.2
	15	6.90	9.3
	25	3.35	8.4
10.2	50	0.90	6.2
	75	0.30	4.5

where h is the thickness of the substrate (m) and f the frequency (Hz).

The maximum relative error in (7.65) as compared to the more accurate expression of Silvester and Benedek [12] is less than 4% for $W/h \geq 0.2$ and $2 \leq \epsilon_r \leq 50$ [10].

As an illustration of the magnitude of the open-end parasitic, the parasitic electrical line length at 10 GHz associated with different width-to-height ratios and dielectric constants $\epsilon_r = 2.5$ and $\epsilon_r = 10.2$ for h = 0.635mm are tabulated in Table 7.9. It is clear from these results that the parasitic influence of open-end parasitics cannot be neglected at the higher frequencies and is more pronounced with higher dielectric constants, low impedance lines, and thicker substrates.

The simplest way to compensate for the increase in line length caused by open-end parasitics is to reduce the length of the designed line by the correct amount.

Steps in Width

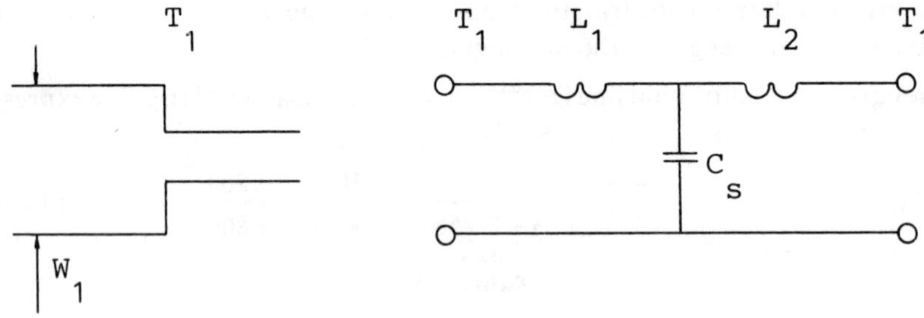

Figure 7.9   The Equivalent Circuit of a Step Discontinuity

Similarly to an open-ended line, the effect of the fringing capacitance associated with the wider line of a step discontinuity is to increase the wide line's length and to decrease that of the short line by the same amount. The change in electrical line length [11] (in degrees) can be approximated by

$$\Delta\theta_{step} = \Delta\theta_{oc} \left[1 - W_2/W_1\right] \tag{7.66}$$

where $\Delta\theta_{oc}$ can be calculated by using (7.65).

An alternative and more accurate approach to characterizing a step discontinuity is to use the equivalent circuit shown in Fig. 7.9b. An approximate expression for the inductance $L_s = L_1 + L_2$ in the circuit ($\pm 5\%$ for $W_1/W_2 \leq 5.0$

and $W_2/h=1.0$) is [10]:

$$L_s \,(\mathrm{nH/m}) = 40.5\, h \left(\frac{W_1}{W_2} -1.0\right) -75 \log\left(\frac{W_1}{W_2} + 0.2\right)\left(\frac{W_1}{W_2} -1\right)^2 \qquad (7.67)$$

The individual inductances are given approximately by [10]:

$$L_1 = L_{W1}/[L_{W1} + L_{W2}] \cdot L_s \qquad (7.68)$$

and

$$L_2 = L_{W2}/[L_{W1} + L_{W2}] \cdot L_s \qquad (7.69)$$

where L_{W1} and L_{W2} are the inductances associated with the characteristic impedances of the two lines.

An approximate closed form expression for the capacitance C_s in Fig. 7.9b ($\pm 10\%$ for $\epsilon_r \leq 10$ and $1.5 \leq W_2/W_1 \leq 3.5$) is [10]:

$$\frac{C_s}{\sqrt{W_1 \, W_2}} \,(\mathrm{pF/m}) =[10.1 \log \epsilon_r +2.33] \,\frac{W_1}{W_2} - 12.6 \log \epsilon_r -3.17 \qquad (7.70)$$

An idea of the magnitude of the parasitic effects associated with step discontinuities can be obtained from the extentions in line length resulting from an open-ended line as given in Table 7.9 and (7.66).

A first-order compensation technique for a step discontinuity would be to decrease the length of the wider line by an appropriate amount and to increase that of the narrower line by the same amount. The phase shift corresponding to a step discontinuity will always be less than that caused by an open-end in a line with the lower characteristic impedance.

Microstrip Bends

The equivalent circuit for a microstrip bend with lines of equal characteristic impedance is shown in Fig. 7.10.

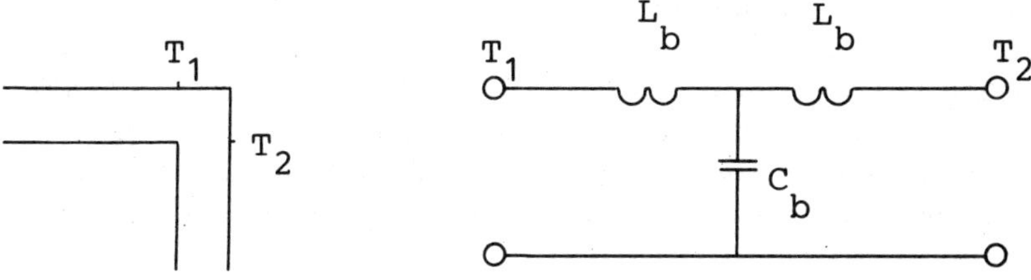

Figure 7.10   Equivalent Circuit for a Microstrip Bend

The appropriate closed form expressions for right-angled bend discontinuity capacitance and inductance are [9]:

$$\frac{C_b}{W} \text{ (pF/m)} = \begin{cases} \dfrac{[14\epsilon_r + 12.5] \, W/h - [1.83\,\epsilon_r - 2.25]}{\sqrt{W/h}} + \dfrac{0.02\,\epsilon_r}{W/h} \quad (W/h \leq 1) \\[4mm] [9.5\,\epsilon_r + 1.25] \, W/h + 5.2\epsilon_r + 7.0, \ (W/h \geq 1) \end{cases} \quad (7.71)$$

$$L_b/h \text{ (nH/m)} = 100 \, [4\sqrt{W/h} - 4.21] \tag{7.72}$$

Equation (7.71) is accurate to within 5% for $2.5 \leq \epsilon_r \leq 15$ and $0.1 \leq W/h \leq 5$. The accuracy of (7.72) is about 3% for $0.5 \leq W/h \leq 2.0$ [10].

Table 7.10

The VSWR (Theoretical) Resulting from an Unchamfered 90° Bend in a 75Ω (ϵ_r = 2.5) and a 50Ω (ϵ_r = 10.2) Line as a Function of the Frequency

ϵ_r	Z_0 (Ω)	f (GHz)	VSWR
2.5	75	2	1.03
		4	1.07
		8	1.15
		10	1.19
10.2	50	2	1.06
		4	1.13
		8	1.27
		10	1.34

An idea of the magnitude of the parasitics associated with a bend discontinuity can be obtained from Table 7.10, in which the theoretical VSWR's corresponding to a single bend and ϵ_r = 2.5, Z_0 = 75Ω and ϵ_r = 10.2, Z_0 = 50Ω are tabulated as a function of frequency. Although the effect of a single bend is small at the lower microwave frequencies for the lines considered, it should be kept in mind that it will increase with frequency, the number of bends, and the line width.

The parasitic effects of bend discontinuities are usually reduced by chamfering the bend as shown in Fig. 7.11. The optimum value of W_c in this figure is about $1.8W$ for 50Ω lines on alumina and rexolite substrates, and seems to be independent of the bend angle [9].

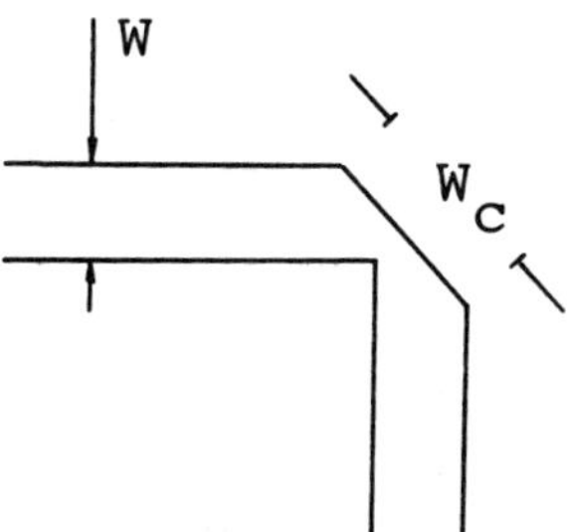

Figure 7.11   Compensation of a Microstrip Bend

T-Junctions

Hammerstad and Bekkadal's approach [11] to characterizing the parasitic effects of a T-junction with constant main-line width is illustrated in Fig. 7.12. The relevant equations are

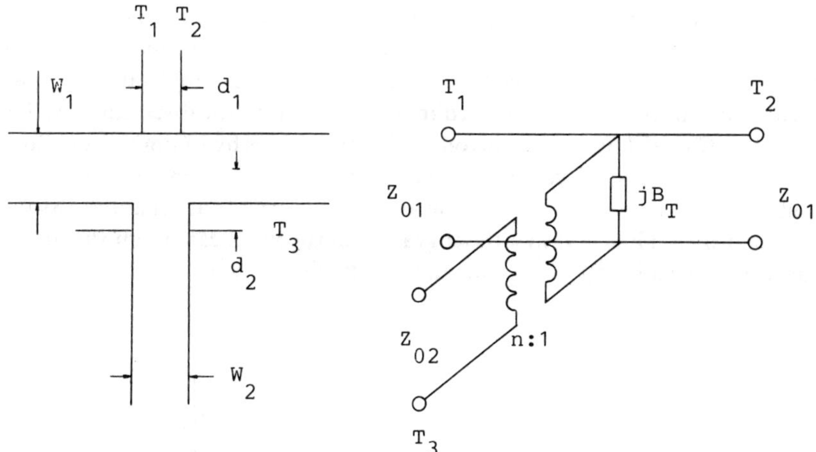

Figure 7.12   Equivalent Circuit for a Microstrip T-Junction

$$D_1 = 120\,\pi\,h/Z_{01}\ \text{(air)},\quad D_2 = 120\,\pi\,h/Z_{02}\ \text{(air)} \tag{7.73}$$

$$n = \frac{\sin\left(\dfrac{\pi}{2}\ \dfrac{2D_1}{\lambda_m}\ \dfrac{Z_{01}}{Z_{02}}\right)}{\dfrac{\pi}{2}\ \dfrac{2D_1}{\lambda_m}\ \dfrac{Z_{01}}{Z_{02}}} \tag{7.74}$$

$$d_1/D_2 = 0.05\,n^2\,Z_{01}/Z_{02} \tag{7.75}$$

$$d_2/D_1 = 0.5 - 0.16 \left[1 + (2D_1/\lambda_m)^2 - 2 \ln(Z_{01}/Z_{02})\right] Z_{01}/Z_{02} \qquad (7.76)$$

$$\frac{B_T \lambda_m}{Y_{01} D_1} = \begin{cases} -[1-2D_1/\lambda_m]\, Z_{01}/Z_{02}, \quad Z_{01}/Z_{02} \leq 0.5 \\[2em] [1-2D_1/\lambda_m]\,[3Z_{01}/Z_{02} - 2], \quad Z_{01}/Z_{02} \geq 0.5 \end{cases} \qquad (7.77)$$

$$\lambda_m = \lambda_0 / \sqrt{\epsilon_{r,\,eff}} \qquad (7.78)$$

When $Z_{01}/Z_{02} \geq 2$, the calculated values of d_2/D_1 are too high. In this range, a better value for d_2 can be obtained by replacing Z_{01}/Z_{02} in (7.76) with its inverse [13].

When open-ended or short-circuited stubs are designed, the stub width or length can be changed to make the shunt reactance equal to that required at the frequency of interest by using (7.73) through (7.78).

7.9   A COMPENSATION TECHNIQUE FOR MICROSTRIP DISCONTINUITIES

It is possible to reduce all the parasitic effects described in the previous sections by tapering the lines correctly. In doing so, fringing capacitance is exchanged for plate capacitance to keep the characteristic impedance constant as the width of the line is decreased in order to reduce the discontinuity. This compensation technique was introduced for striplines by Malherbe and Steyn [5]. The same approach can be followed with microstrip lines and the relevant equations will be derived here. Basic assumptions made are that the mode of propagation is TEM and that the curvature of the fields can be ignored for the purpose of calculating the capacitance of the line.

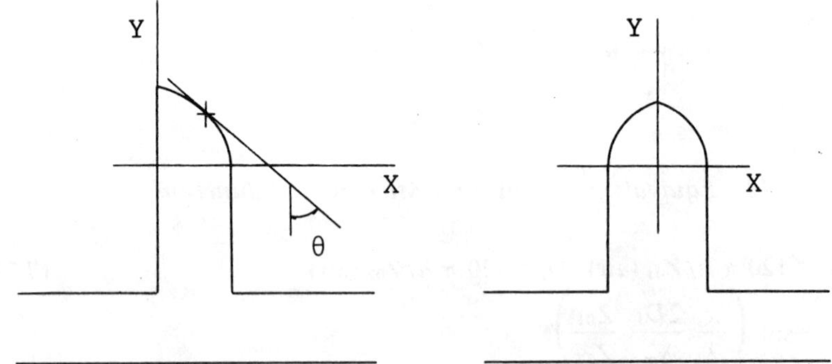

Figure 7.13   Elimination of the Parasitic Effects associated with an Open-Ended Stub by Tapering the Line [a] Asymmetrically, [b] Symmetrically

The characteristic impedance of a microstrip line is given by

$$Z_0 = 1/[v_p\, C] \tag{7.79}$$

$$= \sqrt{\epsilon_r}/[cC] \tag{7.80}$$

where ϵ_r is the relative effective dielectric constant of the line, c is the speed of light, and C is the capacitance per unit length. If the characteristic impedance is to remain constant with narrowing width-to-height ratio, the quotient $C/\sqrt{\epsilon_r}$ in (7.80) must remain constant.

The capacitance corresponding to an incremental distance dy and the untapered width, W, of the line is

$$C_W\, dy = \epsilon_0\, \epsilon_r\, W/h\, dy + \epsilon_0\, W/H_2\, dy + 2C_{FW}\, dy \tag{7.81}$$

where ϵ_r is the dielectric constant of the material used and C_{fw} the fringing capacitance per unit length corresponding to the width W, H_2 is the distance from the conductor to the cover, and h is the dielectric thickness.

The capacitance corresponding to an incremental section of an asymmetrical line with tapered width x is

$$C_x\, dy = \epsilon_0\, \epsilon_r\, x/h + \epsilon_0\, x/H_2 + C_{fx} + C_{fx}\sec\theta \tag{7.82}$$

The fringing capacitance (C_{fx}) can be obtained from the characteristic impedance and effective relative dielectric constant corresponding to an untapered line with width x by using the equation:

$$C_{fx} = 0.5\left[\sqrt{\epsilon_{rx}}/(cZ_{0x}) - \epsilon_0\, \epsilon_r\, x/h - \epsilon_0\, x/H_2\right] \tag{7.83}$$

where ϵ_{rx} is the effective dielectric constant of an untapered line with width x. This equation can be derived by combining (7.80) and (7.81) with W set equal to x in (7.81).

Equating the characteristic impedances at x and W results in

$$1+\sec\theta = \left\{ \frac{\epsilon_{rx}}{\epsilon_{rW}}\left[\frac{\epsilon_0\, \epsilon_r\, W}{h} + \frac{\epsilon_0\, W}{H_2} + 2C_{FW}\right] - \frac{\epsilon_0\, \epsilon_r\, x}{h} - \frac{\epsilon_0\, x}{H_2} - C_{fx}\right\} / C_{fx} \tag{7.84}$$

With θ known as a function of x, it is a simple matter to construct the taper corresponding to it. Alternatively, the exact position y where the taper width is equal to X can be determined numerically by evaluating the integral

$$y = -\int_W^X \cot\theta\, dx \tag{7.85}$$

The length of the taper can be determined by setting the boundary X in (7.85) equal to the final value of x.

The electrical length of the tapered section can be obtained by evaluating the integral

$$\Delta\theta \text{ (degrees)} = 306 f/c \int_{W}^{X} - \sqrt{\epsilon_x}\cot\theta\, dx \tag{7.86}$$

where f is the relevant frequency and X the final value of x.

When a symmetrical taper is required, (7.85) and (7.86) in conjunction with the following equations can be used:

$$C_{fx} = 0.5\left[\sqrt{\epsilon_{rx}}/(cZ_{0x}) - \epsilon_0\,\epsilon_r\,2x/h - \epsilon_0\,2x/H_2\right] \tag{7.87}$$

$$\sec\theta = \left\{\frac{\epsilon_{rx}}{\epsilon_{rW}}\left[\frac{\epsilon_0\,\epsilon_r\,W}{h} + \frac{\epsilon_0\,W}{H_2} + 2C_{FW}\right] - \frac{\epsilon_0\epsilon_r\,2x}{h} - \frac{\epsilon_0\,2x}{H_2}\right\}/\left[2C_{fx}\right]$$

$$\tag{7.88}$$

Application of the tapering technique to reduce the effect of the discontinuities in a design is straightforward and is illustrated in Fig. 7.14.

Example 7.6

In order to check if the characteristic impedance of a line tapered as outlined above indeed remains constant, the reactance of a tapered 25.4Ω stub (Fig. 7.15a; $\epsilon_r = 10.0$ was measured at the frequency where the lengths of the feeding 50Ω lines were equal to one-half of a wavelength (4.6 GHz). The measured reactance of the 50.2° long stub (13.2° tapered + 37.1° untapered) was within 15% of its expected value. The measured error in the reactance of a 1pF untapered stub compensated by using (7.73) to (7.78) was 12%. The ratio of the characteristic impedance of the main line and that of the stub was equal to two, which is in the inaccurate range of (7.73) to (7.78). The open-end capacitance was assumed to be equal to the value given by (7.65) in both cases.

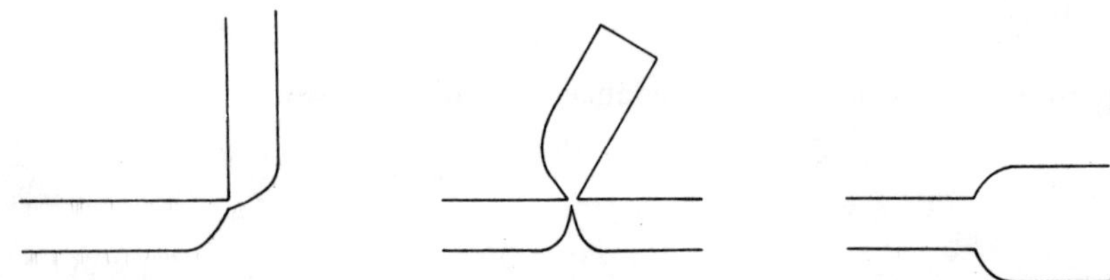

Figure 7.14   The Tapering Technique Applied to Reduce the Discontinuity Effects associated with T-Junctions, Bends, and Steps in Width

The frequency range over which good results can be expected with the taper is illustrated by the measured VSWR of the tapered 50Ω lines of Fig. 7.15b ($\epsilon_r = 2.17$; design frequency 4.6 GHz). The VSWR for the four cascaded tapers was smaller than 1.5 for all frequencies up to 7.3 GHz and smaller than 2.0 below 9.5 GHz. With the same load and connectors, the input VSWR of an untapered 50Ω line was measured to be 1.5 below 9.6 GHz and lower than 2.0 below 16.5 GHz.

Better results can probably be obtained if the taper capacitance is calculated between equipotentials. Analytically, this is, however, not a simple task.

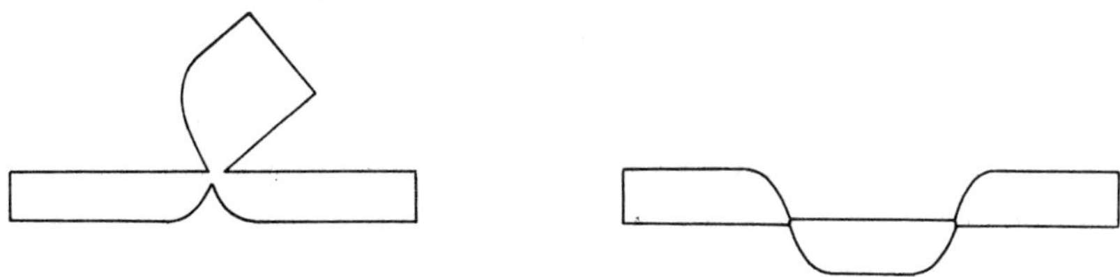

Figure 7.15   [a] The Layout of the Feeding Line and Tapered Stub of Example 7.6, [b] the Tapered 50Ω Lines of Example 7.6

In practice it is usually possible to constrain the characteristic impedance of the lines used to a few standard values in non-commensurate designs. Only a few tapers are then required and these can be reproduced photographically. In following this approach, the cost and effort involved in designing the necessary tapers can be kept low.

Questions and Problems

1. Which two factors determine the frequency above which an inductor can no longer be considered lumped? Which two factors determine the frequency above which a series transmission line can no longer be considered a lumped element?

2. Why is the Q obtainable with a round-wire inductor, with diameter d, higher than that obtainable with a strip inductor of the width $w = d$?

3. Calculate the expected Q-factor of a bonding-wire inductor with a 50μm diameter and length of 2mm at (a) 2 GHz and (b) 10 GHz.

4. Determine the dimensions of a circular 2nH single-loop inductor. The conductor thickness is 35μm.

    If the substrate thickness is 0.635mm (ϵ_r=10.0) and a cavity with the same depth is created beneath the strip to decrease the parasitic capacitance, determine the frequency at which the insertion loss in Fig. 7.1 will be equal to 0.25 dB if the inductor is used at a point in a circuit where the load Q is equal to (a) 0, (b) 2, and (c) 4.

Calculate the expected Q-factor of the inductor at different frequencies below the 0.25 dB frequency corresponding to $Q=0$.

5. Design an air-cored solenoidal coil with reactance as high as possible at 6 GHz (RF choke). Design a square spiral to have the same inductance.

6. Compare the maximum inductance obtainable with the different types of inductors and the same line length in Table 7.3.

7. Up to which frequency will the reactance of a high quality 1pF microwave chip capacitor be within 10% of its expected value?

8. Design a 1pF interdigital capacitor. Up to which frequency can the associated parasitics be ignored?

9. Find a shunt transmission line equivalent for a 1pF shunt capacitor at 4 GHz. Find the optimum (lowest deviation from the expected response) shunt equivalent over the pass band 2-4 GHz.

10. Find a shunt transmission line equivalent for a 2nH inductor at 4 GHz. Find the optimum shunt equivalent over the pass band 2-4 GHz.

11. A 1.5nH inductor is part of a low-pass Π-section. Determine the parameters of the optimum series transmission line equivalent for it at 6 GHz, if the highest realizable characteristic impedance is 75Ω. Determine the shunt capacitance required.

12. A 0.25pF shunt capacitor is part of a low-pass T-section. Determine the parameters of the optimum series transmission line equivalent for it at 8 GHz, if the lowest realizable characteristic impedance is 10Ω. Determine the series inductance required.

13. Determine transmission-line equivalents for the lumped-element circuits shown in Fig. 7.16 over the pass band 2-6 GHz. Compare the frequency response of each circuit before and after the transformation.

14. Calculate the increase in line length associated with an open-ended 25Ω line on a substrate with $\epsilon_r = 10.0$ and $h = 0.635$ mm at (a) 2 GHz and (b) 10 GHz.

15. Calculate the parasitic change in the lengths of a 75Ω to 50Ω step junction on the substrate used in *Problem 7.14* at the same frequencies. Determine the equivalent circuit of the step junction.

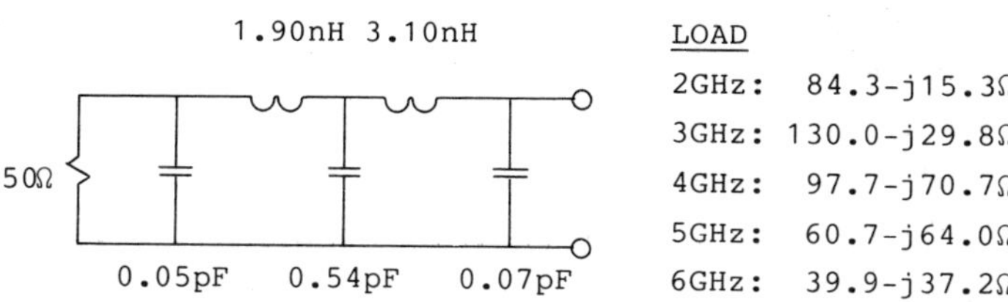

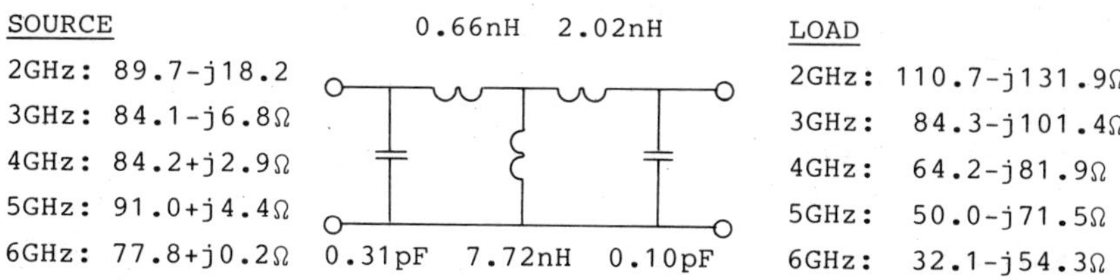

Figure 7.16   The Lumped-Element Circuits Relevant to Problem 7.13

16. Determine the VSWR of a 50Ω line terminated in a 50Ω load and with two unchamfered bends separated by one-half of a wavelength at 8 GHz on the substrate used in the previous problems.

17. Design an open-ended stub to have 0.5pF capacitance at 8 GHz on the substrate used in the previous problems by using (7.73) through (7.78) to compensate for the associated parasitics.

References

1. De Brecht, R., and M. Caulton, "Lumped Elements in Microwave Integrated Circuits in the 1-12 GHz Range," *IEEE G-MTT Internat. Microwave Symp.*, May 1970, p. 14.

2. Hammerstad, E., and O. Jensen, *IEEE MTT-S Symposium Digest,* June 1980, pp. 407-409.

3. March, S., "Microstrip Packaging: Watch The Last Step," *Microwaves,* Dec. 1981.

4. Gupta, K.C., R. Garg, and I.J. Bahl, *Microstrip Lines and Slotlines,* Dedham, MA: Artech House, 1979.

5. Malherbe, J.A.G., and A.F. Steyn, "The Compensation of Step Discontinuities in TEM-Mode Transmission Lines," *IEEE Trans. Microwave Theory Tech.,* Vol. MTT-26, No. 11, Nov. 1978.

6. Caulton, M., S.P. Knight, and D.A. Daly, "Hybrid Integrated Lumped-Element Microwave Amplifiers," *IEEE J. Solid-State Circ.*, Vol. SC-3, No. 2, June 1968.

7. Sobol, H., "Application of Integrated Circuit Technology to Microwave Frequencies," *Proc. IEEE,* Vol. 59, Aug. 1971, pp. 1200-1211.

8. Young, L., *Advances in Microwaves,* New York: Academic Press, 1977.

9. Gupta, K.C., R. Garg, and I.J. Bahl, *Microstrip Lines and Slotlines,* Dedham, MA: Artech House, 1979.

10. Garg, R. and I.J. Bahl, "Microstrip Discontinuities," *Int. J. Electronics,* Vol. 45, July 1978.

11. Hamerstad, E.O., and F. Bekkadal, *Microstrip Handbook,* ELAB Report STF 44/ A74169, University of Trondheim, Norwegian Institute of Technology, 1975.

12. Silvester, P., and P. Benedek, "Equivalent Capacitance of Microstrip Open Circuits," *IEEE Trans. Microwave Theory Tech.,* Vol. MTT-20, 1972, pp. 511-516.

13. Dielectric Laboratories, Inc., Fairfield, CT, USA.

14. Alley, G.D., "Interdigital Capacitors and their Application to Lumped-Element Microwave Integrated Circuits," *IEEE Trans. Microwave Theory Tech.,* Vol. MTT-18, No. 12, December 1970.

15. Terman, F.E., *Radio Engineers Handbook,* New York: McGraw-Hill, 1943.

16. Dill, H.G., "Designing Inductors for Thin-Film Applications," *Electron, Des.,* February 17, 1967, pp. 52-59.

17. Dukes, J.M.C., *Printed Circuits, their Design and Application,* London: MacDonald, 1961, pp. 120-135.

CHAPTER 8

THE DESIGN OF RADIO-FREQUENCY
AND MICROWAVE AMPLIFIERS

8.1 INTRODUCTION

In this chapter the design of radio-frequency and microwave amplifiers will be considered. Basic considerations such as amplifier stability, tunability, unilateralness, and dynamic range will be introduced and procedures will be developed for the design of amplifiers with any realizable transducer power gain *versus* frequency response, low input or output VSWR, or optimum noise performance. Excellent results can be obtained with the procedures outlined here, even when the reverse transfer gain of the transistors used is not negligible, and if the iterative impedance matching techniques discussed in Ch. 6 are used to design the matching networks required, computer-optimization of a designed amplifier is very seldomly if ever required. The design of reflection, balanced, and power amplifiers will also be considered.

The design procedures are developed in terms of the Y-parameters of the transistors used. Equivalent S-parameter expressions for most of the expressions given here are derived in [1] and [2]. The S-parameter equivalents for the expressions derived here are summarized in Appendix E.

8.2 AMPLIFIER STABILITY

In order to prevent an amplifier from oscillating at the design stage or to cure the problem when it occurs, it is necessary to know more about the conditions under which oscillations can occur. These conditions will be derived here. Consider the two-port network in Fig. 8.1. The terminal currents of the new two-port formed by considering the terminations as part of the original two-port can be calculated by using the equation:

$$\begin{bmatrix} I_1 \\ I_2 \end{bmatrix} = \begin{bmatrix} y_{11} + Y_s & y_{12} \\ y_{21} & y_{22} + Y_L \end{bmatrix} \begin{bmatrix} V_1 \\ V_2 \end{bmatrix} \tag{8.1}$$

If the circuit is oscillating, with no signal applied, the terminal voltages will not be equal to zero, while the terminal currents are equal to zero.

So that the voltages in (8.1) are not equal to zero while the currents are equal to zero, the determinant of the extended Y-parameter matrix must be equal to zero. Were this not the case, the Y-parameter matrix would have had an inverse and the only solution corresponding to zero values for I_1 and I_2 would have been zero values for both V_1 and V_2 as well.

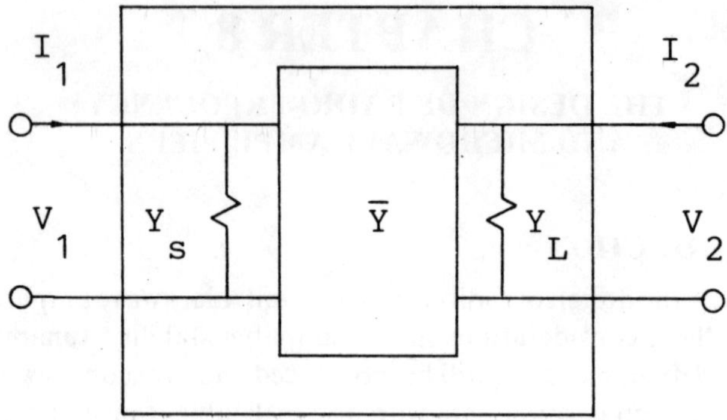

Figure 8.1   *A Two-Port Network with its Terminations*

A zero value for the determinant implies that

$$[y_{11} + Y_s][y_{22} + Y_L] - y_{12}\,y_{21} = 0 \tag{8.2}$$

leading to

$$y_{11} + Y_s - \frac{y_{12}\,y_{21}}{y_{22} + Y_L} = 0 \tag{8.3}$$

and

$$y_{22} + Y_L - \frac{y_{12}\,y_{21}}{y_{11} + Y_s} = 0 \tag{8.4}$$

The last two equations are easily recognized as the equations for the input and output admittances respectively of the augmented two-port.

A necessary condition for any oscillation to occur is, therefore, that the input and output admittance of the two-port, with its terminations considered as part of it, must be equal to zero.

It is clear from (8.2) to (8.4) that if the input admittance of a network is equal to zero, the same will apply to its output admittance as long as the product y_{12} y_{21} is not equal to zero, which will not be the case whenever there is a possibility of an oscillation, and the load and source terminations are passive.

It is also clear from (8.3) and (8.4) that whenever the resistive part of the input or output admittance is greater than zero at any particular frequency, it will not be possible for an oscillation to occur at that particular frequency. Therefore, if care is taken to ensure that the input (or output) conductance of an amplifier is always greater than zero, oscillations will not be possible. The locus of load admittances (source admittances) for which the input (output) conductance of a two-port network will be equal to zero, therefore, is of interest.

Considering only the input admittance for the moment, the locus of load admittance for which the input conductance will be equal to zero can be derived easily be setting $Re\,[\,Y_{IN}\,(\,Y_L)\,] = 0$:

$$Re\,(\,Y_{IN}) = G_s + g_{11} - Re\,\,\frac{y_{12}\,y_{21}}{y_{22} + Y_L}\,\, = 0$$

where

$$g_{11} + jb_{11} = y_{11} \tag{8.5}$$

With

$$g_{22} + jb_{22} = y_{22} \tag{8.6}$$

$$y_{12}\,y_{21} = P + jQ \tag{8.7}$$

and

$$G_L + jB_L = Y_L \tag{8.8}$$

it follows that

$$G_s + g_{11} - Re\,\left\{\frac{(P + jQ)\,[\,g_{22} + G_L - j(b_{22} + B_L)\,]}{(g_{22} + G_L)^2 + (b_{22} + B_L)^2}\right\} = 0$$

This equation can be manipulated into the following form:

$$\left[G_L + g_{22} - \frac{P}{2\,(G_s + g_{11})}\right]^2 + \left[B_L + b_{22} - \frac{Q}{2\,(G_s + g_{11})}\right]^2 = \frac{|y_{12}\,y_{21}|^2}{4\,(G_s + g_{11})^2} \tag{8.9}$$

Equation (8.9) is the equation of a circle in the linear admittance plane with center:

$$G_L + jB_L = \left[\frac{P}{2\,(G_s + g_{11})}\right] - g_{22}\,\, + j\,\left[\frac{Q}{2\,(G_s + g_{11})} - b_{22}\right] \tag{8.10}$$

and radius:

$$R = \left|\frac{y_{12}\,y_{21}}{2\,(G_s + g_{11})}\right| \tag{8.11}$$

For all load admittances falling on the circumference of this circle, the input conductance will be equal to zero, and if the input susceptance is also equal to zero, the amplifier will oscillate. The input conductance is usually negative for all load admittances falling inside the stability circle.

For passive loads, the worst-case condition will clearly be when $G_s = 0$. Under this condition (8.10) and (8.11) simplify to

$$G_L^* + jB_L^* = \frac{P}{2g_{11}} - g_{22} + j\left[\frac{Q}{2g_{11}} - b_{22}\right] \tag{8.12}$$

and

$$R_L^* = \left|\frac{y_{12}\,y_{21}}{2g_{11}}\right| \tag{8.13}$$

Whenever this stability circle lies to the left of the imaginary axis of the admittance plane, it will not be possible for an amplifier to oscillate at that particular frequency as long as its terminations are passive. Such an amplifier is said to be inherently stable.

Proceeding as above, the parameters of the stability circle corresponding to zero values of the output conductance of an amplifier can be determined easily. The resulting equations are

$$G_s + jB_s = \left[\frac{P}{2\,(G_L + g_{22})} - g_{11}\right] + j\left[\frac{Q}{2\,(G_L + g_{22})} - b_{11}\right] \tag{8.14}$$

$$R = \left|\frac{y_{12}\,y_{21}}{2\,(G_L + g_{22})}\right| \tag{8.15}$$

A very useful stability factor, the Linville stability factor, can be defined in terms of the parameters of the stability circle in the following way:

$$C = -\frac{R_L^*}{G_L^*} \tag{8.16}$$

$$= -\frac{|y_{12}\,y_{21}|}{P - 2g_{11}\,g_{22}} \tag{8.17}$$

Whenever $0 \le C < 1$ for a chosen device, the stability circle will lie to the left of the imaginary axis of the admittance plane and the device under consideration will be inherently stable. The stability circle is plotted for different values of C in Fig. 8.2.

Since the Linville stability factor is independent of the amplifier terminations and only a function of the Y-parameters of the device used, it is basically a measure of device stability, and as such it is useful when comparing different devices.

Another very useful measure of stability, which also takes the influence of the resistive loading by the load and source admittance into account and is, therefore, a measure of the amplifier stability, is the Sterne stability factor:

$$K = \frac{g_{22} + G_L}{\dfrac{|y_{12}\,y_{21}|}{2(g_{11} + G_s)} + \dfrac{P}{2(g_{11} + G_s)}} \tag{8.18}$$

An amplifier will be stable at any frequency for which $K \geq 1$.

Whenever either or both of the terminations can change as is the case where an amplifier is connected to a mobile antenna, it is advisable to set G_L or G_s in (8.18) equal to zero, as applicable.

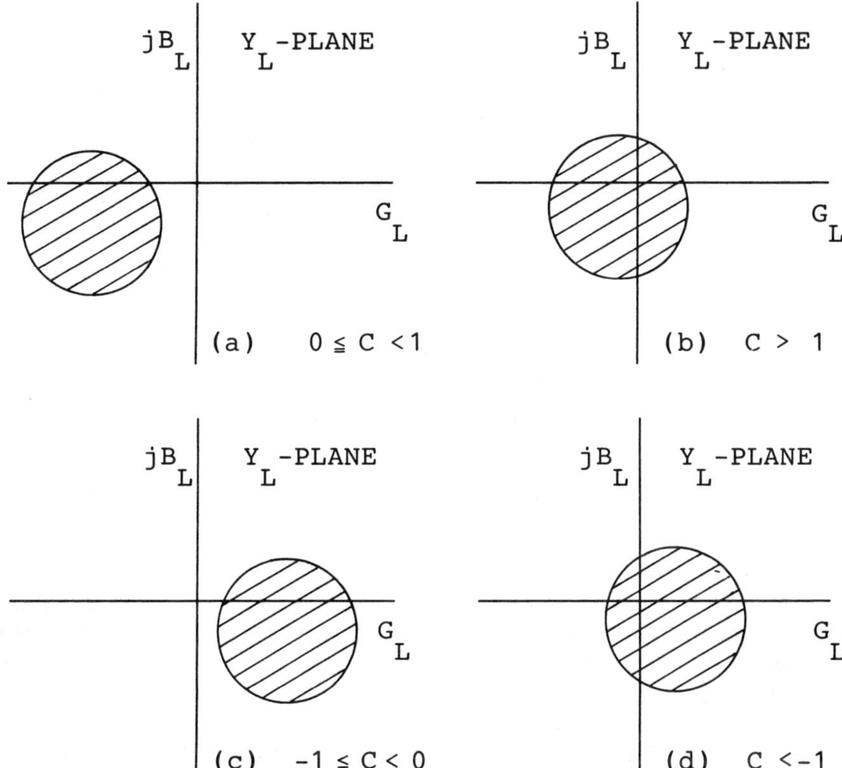

Figure 8.2   The Relationship between the Position of a Stability Circle Relative to the Imaginary Axis of the Admittance Plane and the Linville Stability Factor

The locus of load and source reflection coefficients for which the input or output conductance of an amplifier will be equal to zero are also circles on a Smith chart. These circles are useful when impedance-matching networks are designed graphically, but because much better results can be obtained in considerably less time with a programmable calculator, the parameters of these circles will not be derived here and the interested reader is referred to [1] or [2]. The parameters of these circles are repeated in Appendix E.

Based on the theory derived above, the procedure described below can be followed to ensure that a designed amplifier will be stable.

Procedure For Ensuring that a Designed Amplifier Will Be Stable

1)   Calculate the Linville stability factor at a number of frequencies in the pass band to determine whether the device is inherently stable.

Whenever possible, choose the load or source impedance-matching networks to be of low-pass form to prevent oscillations from occurring at low frequencies outside the pass band. When this is not possible, it is advisable to check the designed amplifier for stability at a few frequencies outside the pass band by calculating the Sterne stability factor at these frequencies.

If the amplifier is not stable at these frequencies, change the element values or topologies of the matching networks and repeat this step.

2)   If the device used is not inherently stable and its terminations are fixed, it is not necessary to be concerned about stability inside the pass band. By designing the amplifier to have a specified gain *versus* frequency-response stability will be ensured automatically inside the pass band.

3)   If the device is not inherently stable inside the pass band, it can be compensated by using the techniques outlined in the next section. Stabilization of a device is advisable in all cases where the terminations are not fixed. When a low-noise amplifier is designed, non-resistive feedback techniques should be used for the stabilization.

4)   If the terminations are not fixed and the device has not been compensated to be inherently stable, check the positions of the load or source impedances as seen from the device terminals relative to that of the stability circles. If the load (source) impedance alone can change, it is only necessary to ensure that the output (input) conductance is positive.

8.3   The Optimal Stabilization of an Amplifier by Resistive Loading

A potentially unstable transistor can always be stabilized by shunting its input or output terminals with enough conductance and often by inductive current series, capacitive voltage shunt, or resistive voltage-shunt feedback. These four methods of stabilizing a device are illustrated in Fig. 8.3.

When resistive voltage-shunt feedback is used, the optimum values of R_f and L_f in Fig. 8.3b can be easily determined iteratively. L_f can be ignored at the lower pass-band frequencies of a wideband amplifier, and R_f can be changed until the device is stable across the pass band. The maximum operating or available power gain obtainable from the device should be monitored during the process. The values of L_f can then be changed to flatten the gain across the pass band by removing some of the feedback at the higher frequencies. Because of the undesirable phase shift across high-inductance lumped inductors, it is usually not possible to use a series inductor with R_f at microwave frequencies.

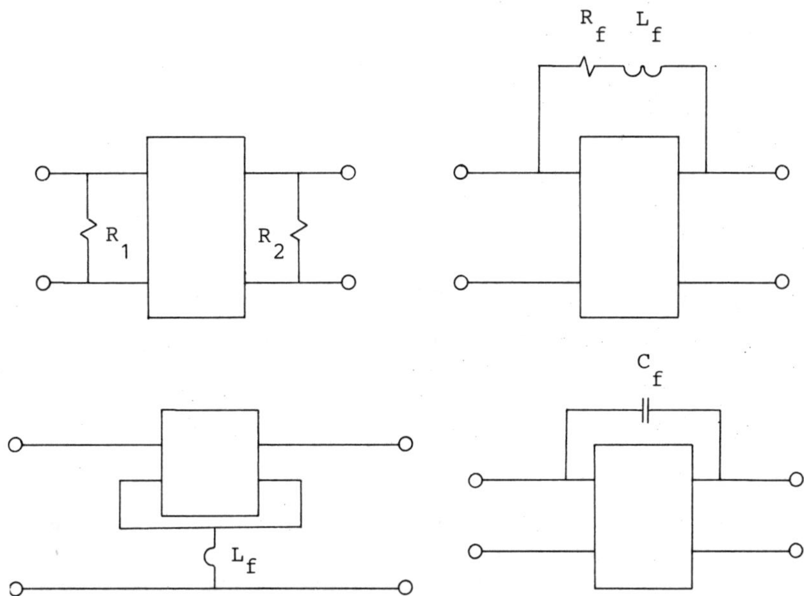

Figure 8.3 Stabilization of a Potentially Unstable Device by [a] Resistively Loading its Input and Output Terminals, [b] Resistive Voltage-Shunt Feedback, [c] Inductive Current-Series Feedback, and [d] Capacitive Voltage-Shunt Feedback.

It is a simple matter to find the optimum value for the required current-series feedback inductor or voltage-shunt feedback capacitor iteratively. Non-resistive feedback is preferable when the noise performance is important.

Any transistor can be stabilized by resistively loading its in- or output terminals. It is sometimes better to load only the output or only the input, or sometimes both terminals, resistively. The optimum values for the loading resistors will be derived here.

By using the Linville stability factor, it follows that a necessary condition for a transistor to be inherently stable is that

$$g_{11}\, g_{22} > 0.5\,[\,|y_{12}\, y_{21}| + P\,] \tag{8.19}$$

When this is not the case and the transistor is resistively loaded, the values of g_{11} or g_{22} are increased and (8.19) can be enforced.

If a stabilizing resistor is used only across the output terminals of the transistor, the minimum value of the required conductance is given by the equation:

$$g_{22C} = \frac{|y_{12}\,y_{21}|}{2g_{11}\,C} + \left[\frac{P}{2g_{11}} - g_{22}\right] \tag{8.20}$$

where C is the required Linville stability factor after compensation.

When the transistor is only compensated across its input terminals, the required conductance is

$$g_{11C} = \frac{|y_{12}\,y_{21}|}{2g_{22}\,C} + \left[\frac{P}{2g_{22}} - g_{11}\right] \tag{8.21}$$

When the loss in gain caused by stabilizing the transistor across either its input or output terminals becomes significant, it becomes necessary to determine whether it will be best if the input or output circuit or both are compensated.

The maximum power gain for an inherently stable transistor in terms of its Linville stability factor at a particular frequency, turns out to be

$$G_{w-MAX} = \left|\frac{y_{21}}{y_{12}}\right| \; (1/C - \sqrt{1/C^2 - 1}) \tag{8.22}$$

Because the influence of g_{11} and g_{22} on the maximum operating power gain are therefore limited to the influence of the Linville factor, which in turn is influenced in exactly the same way by both g_{11} and g_{22}, it follows that as far as the gain at the frequency where the transistor is to be stabilized is concerned, it is immaterial whether g_{11} or g_{22}, or both, are increased.

Although the gain at the frequency (or frequencies) where the transistor is potentially unstable is not influenced by the relative compensation at the input or output of the transistor, the gain at the other frequencies definitely is influenced. The effect is usually most significant at the higher frequencies in the pass band.

It can be shown that if the product $g_{11}\,g_{22}$ at the frequency where the transistor is potentially unstable, e.g., f_u, must be increased by a factor r to achieve the desired stability (C), and if g_{11} is changed by a factor α_s, and therefore g_{22} by a factor r/α_s, in order to do this, then the optimum value for α_s that will ensure the highest maximum operating power gain at another frequency, which is usually the highest frequency in the pass band, f_H, can be obtained from the equation:

$$\alpha_s^2 = r\,\frac{g_{11H}/g_{11u} - 1}{g_{22H}/g_{22u} - 1} \tag{8.23}$$

where g_{11H} and g_{22H} are the real parts of y_{11} and y_{22} at the highest (other) frequency, and g_{11u} and g_{22u} are the real parts of y_{11} and y_{22} at the frequency where the transistor is potentially unstable (f_u).

The optimum compensating conductances are then

$$g_{11C} = [\alpha_s - 1]\, g_{11u} \tag{8.24}$$

and

$$g_{22C} = [r/\alpha_s - 1]\, g_{22u} \tag{8.25}$$

The value of r can be determined by using either (8.16) or (8.17). Independently of which equation is used, it follows that

$$r = \frac{|y_{12}\, y_{21}|}{2g_{11}\, g_{22}\, C} + \frac{P}{2g_{11}\, g_{22}} - 1 \tag{8.26}$$

The parameters used in this equation are those at the frequency where the transistor is potentially unstable (f_u) before the compensation.

If it turns out that α_s or r/α_s is bigger than r, it will be best to compensate the transistor only at the input or output, respectively.

Instead of using only shunt resistors, it is, of course, possible to stabilize a transistor with a frequency-selective loading network (or networks) which provides the required conductance at the frequency where the transistor is potentially unstable and reduces the conductance at the other frequencies. The simplest example of such a network is a series RL combination (series resistance, shorted-stub combination at microwave frequencies).

Example 8.1

Over the pass band 8-12 GHz, the Plessey COD device [1, 6] is potentially unstable at 8 and 9 GHz. As an example of the application of the expressions derived above, consider the stabilization of the transistor over this pass band.

Table 8.1

The S-Parameters of the Plessey COD Device at 8, 9, 10, 11, and 12 GHz

Frequency (GHz)	s_{11} (dB; °)		s_{12} (dB; °)		s_{21} (dB; °)		s_{22} (dB; °)	
8	−4.15	−64	−23.6	134	1.51	150	−1.11	−30
9	−6.74	−87	−21.0	129	2.54	140	−1.51	−43
10	−7.74	−96	−21.7	138	1.73	130	−1.72	−38
11	−7.96	−116	−21.7	138	0.75	125	−1.62	−45
12	−12.0	−107	−21.3	154	0.94	120	−1.31	−42

By applying (8.20) and (8.21), it follows that the transistor can be stabilized over the given frequency range by shunting its input terminals with 323Ω, or its output terminals with 1175Ω.

Because the factor α_s as given by (8.23) is

$$\alpha_s = 2.15$$

while

$$r = 1.42$$

the optimum solution as far as the minimum reduction in gain at 11 GHz, where the operating power gain is a minimum is concerned, is to shunt the input terminals of the transitor with 323Ω; that is, if pure resistances are to be used for the compensation.

If the transistor is stabilized with 323Ω, the maximum operating gain at 11 GHz decreases from 5.8 to 4.4. By carrying out the stabilization with a resistance of 253Ω in series with a short-circuited stub with $f_{\lambda/4} = 12$ GHz, the maximum gain at 11 GHz increases to 5.44, which is very close to the optimum.

8.4   CONSTANT GAIN CIRCLES

In order to design an amplifier, it is necessary to know the possible terminations required to achieve the desired performance. With the terminations known, the impedance-matching techniques considered in the previous chapters can be applied to complete the design.

There is usually more than one load or source impedance which will ensure the desired performance, and the locus of these impedances or the equivalent admittances are therefore of interest. Fortunately, it turns out that all the relevant loci are circles on both the linear admittance plane and the Smith chart. The loci relevant to the admittance plane will be derived here and the expressions relevant to the Smith chart are repeated in Appendix E.

8.4.1   Circles of Constant Mismatch

It was shown in sec. 6.4.3.1 that the locus of load admittances for which the transducer power gain G_T of a passive source with internal admittance $Y_s = G_s + jB_s$ terminated in a passive load $Y_L = G_L + jB_L$ will remain constant is a circle in the linear admittance plane with center:

$$G_0 + jB_0 = [2/G_T - 1] G_s - jB_s \tag{6.91}$$

and radius:

$$R_{Y0} = 2 [1/G_T^2 - 1/G_T]^{1/2} G_s \tag{6.92}$$

Similarly, the locus of constant transducer power gain is also a circle on the Smith chart. The parameters of these circles can be found in Appendix E.

These equations will prove to be useful again in this chapter.

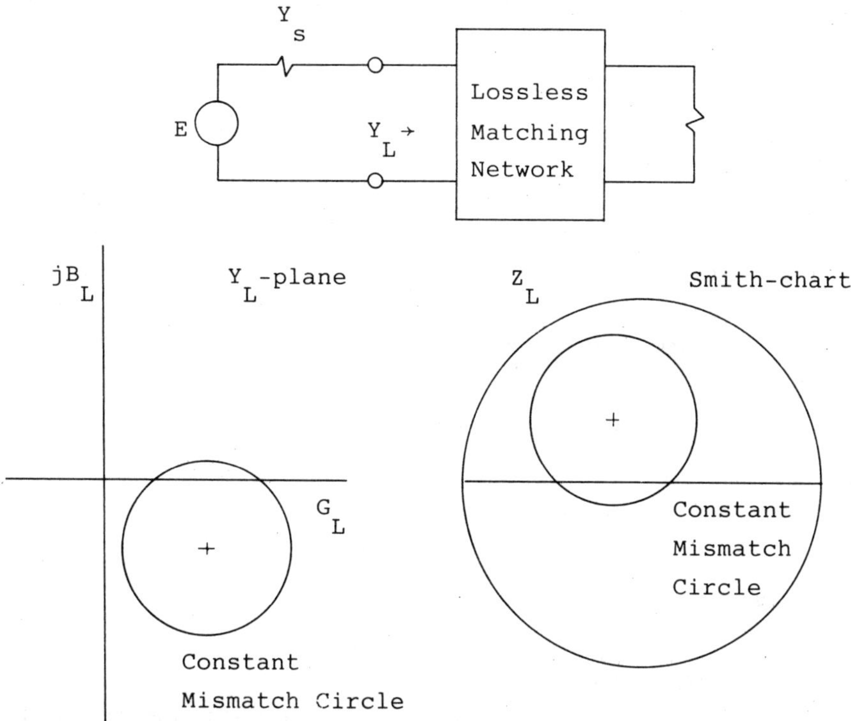

Figure 8.4 [a] The Equivalent Circuit Relevant to the Derivation of the Constant Mismatch Circles and an Example of these Loci on [b] the Admittance Plane and [c] a Smith Chart

8.4.2 Constant Operating Power Gain Circles

In designing small-signal amplifiers, expressions for the parameters of the locus of load admittances for which the operating power gain will remain constant are required. This locus turns out to be a circle in the linear admittance plane, as well as on the Smith chart. The parameters of a constant operating power gain circle on the admittance plane will be derived here, and those for the corresponding circle on a Smith chart [1, 2] are repeated in Appendix E.

The operating power gain of a transistor is given by

$$G_w = P_L / P_{IN} \tag{8.27}$$

$$= \left| \frac{y_{21}}{y_{22} + Y_L} \right|^2 \frac{G_L}{Re\left(y_{11} - \dfrac{y_{12}\, y_{21}}{y_{22} + Y_L}\right)} \tag{8.28}$$

where P_L is the power dissipated in the load, and P_{IN} is the power entering the input terminals of the amplifier.

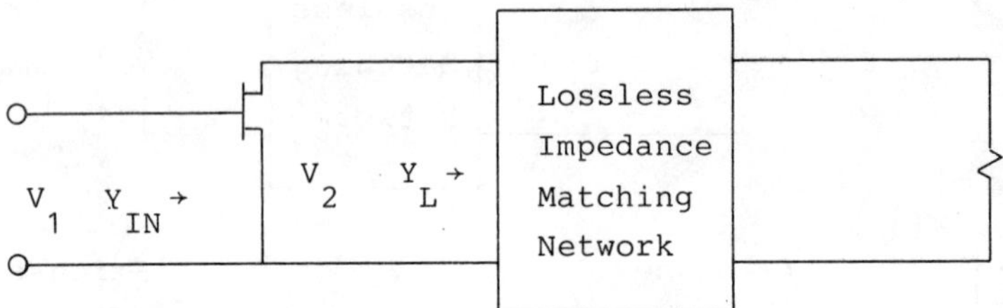

Figure 8.5   *The Circuit Relevant to Calculating the Operating Power Gain of an Amplifier*

With

$$y_{12}\, y_{21} = P + jQ \qquad (8.7)$$

as defined before, and

$$y_{22} + Y_L = [G_L + g_{22}] + j\,[B_L + b_{22}] = G_L' + jB_L' \qquad (8.29)$$

Eq. (8.28) becomes

$$G_w = \frac{|y_{21}|^2\, G_L}{\left| g_{11}\, G_L'^2 + g_{11}\, B_L'^2 - P\, G_L' - Q\, B_L' \right|} \qquad (8.30)$$

By multiplying both sides of this equation with the denominator of the right-hand side and dividing them by $g_{11}\, G_w$, the following equation is obtained:

$$\left[G_L'^2 - \frac{PG_L'}{g_{11}} + B_L'^2 - \frac{QB_L'}{g_{11}} \right] = \frac{|y_{21}|^2\,[G_L' - g_{22}]}{g_{11}\, G_w} \qquad (8.31)$$

This equation can be manipulated into the explicit form of the equation for a circle. The center of this circle is found to be

$$G_{Lw} + jB_{Lw} = \left[\left(\frac{P}{2g_{11}} - g_{22} \right) + \frac{\dfrac{|y_{21}|^2}{2g_{11}}}{G_w} \right] + j \left(\frac{Q}{2g_{11}} - b_{22} \right) \qquad (8.32)$$

and its radius (R_{Lw}) can be obtained from the equation:

$$R_{Lw}^2 = G_{Lw}^2 + \left| \frac{y_{12}\, y_{21}}{2g_{11}} \right|^2 [1 - 1/C^2] \qquad (8.33)$$

where C is the Linville stability factor.

When the transistor is inherently stable $[0 \leq C < 1]$, the operating power gain circles lie entirely in the right-hand side of the admittance plane. When the transistor is potentially unstable, these circles cross over into the left-hand side of the plane, as is illustrated in Fig. 8.6. Note that the gain circles cross the

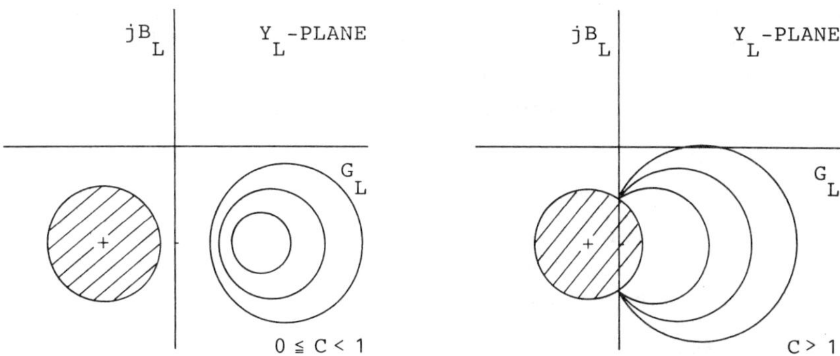

Figure 8.6   The Position of the Constant Operating Power Gain Circles Relative to the Imaginary Axis of the Admittance Plane, Illustrated for [a] an Inherently Stable Transistor and [b] a Transistor for which C>1

imaginary axis in the same two points, and that the gain circle corresponding to an infinite value for the gain is also the stability circle on the load admittance plane, as derived in sec. 8.2.

With the locus of load admittances for which the operating power gain of a device will be equal to any specified known value, an amplifier can be designed to have any realizable gain at any single frequency. This can be done by transforming the physical load of the amplifier (usually 50Ω) to any admittance on the relevant constant gain circle by using the impedance-matching techniques described previously.

The input impedance of an amplifier is usually conjugately matched to the source, and under this condition the transducer power gain will be equal to the operating power gain. In general, the transducer power gain is given by

$$G_T = [1 - |S_{IN}|^2] \, G_w \qquad (8.34)$$

where S_{IN} is the input reflection parameter of the amplifier with the actual source impedance as normalizing impedance.

The input admittance required to design the input matching network can be calculated by using the equations derived in Ch. 1 for this purpose ((1.6) or (1.83)).

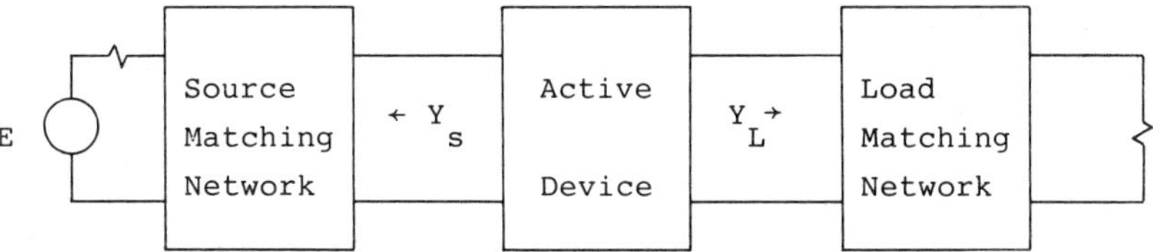

Figure 8.7   Configuration of a Single-Stage Amplifier

When a transistor is inherently stable, the maximum realizable power gain can be derived by determining the operating power gain corresponding to the gain circle with radius equal to zero, or by using (8.22). With R_{Lw} set equal to zero in (8.33), the maximum realizable power gain is found to be

$$G_{w-MAX} = \frac{|y_{21}|^2 / [2g_{11}]}{G_{Lw-OPT} - G_L^*} \tag{8.35}$$

where G_{Lw-OPT} is the real part of the load termination corresponding to the maximum realizable gain, and G_L^* is defined by (8.12). The load termination corresponding to the maximum realizable gain is given by

$$Y_{L-OPT} = G_{Lw-OPT} + jB_{Lw-OPT}$$

$$= \left|\frac{y_{12}\, y_{21}}{2g_{11}}\right| [1/C^2 - 1]^{1/2} + jB_L^* \tag{8.36}$$

with B_L^* as defined in (8.12).

When a transistor is potentially unstable, the maximum gain obtainable is theoretically equal to infinity, but the gain is limited by practical considerations such as stability, the ease by which the amplifier can be tuned, and sensitivity considerations to some finite value. The last two considerations also apply to inherently stable devices and the maximum realizable gain is often not easily realizable at radio frequencies.

8.4.3 Constant Available Power Gain Circles

The available power gain of an amplifier is given by

$$G_A = \frac{P_{AV-O}}{P_{AV-E}} \tag{8.37}$$

$$= \left|\frac{y_{21}}{y_{11} + Y_s}\right|^2 \frac{Re\,[Y_s]}{Re\,[Y_o]} \tag{8.38}$$

where P_{AV-O} is the maximum power available at the output of the amplifier, and P_{AV-E} the maximum power available from the source.

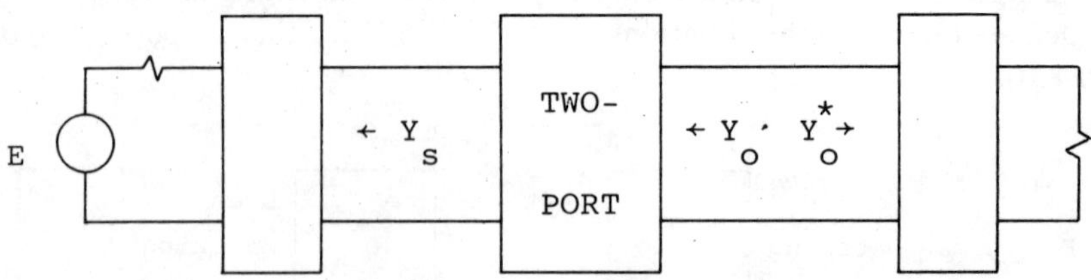

Figure 8.8   The Equivalent Circuit Relevant to Determining the Available Power Gain of an Amplifier

Comparison of (8.38) and the expression for the operating power gain of an amplifier (1.10), yields that if y_{11} is replaced with y_{22}, Y_s with Y_L, and Y_o with Y_{IN}, the two expressions are identical. Because y_{21}, y_{11} and y_{22} are constants, and the relationship between Y_s and Y_o is identical to that between Y_L and Y_{IN}, it is possible to determine the locus of source admittances for which the available power gain of a transistor will be equal to a specified value by using the results obtained for the operating power gain. By following this approach, the center of a constant available power gain circle is found to be located at

$$G_{SA} + jB_{SA} = \left[\left(\frac{P}{2g_{22}} - g_{11} \right) + \frac{\frac{|y_{21}|^2}{2g_{22}}}{G_A} \right] + j \left(\frac{Q}{2g_{22}} - b_{11} \right) \tag{8.39}$$

with its radius (R_{SA}) obtainable from

$$R_{SA}^2 = G_{SA}^2 - \left| \frac{y_{12}\,y_{21}}{2g_{22}} \right|^2 [1/C^2 - 1] \tag{8.40}$$

The parameters of an available power gain circle on a Smith chart [1] are repeated in Appendix E.

When the output match of a transistor is more important than the input match, it can be designed to have any realizable transducer power gain and a conjugately matched load at any single frequency by using the constant gain circle corresponding to the specified available power gain. With its output impedance matched to the load, the transducer power gain of an amplifier is equal to the available power gain. In the case of a mismatch at the output port, the following relationship can be used:

$$G_T = [1 - |S_O|^2]\, G_A \tag{8.41}$$

where S_O is the output reflection parameter of the amplifier with the load as normalizing impedance.

8.5 TUNABILITY

When a designed amplifier is realized, it is usually necessary to tune it slightly to obtain the exact results predicted. This is mainly because of component tolerances. These include tolerances in the passive and active components used.

When the influence of reverse transfer gain of a transistor is not negligible, its input impedance will inevitably be a function of the load termination and, if the load changes, whether because of tuning, temperature drift, or a change in load, the input impedance will also change. The consequent dependence of the input circuit match on the changes in the output circuit (and inversely) is undesirable, especially when an amplifier is tuned for optimum performance.

The tunability factor

$$\delta = \left| \frac{\partial Y_{IN} / Y_{IN}}{\partial Y_L / Y_L} \right| \qquad (8.42)$$

$$= \frac{|y_{12}\, y_{21}\, Y_L|}{|y_{22} + Y_L|^2\, |Y_{IN}|} \qquad (8.43)$$

$$= \left| \frac{y_{12}}{y_{21}} \right|\, G_w \sec \theta_L / \sec \theta_{IN} \qquad (8.44)$$

where

$$\theta_{IN} = \tan^{-1}\left[B_{IN} / G_{IN} \right] \qquad (8.45)$$

and

$$\theta_L = \tan^{-1}\left[B_L / G_L \right] \qquad (8.46)$$

is an excellent measure of the relative dependence of the input circuit match on changes in the output circuit.

It is obvious from (8.44) that the tunability factor δ is a strong function of the gain of an amplifier. If the gain is decreased enough, the output circuit will usually have very little influence on the input circuit and *vice versa*. A tunability factor of less than 0.3 is advisable [2].

The load admittance with the highest gain and $\delta = 0.3$ can be determined easily by using a computer program to calculate δ as a function of the angle around any constant operating power gain circle. Starting with a high value for the gain, it can be decreased progressively in small enough steps until the optimum load is found. If only the optimum load is of interest it can be found even faster by using the fact that the admittance with the lowest δ on any gain circle is almost without exception the one furthest from the imaginary axis. The admittances close to the imaginary axis are seldomly tunable.

When a design is carried out by using the available power gain circles, the optimum source is of interest. The tunability factor

$$\delta' = \frac{\partial Y_{OUT} / Y_{OUT}}{\partial Y_s / Y_s} \qquad (8.47)$$

$$= \frac{|y_{12}\, y_{21}\, Y_s|}{|y_{11} + Y_s|^2\, |Y_{OUT}|} \qquad (8.48)$$

can then be used.

8.6 UNILATERALNESS

When the reverse transfer gain of a transistor is not negligible, its output impedance is a function of the source termination, and its input impedance a function of the load. The output and input impedances of an amplifier, therefore, are not both known before the design is completed when the reverse transfer gain is not negligible.

When a narrow-frequency amplifier is designed, this dilemma is not important when the design is carried out around the operating power gain or the available power gain. The optimum load (or source) can be determined as described in the previous section and, with the load (source) known, the input (output) impedance can be calculated, and the input (output) matching network can be designed.

When a wideband amplifier is designed by following this approach, it is a problem to find a matching network which will transform the physical load to fall on the circumference of each gain circle of interest. The ideal situation would be if the problem could be reduced to an impedance-matching problem. When the influence of the reverse transfer gain of a transistor is negligible, its input and output admittances are always equal to y_{11} (reflection parameter s_{11}) and y_{22} (reflection parameter s_{22}), respectively, and the problem can be changed to an impedance matching problem easily. The reader who is interested in the details of this special case is referred to [1, 2].

It should be noted that most microwave transistors cannot be considered unilateral without introducing significant errors in the gain of the designed amplifier.

It will be shown in the next section that it is possible to transform the problem of finding a network which will transform a given load (source) to fall on the circumferences of a given set of operating (available) power gain circles to an impedance matching problem, without making any approximations whatsoever whenever the transistor used is inherently stable inside the pass band. In so doing the impedance-matching techniques outlined in Ch. 6 can be applied directly to the design of RF and microwave amplifiers.

When a transistor is not inherently stable, it can be compensated by using the techniques outlined above. This is, however, seldom necessary at microwave frequencies, where few transistors are potentially unstable in the pass band. Furthermore, when wideband GaAs amplifiers are designed, feedback is usually essential to decrease the gain-bandwidth constraints imposed by the impedances to be matched. The required feedback also usually stabilizes the transistor.

8.7 A TECHNIQUE FOR DESIGNING AMPLIFIERS WITH NON-UNILATERAL, INHERENTLY STABLE TRANSISTORS

It was shown in sec. 8.4.1 that the locus of load impedances for which the transducer power gain of a voltage or current source terminated in a passive load will remain constant is a circle in the admittance plane. These constant transducer power gain circles always lie in the right-hand side of the admittance plane. By considering an operating or available power gain circle to be the gain circle of a source terminated in a passive load, the problem of finding a network to transform a given load or source to fall on the circumferences of the relevant operating or available power gain circles can be transformed to that of matching a complex source to a complex load with a specified transducer power gain at each of the frequencies of interest whenever the transistor used is inherently stable inside the pass band.

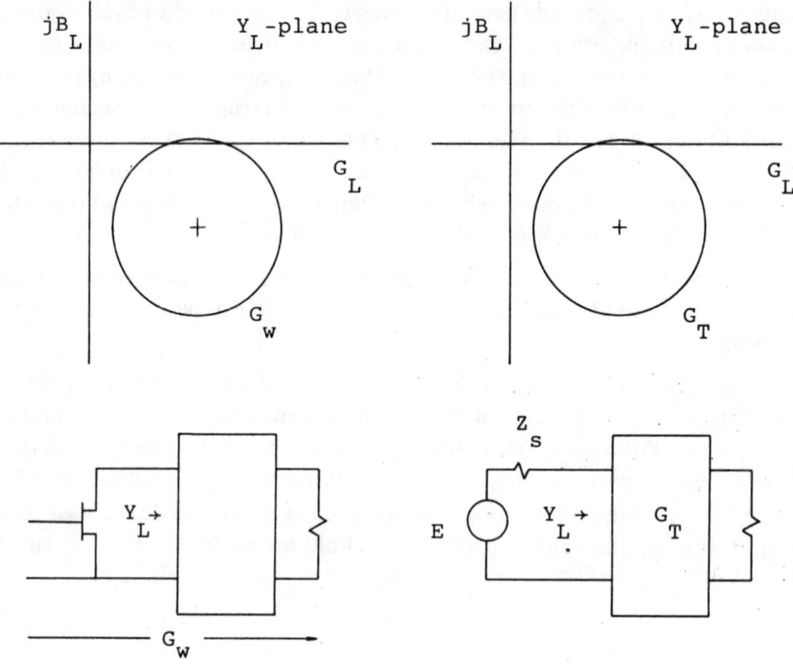

Figure 8.9   *Illustration of the Equivalence between a Constant Operating Power Gain Circle and the Circle Corresponding to Mismatching a Voltage Source with a Fixed Amount to a Passive Load*

The restriction of inherent stability is imposed by the fact the operating or available power gain circles will cross over into the left-hand side of the admittance plane at any frequency where the transistor is potentially unstable and an equivalent transducer power gain circle can then not be found.

The equations relevant to finding the output admittance and transducer power gain equivalent to a given operating power gain circle can be derived by setting

$$G_{Lw} + jB_{Lw} = G_0 + jB_0 \qquad (8.49)$$

(refer to (8.33) and (6.91)) and

$$R_{Lw} = R_{Y0} \qquad (8.50)$$

(refer to (8.33) and (6.92)).

The resulting set of equations is

$$G_{s\text{-}OUT} = R_{Lw}\, G_{T\text{-}OUT}/(2\,[1 - G_{T\text{-}OUT}]^{1/2}) \qquad (8.51)$$

$$B_{s\text{-}OUT} = -B_{Lw} \qquad (8.52)$$

$$G_{T\text{-}OUT} = 1 - \left(\frac{G_{Lw}}{R_{Lw}} - \sqrt{\frac{G_{Lw}^2}{R_{Lw}^2} - 1} \right)^2 \qquad (8.53)$$

where $G_{s\text{-}OUT} + jB_{s\text{-}OUT}$ is the equivalent output admittance of the transistor and $G_{T\text{-}OUT}$ the required transducer power gain.

Similarly, the equations necessary for transforming an available power gain circle to an equivalent load admittance (equivalent input admittance of the transistor) and transducer power gain are

$$G_{L\text{-}IN} = R_{SA}\, G_{T\text{-}IN}\, /\{2\,[1 - G_{T\text{-}IN}]^{1/2}\} \qquad (8.54)$$

$$B_{L\text{-}IN} = -B_{SA} \qquad (8.55)$$

$$G_{T\text{-}IN} = 1 - \left(\frac{G_{SA}}{R_{SA}} - \sqrt{\frac{G_{SA}^2}{R_{SA}^2} - 1} \right)^2 \qquad (8.56)$$

The S-parameter equivalents of (8.51) to (8.56) are given in Appendix E.

8.8 THE DYNAMIC RANGE OF AN AMPLIFIER

The range of acceptable input signals (dynamic range) for an amplifier is determined by its noise performance at low signal levels and linearity considerations at high signal levels. The evaluation and optimization of the noise performance of an amplifier will be discussed in section 8.8.1, while measures for evaluating the linearity of an amplifier will be considered in section 8.8.2.

8.8.1 Evaluation and Optimization of the Noise Performance of an Amplifier

The noise performance of an amplifier is evaluated by calculating its noise figure at the frequencies of interest. The noise figure of a device is defined by

$$F = P_{n\text{-}AO}\,/\,P_{n\text{-}AO\text{-}I} \qquad (8.57)$$

where $P_{n\text{-}AO}$ is the noise power available at the output of the device and $P_{n\text{-}AO\text{-}I}$ the noise power which would have been available if the device under consideration was replaced with an identical but noiseless equivalent. In the latter case, the only noise present at the output would be the amplified thermal noise of the relative source. The thermal noise power available from any purely resistive source is the same over the same bandwidth and at the same temperature, and is given by

$$P_{n-th} = kTB \tag{8.58}$$

where B is the bandwidth (ideal filter response assumed) is hertz, k is Boltzman's constant (1.37E-23 Joule/$^\circ$K at 290°K), and T the absolute temperature ($^\circ$K). The noise power available at the output of a noiseless equivalent of the device, therefore, is given by

$$P_{n-AO-I} = kTB \, G_A \tag{8.59}$$

where G_A is the available power gain of the device.

The gain response of a practical amplifier is seldom of the same form as that of an ideal filter. In order to calculate the noise power at the output of a practical amplifier, it is useful to define an equivalent noise bandwidth to be used in such cases. The equivalent noise bandwidth for any system is defined as the bandwidth of an ideal filter which will result in exactly the same noise power at the output and can be calculated by using the equation:

$$B_{ne} = \frac{1}{G_{A-max}} \int_0^\infty G_A(f) \, df \tag{8.60}$$

where G_{A-max} is the maximum value of the available power gain in the pass band which should be used with the effective bandwidth in (8.59). It should be noted that the equivalent noise bandwidth is only meaningful when the spectrum of the noise power density can be considered flat before filtering as long as the output power density is significant.

The equivalent noise bandwidth of a single-pole low-pass filter as well as that of a single-tuned filter, is equal to $(\pi/2)$ times its 3 dB bandwidth [3].

The bandwidth over which the available noise power is measured or calculated is usually taken to be very narrow. Because of the narrow bandwidth, the noise density can be considered constant as long as it is of interest and the following manipulations on (8.57) are possible:

$$
\begin{aligned}
F &= P_{n-AO}/P_{n-AO-I} \\
&= \int P'_{n-AI} \, G_A \, df \Big/ \left[\int P'_{n-AI-I} \, G_A \, df \right] \\
&= P'_{n-AI} \int G_A \, df \Big/ \left[P'_{n-AI-I} \int G_A \, df \right] \\
&= P'_{n-AI}/P'_{n-AI-I} \\
&= P'_{n-AI} \int G_T \, df \Big/ \left[P'_{n-AI-I} \int G_T \, df \right]
\end{aligned}
$$

$$= \int P'_{n-AI} \, G_T \, df \Big/ \left[\int P'_{n-AI-I} \, G_T \, df \right]$$

$$= P_{n-O} / P_{n-OI} \tag{8.61}$$

where P'_{n-AI} is the available noise power density at the output referred to the input of the amplifier, P'_{n-AI-I} the available noise power density at the output if the amplifier were replaced with a noiseless equivalent referred to the input, P_{n-O} the actual noise power dissipated in the load (not necessary the available noise power), and P_{n-OI} the actual noise power which would have been dissipated in the same load if the device was replaced with a noiseless equivalent.

It follows from (8.61) that is is not necessary to find the available power at the output of a device when measuring or calculating its noise figure. As long as the bandwidth is narrow enough to ensure that the noise power density will remain approximately constant as long as it matters, the actual power dissipated in any load can be measured or calculated.

It can be shown easily that as long as the noise figure is only applicable to a very narrow bandwidth (spot noise figure), (8.57) is also equivalent to

$$F = (S/N)_i / (S/N)_o \tag{8.62}$$

where $(S/N)_i$ is the ratio of the available signal-to-noise power at the input and $(S/N)_o$ the ratio of the available signal to noise power at the output terminals of the device. Apart from comparing the noise performance of a device with that of a noiseless equivalent, the noise figure is therefore also a measure of the degradation in the signal-to-noise ratio from the input to the output of a device.

The noise figure of a transistor is a function of the bias current, temperature, the frequency of interest, and the source impedance as viewed from the input terminals of the transistor. Optimizing the noise performance of a chosen transistor is therefore a matter of finding the optimum bias point, which is often specified by the manufacturer, and determining the optimum source impedance.

The noise figure of a transistor is related to its source admittance in the following way:

$$F = F_{min} + \frac{R_n}{G_s} \left[(G_s - G_{no})^2 + (B_s - B_{no})^2 \right] \tag{8.63}$$

where F_{min} is the lowest possible noise figure for the transistor, $G_{no} + jB_{no}$ the source admittance corresponding to this minimum noise figure and R_n a constant.

Because (8.63) is clearly the equation of a circle in the admittance plane, it follows that the locus of load admittances for which the noise figure of a transistor will remain constant is a circle in the admittance plane. The center of each constant noise figure circle is given by

$$G_{sF} + jB_{sF} = G_{no} + \frac{F - F_{min}}{2R_n} + jB_{no} \tag{8.64}$$

and its radius (R_{YF}) can be obtained from the equation:

$$R_{YF}^2 = G_{sF}^2 - G_{no}^2 \tag{8.65}$$

The expressions for the equivalent circle on a Smith chart is repeated in Appendix E.

With more than one device in cascade, the equivalent noise figure for the cascade is given by the following equation [3]:

$$F_T = F_1 + \frac{F_2^{-1}}{G_{A1}} + \frac{F_3^{-1}}{G_{A1}\,G_{A2}} + \dots \tag{8.66}$$

where F_i is the noise figure of the ith stage in the cascade (counted from input to output) and G_{Ai} its available power gain.

Inspection of (8.66) yields that if its gain is high enough, the noise performance of a cascade amplifier will mainly be determined by the first stage. When this is the case, the noise performance of the cascade can be optimized by optimizing the noise figure of the first stage.

When the available power gain of the first stage is not high enough where the source impedance for optimum noise performance is used, a trade-off between the noise performance of each stage and its available power gain becomes possible.

The optimum combination of available power gain and noise figure for each stage can be determined iteratively at any single frequency. Before a procedure for this purpose can be outlined it is necessary to find a way of determining the admittance with the lowest noise figure on a constant available power-gain circle.

Graphically, the optimum can be determined easily on a Smith chart [1] or on the admittance plane. In both cases, finding the optimum admittance on each circle is simply a matter of finding the noise circle tangent to the available power gain circle, as is illustrated for the admittance plane in Fig. 8.10.

The easiest and fastest way to find the optimum is to calculate the noise figure at different angles around the gain circle.

The optimum combination of available gain and noise figure can now be determined at any single frequency by following the procedure outlined below. The same procedure can be followed to determine the optimum specifications at each relevant frequency for a wideband amplifier. In the latter case, it may not be possible to realize these specifications to great accuracy because of

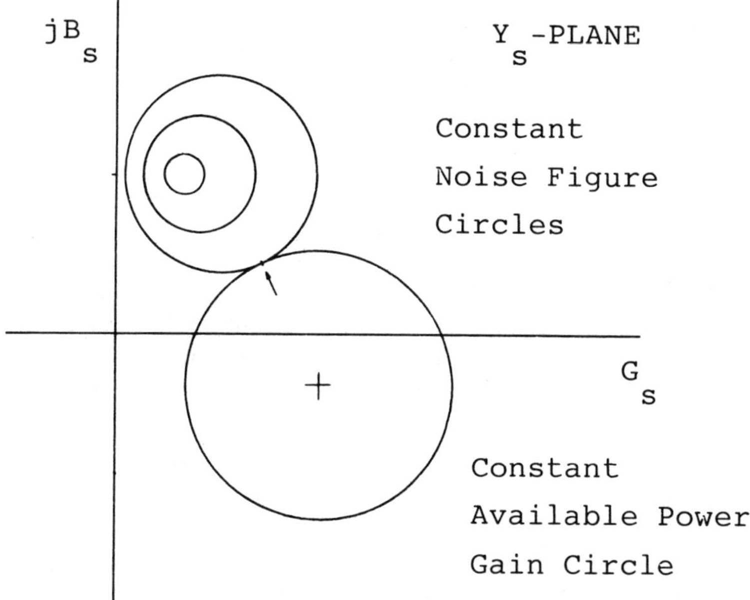

Figure 8.10   A Graphical Approach to Finding the Admittance on a Constant Available Power-Gain Circle with the Lowest Noise Figure

the gain-bandwidth constraints associated with the specifications.

Procedure for Determining the Optimum Combination of Available Power Gain and Noise Figure for a Multistage Amplifier

1)   Start with the available power gain equal to the value corresponding to the minimum noise figure for each stage, and calculate the noise figure corresponding to this combination of gains and noise figures.

2)   Calculate the maximum available power gain for each stage and the noise figure corresponding to it. Calculate the noise figure corresponding to this combination. Where any stage is not inherently stable, increase the gain corresponding to the minimum noise figure of that stage by a factor of 2, for example, and find the admittance with the smallest noise figure as described above, and use this noise figure and gain value to perform the calculations at this step.

3)   Decide on an incremental value for the available power gain (e.g., 1.059) and increase the gain of each stage progressively from its minimum noise value to its maximum value. Generate all possible combinations of available power gain with these different gain values for each stage and calculate the optimum noise figure corresponding to each.

4)   Select the optimum combination of the available power gains and the corresponding noise figures from the results in step 3.

5)   With the optimum source impedance for each stage known, the necessary impedance-matching networks can be designed. This should be done by starting with the first stage.

6)   The required transducer power gain can be obtained by tapering the response of the load matching network.

8.8.2   Evaluation of the Linearity of an Amplifier

Because of the nonlinearities in an amplifier, the output signal at any particular frequency will not increase linearly with the input signal when it becomes too large. In practical amplifiers, the output signal then becomes progressively smaller than expected.

The input signal for which the fundamental component of the output signal is compressed by 1 dB below its expected value (or equivalently, where the gain has decreased with 1 dB from its small signal value) is usually taken as the upper limit of the input signal for the purpose of calculating the dynamic range of the amplifier. This particular signal level is referred to as the 1 dB compression point.

This particular phenomenon is clearly illustrated by the output signal of a memoryless third-order system with input signal $A \sin \omega t$:

$$e_0(t) = a_1 e_i(t) + a_3 e_i^3(t) \tag{8.67}$$

$$= a_1 (A \sin \omega t) + a_3 (A \sin \omega t)^3$$

$$= \left[a_1 A + \left(\frac{3}{4}\right) a_3 A^3 \right] \sin \omega t + \left(\frac{1}{4}\right) a_3 A^3 \sin(3\omega t) \tag{8.68}$$

Because a_3 is usually negative, the fundamental component of the output signal will be smaller than its expected value ($a_1 A \sin \omega t$).

The amount of third-order distortion present in a system is often also evaluated by calculating a third-order intercept point for it. This point is defined as that level of the input signal for which the third harmonic at the output will be equal in amplitude to the uncompressed fundamental component. For a simple third-order system this would mean that value of A for which

$$a_1 A = \left(\frac{1}{4}\right) a_3 A^3 \tag{8.69}$$

Intercept points of other orders and types are also frequently used.

Apart from distorting the amplitude of the input signal and generating harmonics of its components, the nonlinearities in an amplifier has the further

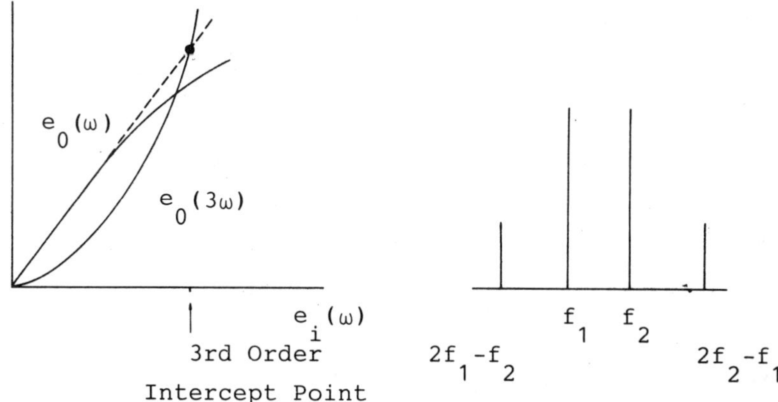

Figure 8.11 [a] *Illustration of the Definition of the Third-Order Intercept Point;* [b] *the In-Band Intermodulation Products of a Third-Order System*

disadvantage that new frequencies (intermodulation products) are also created by the interaction of the different components of the input signal when it consists of more than one tone.

The new frequencies are of the from

$$f_{IMD} = \pm\, m_1 f_1 + m_2 f_2 \pm \ldots \pm m_n f_n \tag{8.70}$$

where f_i is one of the components of the input signal, and m_i any integer greater or equal to zero within the constraint that

$$m_1 + m_2 + \ldots + m_n = p \tag{8.71}$$

where p is the order of the nonlinearity under consideration, and m_i any integer greater than or equal to zero.

These new frequencies are particularly troublesome when they fall inside the pass band and, therefore, cannot be removed from the amplified input signal by filtering.

As an illustration of the new frequency components generated by the nonlinearities, consider the influence of the third-order characteristics of a system on a signal consisting of two frequencies (f_1 and f_2) which are very close to each other. The new components generated are

$$f_1 + 2f_2,\ 2f_2 - f_1,\ 2f_1 + f_2 \text{ and } 2f_1 - f_2\ (m_1 + m_2 = 3)$$

Of these products, the $2f_2 - f_1$ and $2f_1 - f_2$ components will fall inside the pass band and cannot be removed at all.

The maximum level of the in-band intermodulation products compared to that of each component of a two-tone signal consisting of two frequencies, which are very close to each other (usually 1 kHz apart), on a spectrum analyzer are often used as a measure of the nonlinearities present in a power amplifier. Intermodulation distortion (IMD) of less than –40 dB are typical for a class A power amplifier, and –30 dB is typical for a class AB amplifier.

8.9 THE DESIGN OF MULTISTAGE AMPLIFIERS

At this stage, three general procedures can be outlined for the design of small signal amplifiers. All of these procedures are applicable to unilateral, as well as non-unilateral, transistors.

The first procedure outlined below is based on the operating power gain and can be used in most cases. The second procedure is based on the available power gain. The third procedure should be followed when an amplifier is designed to have a specified noise figure.

The operating power gain procedure proceeds from the load to the source and the available gain procedure from the source to the load. Excellent control over the transducer power-gain *versus* frequency-response is ensured by both procedures.

When a single-stage amplifier is designed by following the operating power gain procedure, the input impedance can be matched or mismatched to the source as required and the output matching network is used to taper the gain. When the available power gain procedure is followed, the output impedance can be matched or mismatched to the load and the input matching network is used to taper the gain as required.

When a multistage amplifier is designed, the required transducer power gain can be obtained, and both the input and output impedances can be matched. In order to do this when the transistors used are non-unilateral, the operating power gain of at least the last stage in the chain should be kept as high as possible. When the available power gain approach is followed, the gain of at least the first stage should be maximized.

Ideally, the tapering required for flattening the gain should be concentrated in the first interstage matching network (as counted from the source) when the operating power gain procedure is followed, and in the last interstage matching network when the design is carried out around the available power gain.

When the transistors used are highly non-unilateral, the first matching network designed must usually be re-designed in order to improve the output or input match as applicable.

8.9.1 A Procedure for Designing Small-Signal Amplifiers Based on the Operating Power Gain

1) Calculate the Linville stability factors for each of the transistors to be

used in the amplifier chain by using (8.17). If the transistors are not inherently stable, compensate them by using one of the techniques outlined in sec. 8.2. For the purpose of this procedure it is only necessary for $C < 1$ and a value of 0.99, for example, is adequate. If higher stability is required a lower stability factor can, of course, be chosen.

2)   Calculate the maximum operating gain available for each transistor at each of the frequencies of interest by using (8.22). Determine the maximum tunable ($\delta = 0.3$) operating power gain for each transistor as described in sec. 8.5. It is advisable to keep the operating power gain of each transistor below this value.

3)   Decide on the distribution of the gain between the different stages if more than one transistor is to be used. Find the lowest (tunable) gain in the pass band and use this gain as the design goal. The gain at the other frequencies should be chosen to ensure a flat response over the pass band, that is, of course, if a flat response is required.

When a multistage amplifier is designed, the operating power gain of the stages close to the load should be kept as high as possible (equal to the maximum available gain) if the output impedance of the amplifier should be matched to the load. It is usually only necessary to do this for the last stage in the chain, but when the transistors used are highly non-unilateral a better output match can be expected initially if the gain of the next stage is also maximized. From the viewpoint of tunability it is better to maximize only the gain of the last stage.

It should be noted that when the designed networks can be realized to good accuracy, whether by optimizing them individually or directly, tunability is often not a problem and the maximum realizable gain can often be obtained.

4)   Start with the load matching network and design the required matching networks by using the impedance-matching techniques outlined in Ch. 6. The source impedance and transducer power gain for each network can be obtained by using (8.32), (8.33), and (8.51) to (8.53), while the load impedance of each matching network, excluding the load matching network, is the input impedance of the transistor following it. The input admittance of the relevant transistor can be calculated by using (1.6):

$$Y_{IN} = y_{11} - \frac{y_{12}\, y_{21}}{y_{22} + Y_L} \tag{1.6}$$

As each matching network is designed, the operating power gain obtained can be compared to the specifications, and a corresponding adjustment in the operating power gain assigned to the transistor determining the source specifications of the next network to be designed can be made to compensate for the deviation in gain. By doing this, the error in transducer power gain will not accumulate and the ripple in the transducer power gain of the designed amplifier will be very small.

5)   Calculate the transducer power gain and the input and output impedances of the designed amplifier. If a multistage amplifier were designed and the output VSWR was not small enough, the output matching network can be re-designed with the actual output impedance as specification.

The stability of different designs can be compared by calculating the Sterne stability factor at each frequency of interest (8.18).

Excellent results can be obtained by following this procedure, as will be illustrated below.

8.9.2   A Procedure for Designing Small-Signal Amplifiers based on the Available Power Gain

1)   Calculate the Linville stability factors for each of the transistors to be used in the amplifier chain by using (8.17). If the transistors are not inherently stable, compensate them by using one of the techniques outlined in sec. 8.2. For the purpose of this procedure it is only necessary for $C<1$ and a value of 0.99, for example, is adequate for this purpose. If higher stability is required, a lower stability factor can, of course, be chosen.

2)   Calculate the maximum available power gain for each transistor at each frequency of interest by using the equation:

$$G_{A-MAX} = \frac{\dfrac{|y_{21}|^2}{2g_{22}}}{G_{sA-OPT} - \left(\dfrac{P}{2g_{22}} - g_{11}\right)} \tag{8.72}$$

where

$$Y_{sA-OPT} = G_{sA-OPT} + jB_{sA-OPT}$$

$$= \left|\frac{y_{21}\,y_{21}}{2g_{22}}\right| \sqrt{1/C^2 - 1} + j\left(\frac{Q}{2g_{22}} - g_{11}\right) \tag{8.73}$$

and Y_{sA-OPT} is the source admittance corresponding to the maximum available power gain.

Determine the maximum tunable ($\delta' = 0.3$) available power gain for each transistor as described in sec. 8.4. It is advisable to keep the available power gain of the transistor below this value.

3)   Decide on the distribution of the gain between the different stages if more than one transistor is to be used. The lowest (tunable) available gain in the pass band should be used as the design goal and the gain at the other frequen-

cies should be chosen to ensure a flat response over the pass band, that is, of course, if a flat response is required.

When a multistage amplifier is designed, the available power gain of the stages close to the source should be kept as high as possible (equal to the maximum available gain) if the input impedance of the amplifier should be matched to the source. It is usually only necessary to do this for the first stage in the chain, but when the transistors used are highly non-unilateral, a better input match can be expected initially if the gain of the next stage is also maximized. From the viewpoint of tunability it is better to maximize only the gain of the first stage.

It should be noted that when the designed networks can be realized to good accuracy whether by optimizing them individually or directly, tunability is often not a problem and the maximum realizable gain can often be obtained.

4)   Start with the source matching network and design the required matching networks by using the impedance matching techniques outlined in Ch. 6. The load impedance and transducer power gain for each network can be obtained by using (8.39), (8.40) and (8.54) to (8.56), while the source impedance of each matching network, excluding the source matching network, is the output impedance of the transistor driving it. The output admittance of the revelant transistor can be calculated by using (1.7).

$$Y_{OUT} = y_{22} - \frac{y_{12}\,y_{21}}{y_{11} + Y_s} \qquad (1.7)$$

As each matching network is designed, the available power gain obtained can be compared to the specifications, and a corresponding adjustment in the available power gain assigned to the transistor determing the load specifications of the next network to be designed can be made to compensate for that deviation in gain. By doing this, the error in transistor determining the load specifications of the next network to be designed can be made to compensate for the deviation in gain. By doing this, the error in transducer power gain will not accumulate and the ripple in the transducer power gain of the designed amplifier will be very small.

5)   Calculate the transducer power gain and the input and output impedances of the designed amplifier. If a multistage amplifier was designed and the input VSWR is not small enough, the input matching network can be re-designed with the actual input impedances as specification.

The stability of different designs can be compared by calculating the Sterne stability factor at each frequency of interest (8.18).

Excellent results can be obtained by following this procedure.

8.9.3   A Procedure for Designing a Wideband Amplifier for a Specified Noise-Figure and Transducer Power Gain

In this design procedure the first stage of the amplifier is designed to have a specified noise figure and the subsequent stages for a specified noise figure or maximum (tunable) available gain, depending on whether the noise figure or maximum available gain, depending on whether the noise figure resulting from doing the latter is acceptable. Alternatively, the design procedure outlined for narrowband low-noise amplifiers in sec. 8.8.1 can be followed to determine the optimum specifications for each stage (tunability ignored).

The gain tapering necessary to flatten the gain is concentrated in the last matching network. If a low output VSWR is required, the gain tapering must be concentrated in the last interstage matching network.

1)   Calculate the parameters of the noise circle corresponding to the noise figure ($G_{sF} + jB_{sF}$; R_{YF}) of the first stage at each relevant frequency by using (8.64) and (8.65), and find the load admittance ($G_L + jB_L$) and transducer power gain (G_{TN}) corresponding to each circle by using the following set of equations:

$$G_L = R_{YF} \, G_{TN}/[2 \,(1 - G_{TN})^{1/2}] \tag{8.74}$$

$$B_L = -B_{sF} \tag{8.75}$$

$$G_{TN} = 1 - \left(\frac{G_{sF}}{R_{YF}} - \sqrt{\frac{G_{sF}^2}{R_{YF}^2} - 1} \right)^2 \tag{8.76}$$

Match the load determined in this way to the source of the amplifier.

It should be kept in mind that the available power gain is not the same at different positions around a finite constant noise-figure circle. In narrowband designs, the source can be transformed to that admittance on the circle with the highest tunable available power gain.

2)   Calculate the output admittance of the first stage by using (1.7):

$$Y_{OUT} = y_{22} - \frac{y_{12}\, y_{21}}{y_{11} + Y_s} \tag{1.7}$$

and calculate the available power gain by using the equation:

$$G_A = \left| \frac{y_{21}}{y_{11} + Y_s} \right|^2 \frac{Re\,[Y_s]}{Re\,[Y_{OUT}]} \tag{8.77}$$

where Y_s is the source admittance as seen from the input terminals of the transistor and Y_{OUT} is given by (1.7).

3)   If only a single-stage amplifier or the load matching network of a multistage amplifier (output mismatched) is designed, calculate the transducer

power gain required for the design of the last impedance-matching network by using the equation:

$$G_T = G_{A\text{-}MIN} / G_A \tag{8.78}$$

where $G_{A\text{-}MIN}$ is the lowest available power gain in the pass band. The required source admittance is the output admittance of the (last) transistor and can be calculated by using (1.7).

If an interstage matching network is designed, calculate the noise figure corresponding to the maximum available power gain of the second stage by using (8.72), (8.73), and (8.63). If the maximum tunable gain is used, find the noise-figure corresponding to this gain by calculating the noise-figure at different positions around the gain circle and use this value (pessimistic) in (8.66). In narrowband designs, the source can be transformed to the (tunable) admittance on the gain circle with the lowest noise figure.

If a low output VSWR is required, the last interstage matching-network must be designed to provide the required gain tapering. The required available power gain specification can be obtained by taking this gain to be equal to its maximum possible value at the frequency where the overall gain is the lowest.

Calculate the expected noise figure for the first two stages if the noise figure of the second stage is equal to that calculated by using the cascade noise figure:

$$F_T = F_1 + \frac{F_2 - 1}{G_{A1}} + \ldots \tag{8.66}$$

If the results are acceptable, design the second stage for maximum available power gain. This can be done by using the calculated output admittance of the first stage as source admittance. The load admittance and required transducer power gain can be obtained from equations (8.39), (8.40), and (8.54) to (8.56), or in the special case where the available power gain is equal to its maximum value by setting the transducer power gain equal to one and using the following admittance as load admittance:

$$G_L + jB_L = \left| \frac{y_{12}\, y_{21}}{2g_{22}} \right| \sqrt{1/C^2 - 1} \; - j \left(\frac{Q}{2g_{22}} - g_{11} \right) \tag{8.79}$$

If the noise figure with the gain set equal to its maximum (tunable) value is not acceptable, design the second stage in the same way as the first stage (that is, for minimum noise figure).

Proceed in this way until the design is completed.

4)   Calculate the transducer power gain, noise figure, and Sterne stability factor for the designed amplifier.

Example 8.2

As an example of the design of a single-stage amplifier, the operating power

Table 8.2

The Specification for the Output Matching Network of the Plessey COD Device

Frequency (GHz)	Source Impedance (Ω)	Load Impedance (Ω)	Transducer Power Gain
8	$4.094 - j196.8$	$50.00 + j0.00$	0.0870
9	$8.240 - j136.6$	$50.00 + j0.00$	0.2669
10	$31.010 - j149.8$	$50.00 + j0.00$	0.8152
11	$23.050 - j125.3$	$50.00 + j0.00$	0.9906
12	$16.610 - j133.8$	$50.00 + j0.00$	0.7422

The specifications for the input matching network are shown in Table 8.3 and the designed input matching network is shown in Fig. 8.12.

Table 8.3

The Specifications for the Input Matching Network of the Plessey COD Device

Frequency (GHz)	Source Impedance (Ω)	Load Impedance (Ω)	Transducer Power Gain
8	$50.0 + j0.00$	$47.65 - j66.47$	1.0000
9	$50.0 + j0.00$	$36.26 - j45.66$	1.0000
10	$50.0 + j0.00$	$38.78 - j42.34$	1.0000
11	$50.0 + j0.00$	$26.14 - j36.25$	1.0000
12	$50.0 + j0.00$	$32.22 - j43.57$	1.0000

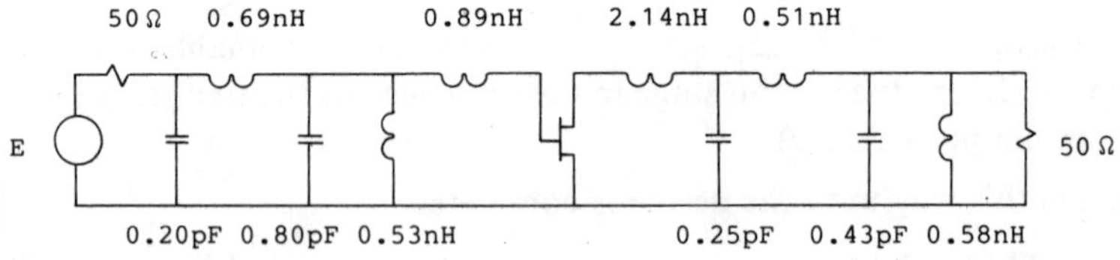

Figure 8.12   The Amplifier Designed in Example 8.2 [stabilized Plessey COD device; 7.22 ± 0.35 dB; Input VSWR ≤ 1.3]

gain procedure outlined above will be followed to design an 8-12 GHz amplifier with the highly non-unilateral Plessey transistor stabilized in *Example 8.1*. The design goal will be to achieve a flat gain of 5.4 (7.3 dB) across the pass band and a low input VSWR, and lumped element matching networks will be used. The optimum networks as determined by using the transformation Q technique outlined in Ch. 6 will be used. The realizability of the networks will not be considered in this example.

The specifications obtained for the output matching network are shown in Table 8.2. The designed output matching network is shown in Fig. 8.12. The resulting operating power gain is 7.245 ± 0.225 dB. If less ripple is required, the operating power gain can be decreased ($G_w = 6.945 \pm 0.115$ dB) or the number of matching elements can be increased.

The transducer power of the designed amplifier is 7.22 ± 0.35 dB and the input VSWR is smaller than 1.3.

Example 8.3

As an example of the design of a multistage amplifier, a distributed two-stage amplifier will be designed over the pass band 2-6 GHz by designing a lumped-element network and using the Π-section transformation described in Ch. 7. In order to use this technique, the impedance-matching networks designed will be constrained to contain low-pass Π-sections whenever possible. The S-parameters of the GaAs Dexcel 1503A transistor used are repeated in Table 8.4.

Because the gain-bandwidth constraints resulting from the input and output impedances of the transistor are too severe, it was decided to use voltage-shunt feedback in order to reduce these constraints. The values of the feedback components were determined iteratively as described in sec. 8.3. More feedback was used on the transistor of the first stage because a low input VSWR is

Table 8.4

The S-Parameters of the Dexcel 1503A Chip GaAs Transistor

Frequency (GHz)	s_{11} (dB; °)		s_{12} (dB; °)		s_{21} (dB; °)		s_{22} (dB; °)	
2	−0.265	−22	−30.5	78	9.99	159	−2.270	−10
3	−0.630	−31	−28.0	76	9.48	150	−2.384	−13
4	−1.012	−42	−24.4	69	9.40	143	−2.734	−16
5	−1.412	−53	−23.1	66	9.48	134	−2.975	−19
6	−1.938	−68	−21.9	56	9.25	122	−4.013	−22

Table 8.5

**The Specifications for the Initial Output Matching Network of the
Two-Stage Amplifier Designed**

Frequency (GHz)	Source Impedance (Ω)	Load Impedance (Ω)	Transducer Power Gain
2	$86.98 - j22.13$	$50.0 + j0.00$	1.000
3	$95.97 - j28.85$	$50.0 + j0.00$	1.000
4	$88.90 - j44.97$	$50.0 + j0.00$	1.000
5	$88.33 - j52.29$	$50.0 + j0.00$	1.000
6	$79.85 - j48.70$	$50.0 + j0.00$	1.000

Table 8.6

**The Specifications of the Interstage Matching Network of the Two-Stage
Amplifier Designed**

Frequency (GHz)	Source Impedance (Ω)	Load Impedance (Ω)	Transducer Power Gain
2	$75.08 + j0.84$	$83.16 - j135.9$	0.7462
3	$81.22 + j2.98$	$53.02 - j102.9$	0.8874
4	$81.94 - j1.52$	$35.56 - j77.55$	0.8802
5	$85.15 - j1.40$	$39.93 - j68.64$	1.0000
6	$81.44 - j1.19$	$22.69 - j46.11$	0.8605

required and the constraints associated with the input impedances of the FET
are more severe than those associated with its output impedance (this is usually the case). Although the gain can theoretically be flattened very effectively,
the feedback should not be removed at the higher frequencies in the pass band
because the required inductance is too large to be realized with negligible
phase shift across it. The feedback components are shown in Fig. 8.13a.

The specifications of the output matching network are shown in Table 8.4.
Because a good output match is required, the operating power gain was
chosen to be as high as possible. The minimum gain of the five-element output
matching network designed is 0.955 and the deviation from the desired response is, therefore, very small.

The specifications of the interstage matching network are shown in Table 8.6.
The designed network is shown in Figure 8.13a. The maximum deviation from
the specified gain response was 0.25 dB.

The specifications of the input matching network are shown in Table 8.7. The designed network is shown in Fig. 8.13a. The calculated transducer power gain of the amplifier is 18.65 ± 0.35 dB, and the input and output VSWRs are smaller than 1.81 and 2.24, respectively. Because the output VSWR is too

Table 8.7

The Specifications for the Input Matching Network of the Two-Stage Amplifier Designed

Frequency (GHz)	Source Impedance (Ω)	Load Impedance (Ω)	Transducer Power Gain
2	$49.95 - j1.57$	$80.13 - j13.83$	0.9383
3	$49.89 - j2.35$	$139.00 - j21.11$	0.9685
4	$49.80 - j3.13$	$102.50 - j79.36$	0.9672
5	$46.69 - j3.90$	$68.13 - j64.62$	1.0000
6	$49.56 - j4.67$	$41.80 - j37.01$	0.9783

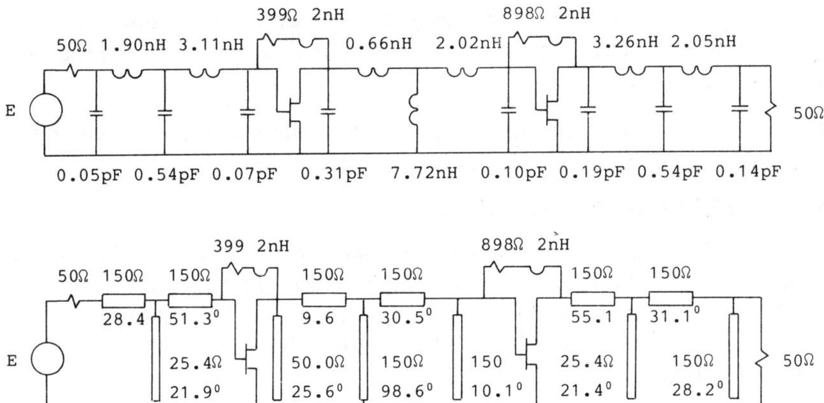

Figure 8.13 [a] *The Lumped Element Two-Stage Amplifier Designed in Example 8.3* [$G_T = 18.60 \pm 0.34$ dB; input VSWR ≤ 1.69, output VSWR ≤ 1.72] *and* [b] *a Distributed Equivalent* [$G_T = 18.65 \pm 0.19$ dB; input VSWR ≤ 1.81; output VSWR ≤ 1.86]

high, the specifications in Table 8.7 were used to re-design the output matching network. The source impedance shown is the actual output impedance of the designed two-stage amplifier. The designed output matching network is

Table 8.8

The Specifications for the Final Output Matching Network of the Two-Stage Amplifier Designed

Frequency	Source Impedance	Load Impedance	Transducer Power Gain
(GHz)	(Ω)	(Ω)	
2	$122.6 - j42.09$	$50.0 + j0.00$	0.944
3	$131.6 - j36.89$	$50.0 + j0.00$	1.000
4	$120.4 - j35.73$	$50.0 + j0.00$	1.000
5	$117.0 - j34.32$	$50.0 + j0.00$	1.000
6	$93.06 - j16.88$	$50.0 + j0.00$	1.000

shown in Fig. 8.13a. The transducer power gain of the final amplifier is 18.60 $\pm$ 0.34 dB, the input VSWR is smaller than 1.69 and the output VSWR is smaller than 1.72.

At this stage the required distributed equivalent can be determined by using the techniques outlined in Ch. 7. The resulting network is shown in Fig. 8.13b. An equivalent for the interstage matching network was determined by ignoring the high inductance shunt inductor. The corresponding transducer power gain is 18.60 $\pm$ 0.34 dB, and its input and output VSWRs are smaller than 1.81 and 1.86, respectively.

The high impedance capacitance stubs in the designed amplifier can be neglected without significantly degrading the performance.

8.10 REFLECTION AMPLIFIERS

At the higher gigahertz frequencies, Impatt, Gunn, and tunnel diodes are usually used to provide the necessary amplification. These negative resistance, single-port devices are usually used in combination with circulators and occasionally with 3 dB hybrid couplers [1, p. 279]. Only the circulator-type will be considered here.

The S-parameter matrix of an ideal circulator is given by

$$\overline{S} = \begin{bmatrix} 0 & 0 & 1 \\ 1 & 0 & 0 \\ 0 & 1 & 0 \end{bmatrix} \tag{8.80}$$

This implies that

$$\begin{bmatrix} b_1 \\ b_2 \\ b_3 \end{bmatrix} = \begin{bmatrix} a_3 \\ a_1 \\ a_2 \end{bmatrix} \tag{8.81}$$

and, therefore, the energy incident at port 1 is always delivered to the load

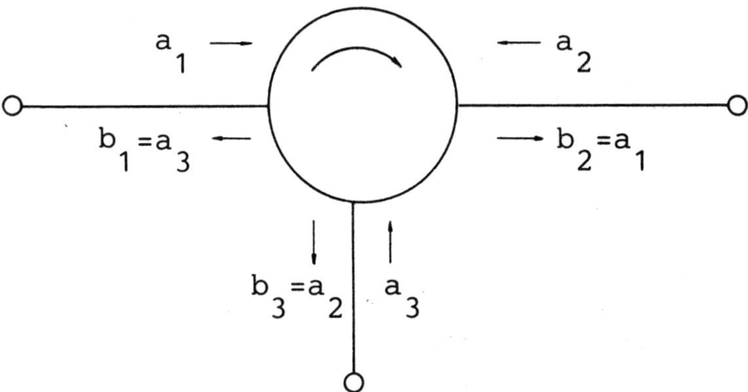

Figure 8.14 The Relationships between the Normalized Incident and Reflected Components of an Ideal Circulator

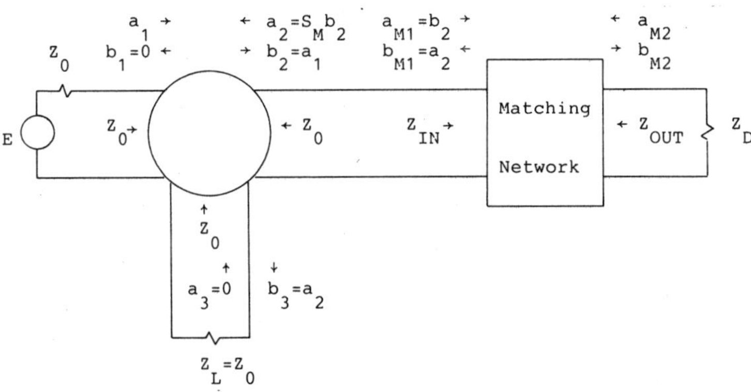

Figure 8.15 The Configuration of a Circulator-Type Reflection Amplifier

connected to port 2, the energy incident at port 2 to the load connected to port 3, and the energy incident at port 3 to the load connected to port 1. Consequently, the energy is propagated in a circular fashion around the circulator; hence the name circulator. These relationships are illustrated in Fig. 8.14.

The configuration of a circulator-type reflection amplifier is shown in Fig. 8.15.

The transducer power gain of the amplifier is defined by

$$G_T = \frac{P_L}{P_{AV\text{-}E}} \qquad (8.82)$$

where P_{AV-E} is the power available from the source. By using the relationships shown in Fig. 8.15 it follows that

$$G_T = \frac{|b_3|^2}{|a_1|^2}$$

$$= \frac{|a_2|^2}{|b_2|^2}$$

$$= \frac{|b_{M1}|^2}{|a_{M1}|^2}$$

$$= \frac{|b_{M2}|^2}{|a_{M2}|^2}$$

$$= \left| \frac{Z_{OUT} - Z_D^*}{Z_{OUT} + Z_D} \right|^2 \qquad\qquad (8.83)$$

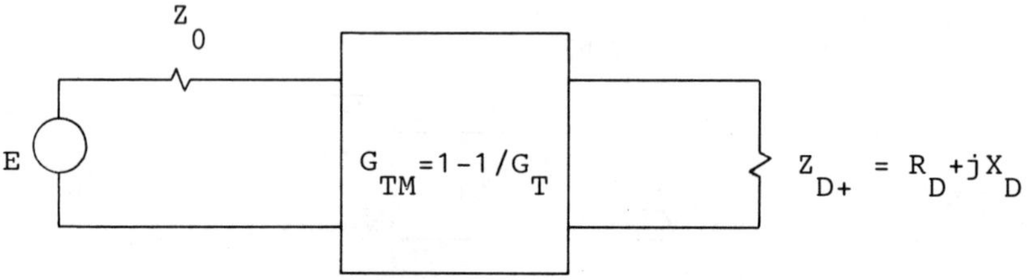

Figure 8.16   The Matching Problem to be Solved when the Amplifier in Fig. 8.15 is Designed

where $Z_D = -R_D + jX_D$ is the impedance of the negative resistance diode and Z_D^* indicates its conjugate.

Equation (8.83) can be manipulated in the following way:

$$G_T = \left| \frac{Z_{OUT} - [-R_D - jX_D]^*}{Z_{OUT} + [-R_D + jX_D]} \right|^2$$

$$= \left| \frac{Z_{OUT} + [R_D + jX_D]}{Z_{OUT} - [R_D + jX_D]^*} \right|^2$$

$$= 1 \bigg/ \left| \frac{Z_{OUT} - [R_D + jX_D]^*}{Z_{OUT} + [R_D + jX_D]} \right|^2$$

$$= 1 \big/ |S_{D+}|^2 \qquad\qquad (8.84)$$

where S_{D+} is the reflection parameter of the network shown in Fig. 8.16 with the source and load impedances shown as normalizing impedances.

The problem of maximizing the gain of a circulator-type reflection amplifier is, therefore, equivalent to that of matching the load in Fig. 8.16 as well as possible to the source shown.

When the amplifier is designed to have a specified gain *versus* frequency-response, the gain of the equivalent matching network should be

$$G_{TM} = 1 - 1/G_T \qquad (8.85)$$

where G_T is the transducer power gain specification for the reflection amplifier.

Table 8.9

The Specifications for the Output Matching Network of the Power Amplifier of Example 8.5

Frequency	Source Impedance	Load Impedance	Transducer Power Gain
(GHz)	(Ω)	(Ω)	
7.0	$50.0 + j0.00$	$10 + j3$	0.900
7.5	$50.0 + j0.00$	$12 + j7$	0.900
8.0	$50.0 + j0.00$	$15 + j10$	0.900
8.5	$50.0 + j0.00$	$19 + j13$	0.900
9.0	$50.0 + j0.00$	$25 + j15$	0.900

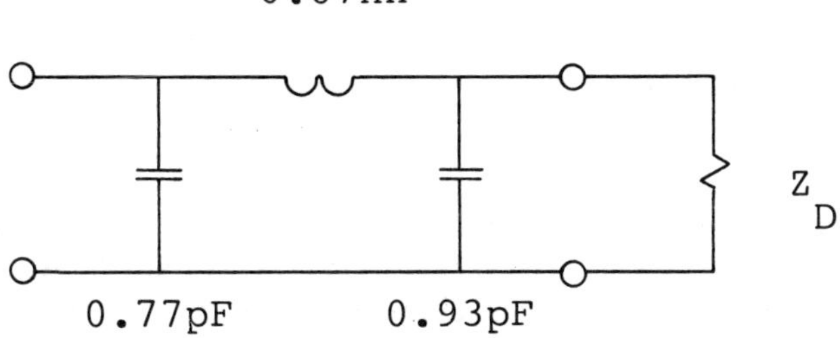

Figure 8.17   The Designed Matching Network for the Reflection Amplifier of Example 8.4

Example 8.4

As an example of designing the matching network of a reflection amplifier, a matching network will be designed for a Gunn diode (M/A-COM, MA-49110) with input impedance corresponding to the load impedance of the corresponding equivalent matching problem as given in Table 8.9 and a gain of 10 dB across the pass band 7-9 GHz.

With the required transducer power gain equal to 10 dB, the transducer power gain of the equivalent matching problem is found to be

$$G_{TM} = 1 - 1/G_T \tag{8.85}$$

$$= 1 - 1/10.0$$

$$= 0.90$$

The specifications of the equivalent matching problem is shown in Table 8.9.

The designed matching network is shown in Fig. 8.17. The maximum deviation from the specified gian response is 0.16 dB and the transformation Q-factors corresponding to the solution are 1.183, 1.506, and 0.511, respectively.

8.11   BALANCED AMPLIFIERS

In a balanced amplifier, the input signal is split to two or more amplifiers and the output signal of these amplifiers are combined to a single load, with isolation between the individual amplifier ports in both cases. The most commonly used configuration is shown in Fig. 8.18.

The S-parameter matrix of a 3 dB 90° hybrid divider is given by [7]:

$$\overline{S}_{Hd} = 0.707 \begin{bmatrix} 0 & j & 1 \\ j & 0 & 0 \\ 1 & 0 & 0 \end{bmatrix} \tag{8.86}$$

and that for a 3 dB 90° hybrid combiner by

$$\overline{S}_{Hd} = 0.707 \begin{bmatrix} 0 & 0 & j \\ 0 & 0 & 1 \\ j & 1 & 0 \end{bmatrix} \tag{8.87}$$

with the ports numbered as in Fig. 8.18.

For the divider, the energy incident at port 1 is, therefore, delivered to the loads connected to ports 2 and 3 with a 90° phase shift between the two components, while the energy incident at ports 2 and 3 in the combiner is routed to port 1, again with a 90° phase shift between the two components.

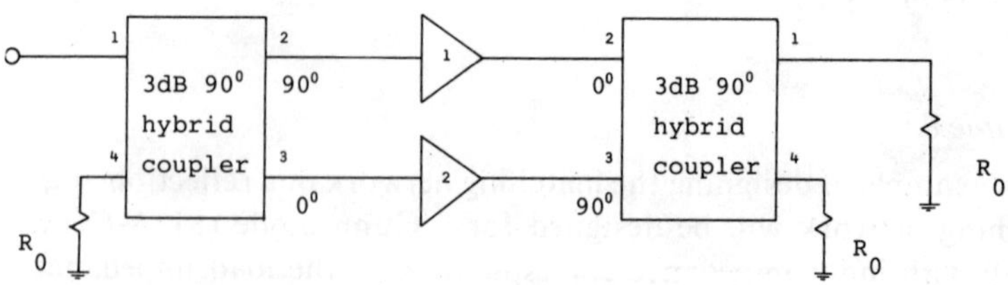

Figure 8.18   The Most Commonly Used Balanced Amplifier Configuration

The S-parameter matrix of the amplifier is given in terms of the S-parameters of the two individual amplifiers by [7]:

$$\overline{S}_T = 0.5 \begin{vmatrix} s_{11,1} - s_{11,2} & j\,(s_{12,1} + s_{12,2}) \\ j\,(s_{21,1} + s_{21,2}) & -s_{22,1} + s_{22,2} \end{vmatrix}$$

It is clear from this equation that if amplifiers 1 and 2 are identical, the input and output reflection parameters of the balanced amplifier will be equal to zero, even when the reflection parameters of the individual amplifiers are not equal to zero. As long as the individual amplifiers are almost identical, the input and output VSWRs of a balanced amplifier will, therefore, be very low, independent of the VSWRs of the individual amplifiers.

The transducer power gain of the balanced amplifier is given by

$$G_T = 0.25 \,|s_{21,1} + s_{21,2}|^2 \tag{8.89}$$

When the individual amplifiers are identical, this reduces to

$$G_T = |s_{21,1}|^2 \tag{8.90}$$

which is identical to the gain of a single amplifier.

Although the gain of the balanced amplifier is, therefore, identical to that of each individual amplifier in the ideal case, the output power is twice that obtainable by using only a single-ended stage.

Should one of the amplifiers comprising the balanced amplifier fail, the gain will be reduced to one-fourth of its original value. This can be proved easily by setting $s_{21,1}$ in (8.88) equal to zero. In some applications this advantage can be an important factor when it is decided whether a balanced or single-ended amplifier should be used.

8.12 CONSIDERATIONS APPLYING TO POWER AMPLIFIERS

The design of RF and microwave power amplifiers differ from the design of small-signal amplifiers in the design of the output circuit. Where the output circuit in small-signal amplifiers is either conjugately matched to the load or used for tapering the gain response, the load impedance of a power amplifier must be chosen in such a way that the required power can be obtained and the efficiency is as high as possible.

The output power obtainable from an amplifier is limited by the limitations of the device used or the output circuit designed. The device limitations stem from the finite voltage, current and power ratings of the device (usually a transistor), and its saturation voltage or saturation resistance. The saturation voltage or resistance of a device determines the lowest peak of the voltage across the device. Saturation voltages of a few volts for bipolar transistors, and saturations resistances of fractions of an ohm up to a few ohms are typical for FETs. At the lower frequencies, the maximum output power corresponding to a load R_L, with the influence of the output susceptance of the device removed by the output matching network when necessary, is given in terms of the supply voltage V_{ss}, the saturation voltage (V_{sat}), and saturation resistance (R_{sat}) by the following equation:

$$P_L = \frac{[V_{ss} - V_{sat}]^2}{[R_L + \alpha R_{sat}]} \frac{R_L}{R_L + \alpha R_{sat}} \tag{8.91}$$

where α is equal to two for class A amplifiers, and equal to one for class B amplifiers. The saturation resistance of bipolar transistors are usually negligibly small, while the saturation voltage for FETs can be neglected.

In order to obtain the maximum possible output power from a device it is, therefore, necessary to use the highest supply voltage possible, and to choose the load resistance as small as possible. (The saturation resistance is usually significantly smaller than the resistance required.) The minimum value of the

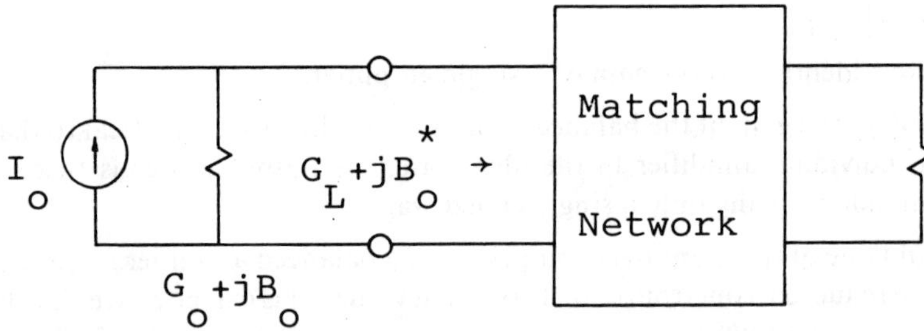

Figure 8.19   The Output Circuit of a High Efficiency Power Amplifier (G_o is usually negligibly small)

load resistance is determined by the maximum dc and RF currents which can be tolerated through the device. The optimum load for a power device is often specified by the manufacturer. Where it is not done, the optimum terminations of the device or terminations resulting in the same performance can be determined practically at each relevant frequency by using stub tuners.

The efficiency of a power amplifier is a function of the class of operation and the effective shunt susceptance in the output circuit (the device susceptance included). With the voltage across the transistor output terminals of sinusoidal form, the efficiency will always be smaller than 50% for class A amplifiers (the conduction angle of the current through the transistor is then 360°), while that for class B amplifiers (180° conduction angle) is constrained to below 78.5%. Higher efficiencies can be obtained with class C amplifiers, but because the same power must be concentrated in a narrower pulse, the peak current though the transistor increases as the efficiency increases. The device specifications for a class C amplifier are, therefore, more severe than those for class A or B amplifiers of the same output power with the same supply voltage. A class C amplifier, of course, cannot be used directly for linear applications.

When the effective load (transistor output admittance considered as part of the load) of a power amplifier is reactive, the efficiency decrease by a factor

$$\eta_r = 1 \ / \ \sqrt{1 + [B_L/G_L]^2} \tag{8.92}$$

because of the increase in the supply current caused by the shunt susceptance. Therefore, in optimizing the efficiency of an amplifier it is essential to remove the influence of the output susceptance of a device. This is often not necessary at the lower RF frequencies where the output susceptance of most RF devices are relatively small.

In order to achieve the required power, the physical load of a power amplifier (usually 50Ω) must be transformed to a lower value. It was shown in Ch. 6 that this can often not be done with LC networks at radio frequencies, and therefore transmission-line transformers are used for this purpose in wideband

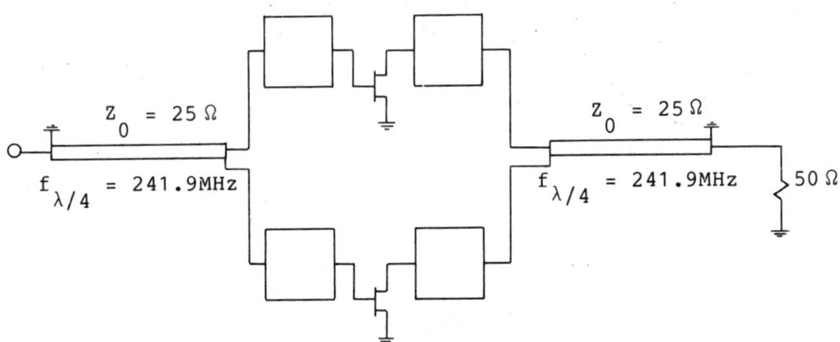

Figure 8.20   The Configuration of the Power Amplifier of Example 8.5

amplifiers. These transformers are also used for various combining and splitting functions. Combiners and splitters are required to connect several devices in parallel or in a balanced or push-pull configuration for higher output power. The cancellation of the output susceptance of a power device is carried out with an LC network between these transformers and the device.

The gain tapering required in a power amplifier is usually done at the input of one of the low-power drivers of the amplifier.

Example 8.5

As an example of the design of a radio-frequency power amplifier, an output matching network will be designed for the balanced amplifier shown in Fig.

8.20 over the pass band 225-260 MHz. The network will be designed for an output power of 165W. The supply voltage will be taken as 28V, and the output capacitance of each transistor as 130pF.

An approximate value for the required load impedance for each transistor can be obtained from (8.91). The saturation voltage will be taken as 3V and the saturation resistance is assumed to be negligible. Application of (8.91) yields that

$$165/2 = P_L = \frac{[28-3]^2}{2 R_L}$$

leading to

$$R_L = 3.79$$

The quarter-wavelength transformer in the output (input) circuit is used to transform the load (source) impedance to approximately $(12.5/2)\ \Omega$ for each transistor, and also serves as a combiner (splitter) for the output (input) power of the two transistors. The exact impedances can be obtained easily by using the standard equation for the input impedance of a transmission line and dividing the results by two to get the load for each transistor. The load impedance thus obtained is the load specification for the impedance-matching network to be designed.

The source impedance for the output matching network to be designed is simply equal to the load resistance corresponding to the power required in parallel with the output capacitance of each transistor. The transducer power gain required is, of course, equal to one. The specifications for the matching network to be designed are summarized in Table 8.10.

Table 8.10

The Specifications for the Output Matching Network of the Power Amplifier of Example 8.5

Frequency (MHz)	Source Impedance (Ω)	Load Impedance (Ω)	Transducer Power Gain (–)
225	$2.55 - j1.78$	$6.31 - j1.03$	1.000
230	$2.51 - j1.79$	$6.28 - j0.72$	1.000
235	$2.48 - j1.80$	$6.27 - j0.43$	1.000
240	$2.44 - j1.81$	$6.25 - j0.16$	1.000
245	$2.41 - j1.82$	$6.25 + j0.20$	1.000
250	$2.38 - j1.83$	$6.27 + j0.49$	1.000
255	$2.33 - j1.84$	$6.29 + j0.80$	1.000
260	$2.30 - j1.85$	$6.32 + j1.10$	1.000

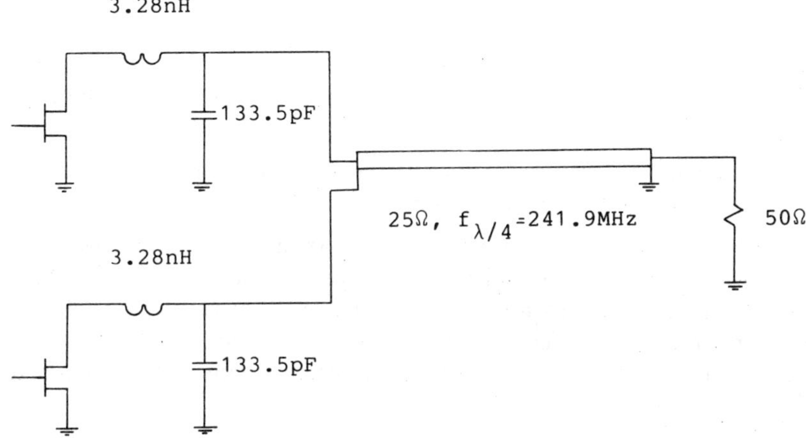

Figure 8.21   The Output Matching Network Designed for the Power Amplifier of Example 8.5

Table 8.11

The S-Parameters of the M/A-COM MA-4F001-500 GaAs FET as Specified by the Manufacturer

Frequency	s_{11}		s_{12}		s_{21}		s_{22}	
(GHz)	(dB; °)		(dB; °)		(dB; °)		(dB; °)	
2.0	−0.355	−51	−28.6	56.9	7.89	137.2	−1.938	−24
3.0	−0.630	−74	−26.2	42.3	7.16	117.6	−2.158	−35
4.0	−1.012	−94	−24.9	29.4	6.32	99.6	−2.384	−45

The designed output matching network is shown in Fig. 8.21. The deviation from the specified performance is negligibly small.

Problems

1. Calculate the parameters of the input and output stability circles on the admittance-plane for the transistor of *Example 8.1*. Also calculate the Linville and Sterne stability factors.

2. Stabilize the transistor with parameters as given in Table 8.11 by (a) resistive loading, (b) resistive voltage shunt feedback, (c) current series feedback and (d) capacitive voltage shunt feedback, if possible. Then (e) calculate the maximum operating power gain, as well as the maximum available power

gain obtainable from the transistor before and after stabilization in each case.

3. Stabilize the transistor of *Example 8.1* by using different feedback techniques and compare the operating gains corresponding to the different possibilities.

4. Derive the parameters of a constant mismatch circle for a source terminated in a passive load on a Smith chart by using (1.92).

5. Find (a) the load corresponding to the maximum operating power gain obtainable from the transistor in *Problem 8.2* and (b) the source corresponding to the maximum available power gain ($G_{A,max}$) at each relevant frequency. Then (c) find the parameters of the operating and available power gain circles on the admittance plane with gain equal to $0.79 \cdot G_{A,max}$.

6. Show that the maximum available power gain of a transistor is given by (8.72) and the optimum source admittance by the conjugate of the load admittance as given by (8.79).

7. Find the maximum tunable operating gain and available power gain by the transistor in *Problem 8.3*.

8. Find (a) the source impedance and transducer power gain, which is equivalent to the operating power-gain circle of *Problem 8.5c*, and (b) the load admittance and transducer power gain which is equivalent to the available power gain circle.

9. (a) Calculate the thermal noise available per unit bandwidth from a resistor at a temperature of $290°$ K. (b) Calculate the noise power available at the output terminals of a transistor with a 1.5 dB noise figure and available power gain of 15 dB if the bandwidth (ideal filter response) is equal to 1 MHz. (c) Repeat the last calculation for the case where the gain response is of the same form as that of a single-tuned resonant circuit.

10. (a) Calculate the parameters of the 2.0 dB constant noise-figure circle of a transistor on the admittance plane, if $F_{min} = 1.2$ dB, $R_n = 100\Omega$, and $G_{on} + jB_{on} = (20.0 + j10.0)$ mS. (b) Calculate the noise figure of the transistor if its source admittance is equal to $(30.0 + j0.0)$ mS.

11. Derive the equation for the noise figure of a cascade of N amplifiers (8.63).

12. (a) Use the material in Ch. 7 to determine if the amplifier of *Example 8.2* is realizable in lumped form.

13. Use the transistor of *Example 8.3* to design a single-stage 2-4 GHz

amplifier which is realizable in (a) lumped form and (b) distributed form.

14. Optimize the noise performance of a two-stage low-noise amplifier at 500 MHz, if the Motorola MRF966 GaAs FET (5V; 10mA) is to be used in the first stage and a BRF91 bipolar transistors (5V; 2mA) in the second. The S-parameters of the two transistors are respectively.

$s_{11} = -0.265$ dB $-14°$
$s_{12} = -47.96$ dB $76°$
$s_{11} = 4.24$ dB $156°$
$s_{22} = -0.3545$ dB $-9.0°$

and

$s_{11} = -5.849$ dB $-125°$
$s_{12} = -19.17$ dB $45°$
$s_{21} = 9.71$ dB $95°$
$s_{22} = -4.15$ dB $-35°$

Table 8.12

The S-Parameters of the Transistor in Problem 8.15 (Hewlett-Packard HFET-1102; $V_{DS} = 3.5V$, $I_{DS} = 15\% I_{DSS}$)

Frequency (GHz)	s_{11} (dB; °)		s_{12} (dB; °)		s_{21} (dB; °)		s_{22} (dB; °)	
3	−1.290	−77.1	−24.4	39.3	6.32	104.4	−3.135	−45.4
4	−2.025	−102.4	−23.1	26.0	5.82	81.1	−3.622	−60.8
5	−2.698	−127.2	−22.6	14.8	5.39	58.7	−4.033	−76.0
6	−3.427	−152.0	−22.5	6.2	4.81	36.4	−4.437	−92.0

Table 8.13

The Noise Parameters of the HFET-1102

Frequency (GHz)	S_{s-opt} (dB; °)		F_{min} (dB)	R_n (Ω)
2	−2.734	60	1.25	19.40
4	−4.180	98	1.60	23.14
6	−4.807	138	2.20	6.64
8	−4.194	−170	2.80	1.88

Table 8.14

**The Impedance of the Gunn Diode relevant to Problem 8.16 (M/A-COM
MA 49139)**

Frequency	Diode Impedance
(GHz)	(Ω)
5	$-10.0 - j13.0$
6	$-9.0 - j2.5$
7	$-9.0 + j5.0$

and the noise parameters

$F_{min} = 1.0$ dB, $S_{s-opt} = -2.384$ dB $22°$, $R_n = 101.5\Omega$

and

$F_{min} = 1.9$ dB, $S_{s-opt} = -10.8$ dB $83.7°$, $R_n = 39.6\Omega$

15. Design (a) a single-stage and (b) a two-stage low-noise amplifier over the pass band 3-6 GHz with the transistor with specifications as given in Tables 8.12 and 8.13.

16. Design a reflection amplifier with a gain of 12 dB by using a Gunn diode with the specifications given in Table 8.14 over the pass band 5-7 GHz.

17. Design a 3-6 GHz two-stage amplifier with a flat transducer power-gain *versus* frequency-response and good input and output VSWRs with the HFET-1102, the parameters of which are given in *Problem 8.14.*

18. If the output capacitance of the transistor to be used is 100pF, the supply voltage 28V, and the saturation voltage 2.5V, design the output circuit of a balanced class B power amplifier with a maximum output power capability of approximately 100W, and efficiency as high as possible over the pass band 30-400 MHz, by using the techniques outlined in Ch. 5 and 6.

References and Additional Reading

1. Ha, T.T., *Solid-State Microwave Amplifier Design,* New York: John Wiley and Sons, 1981.

2. Carson, R.S., *High Frequency Amplifiers,* New York: John Wiley and Sons, 1979.

3. Roddy, D., and J. Coolen, *Electronic Communications,* Reston, VA: Reston Publishing Company, 1981.

4. Krauss, H.L., W.B. Bostian, and F.H. Raab, *Solid State Radio Engi-*

neering, New York: John Wiley and Sons, 1980.

5.   Bodway, G.E., "Two Port Power Flow Analysis Using Generalized Scattering Parameters," *Microwave Journal,* May 1967, pp. 61-69.

6.   Tajima, Y., and P.D. Miller, "Design of Broad-Band Power GaAs Fet Amplifiers," *IEEE Trans. Microwave Theory Tech.,* Vol. MTT-32, No. 3, March 1984.

7.   Russel, K.J., "Microwave Power Combining Techniques," *IEEE Trans. Microwave Theory Tech.,* MTT-27, NO. 5, pp. 472-478. 1979.

8.   Kurokawa, K., "Design Theory. of Balanced Transistor Amplifiers," *Bell Syst. Tech. J.,* Vol. 44, No. 10, 1965, pp. 1675-1698.

9.   Young, G.P., and S.O. Scanlan, "Matching Network Design Studies for Microwave Transistor Amplifiers," *IEEE Trans. Microwave Theory Tech.,* Vol. MTT-29, No. 10, October 1981.

10. Carlin, H.J., and J.J. Komiak, "A New Method of Broad-Band Equalization Applied to Microwave Amplifiers," *IEEE Trans. Microwave Theory Tech.,* Vol. MTT-27, No. 2, February 1979.

11. Yarman, B.S., and H.J. Carlin, "A Simplified 'Real-Frequency' Technique Applied to Broad-Band Multistage Microwave Amplifiers," *IEEE Trans. Microwave Theory Tech.,* Vol. MTT-30, No. 12, December 1982.

Appendix A

PLNM FORTRAN

```
C
C
C      THIS PROGRAM DETERMINES THE COEFFICIENTS OF A POLINOMIAL CORRES-
C      PONDING TO A SET OF (W**2, T(W**2)) CO-ORDINATES WITH LEAST SQUARE
C      ERROR.  THE DEGREE (M) OF THE POLINOMIAL REQUIRED, THE NUMBER
C      OF CO-ORDINATES TO BE SPECIFIED AND THE CO-ORDINATES THEMSELVES
C      MUST BE SPECIFIED.  THE CO-ORDINATES MUST BE ENTERED IN THE
C      FORM (F, (T(W**2)) WHERE F = W/(2*PI).
C
C      THE RESULTS OF THIS PROGRAM CONSIST OF THE REQUIRED COEFFICIENTS,
C      A COMPARISON OF THE SPECIFIED VALUES OF T(W**2) AND THOSE
C      GIVEN BY THE DETERMINED POLINOMIAL AT THE FREQUENCIES (F) SPECI-
C      FIED AND THE ZEROS OF THE POLYNOMIAL. THE COEFFICIENTS ARE PRINTED
C      IN THE SEQUENCE R0 / R1 / R2 ... /RN WHERE T(W**2) = R0 + R1*W**2
C      + R2*W**4 + ...
C
C      WHEN PROBLEMS ARE EXPERIENCED WITH ZEROS ON THE REAL AXIS OF THE
C      W-PLANE , AN EXTRA CO-ORDINATE CAN BE ADDED TO THE GIVEN DATA SET
C      AT A FREQUENCY EQUAL TO 1.5 TIMES THE HIGHEST SPECIFIED FREQUENCY.
C      THE VALUE OF THE DEPENDENT VARIABLE (T(W**2)) AT THIS FREQUENCY
C      WILL BE INCREASED FROM A SPECIFIED INITIAL VALUE IN SPECIFIED
C      INCREMENTS FOR A SPECIFIED NUMBER OF TIMES.   IF THE REASON FOR
C      THE ZEROS ON THE REAL AXIS OF THE W-PLANE IS THAT THE INCREASE
C      IN THE DEPENDENT VARIABLE IS TOO SLOW AT HIGHER FREQUENCIES,
C      THIS PROCEDURE WILL USUALLY SOLVE THE PROBLEM.
C
C      IN ORDER TO USE THIS FACILITY, A ONE MUST BE ENTERED INTO THE
C      PROGRAM AT THE APPROPRIATE POINT (ZERO-PROBLEM?).   THE INITIAL
C      VALUE OF THE DEPENDENT VALUABLE OF THE ADDED CO-ORDINATE RELATIVE
C      TO THAT OF THE LAST CO-ORDINATE GIVEN, THE MULTIPLICATION FACTOR
C      WHEREBY IT IS TO INCREASED AND THE NUMBER OF TIMES THIS MUST BE
C      DONE, MUST ALSO BE ENTERED INTO THE PROGRAM WHEN ASKED FOR.
C
C      OTHER REASONS FOR AN OVERSHOOT IN THE DEPENDENT VARIABLE (REAL
C      ZEROS) ARE THAT THE VALUE OF THE DEPENDENT VARIABLE IS TOO HIGH
C      AT THE LOWER FREQUENCIES (THE DECREASE IN THE DEPENDENT VARIABLE
C      IS THEN TOO SHARP) OR THAT THE NUMBER OF CO-ORDINATES SPECIFIED
C      AT POINTS WHERE THE VALUE OF THE DEPENDENT VARIABLE ARE SMALL,
C      IS INADEQUATE.
C
       DIMENSION W(20),R(20),T(20,20)
       DIMENSION S(40),U(20),Q(20)
       DIMENSION A(16)
       REAL P,F,TMULT
       COMPLEX Z(12)
C
```

```
        WRITE(2,10)
10      FORMAT(/' DEGREE OF POLINOMIAL REQUIRED, NUMBER OF CO-ORDINATES TO
       *BE SPECIFIED; ZERO-'/' PROBLEM? (1:YES, 0:NO)'/)
        I=3
        READ(1,*) M,N,M9
        WRITE(2,15) M,N,M9
15      FORMAT(I3,3X,I3,3X,I3)
        WRITE(2,20)
20      FORMAT(/' CO-ORDINATES:  F, T(W**2) /...'/)
        DO 30 I=1,N
        READ(1,*) W(I),R(I)
        WRITE(2,25) W(I),R(I)
25      FORMAT(E15.8,3X,E15.8)
        W(I)=4.000*3.1415900**2*W(I)**2
30      CONTINUE
C
        II6=1
        IF (M9.EQ.0) GOTO 40
        WRITE(2,35)
35      FORMAT(/' ZERO-PROBLEM:'/' ENTER NUMBER OF ITERATIONS DO BE DONE '
       */'          INCREMENT FACTOR'/'          INITIAL MULTIPLICATION FACTOR')
        READ(1,*) II6,DD,EE
        WRITE(2,37) II6,DD,EE
37      FORMAT(I3,3X,F10.5,3X,F10.5)
        N=N+1
        W(N)=W(N-1)*1.500
        R(N)=R(N-1)*EE
C
40      M1=M+1
        M2=M+2
        M3=2*M+1
C
        DO 3600 I8=1,II6
C
        IF (M9.EQ.0) GO TO 90
        R(N)=R(N)*DD
C
90      U(1)=0.000
        S(1)=0.000
C
        DO 100 I=1,N
        U(1)=U(1)+R(I)
        S(1)=S(1)+1.000
100     CONTINUE
C
        DO 125 I=2,M1
        S(I)=0.000
        U(I)=0.000
125     CONTINUE
C
        DO 130 I=M1,M3
        S(I)=0.000
130     CONTINUE
C
        DO 200 I=1,N
        DO 150 J=2,M3
        S(J)=S(J)+W(I)**(J-1)
        IF (J.GT.M1) GO TO 150
        U(J)=U(J)+R(I)*W(I)**(J-1)
150     CONTINUE
200     CONTINUE
C
        K=0
```

```fortran
      DO 500  J=1,M1
      DO 400  I=1,M1
      T(J,I)=S(K+I)
400   CONTINUE
      K=K+1
500   CONTINUE
C
      DO 600  I=1,M1
      T(I,M2)=U(I)
600   CONTINUE
C
      DO 2000 L=1,M
C
      D1=T(L,L)
      DO 700  I=L,M2
      T(L,I)=T(L,I)/D1
700   CONTINUE
C
      L1=L+1
      DO 900  J=L1,M1
      D2=T(J,L)
      DO 800  I=L,M2
      T(J,I)=T(J,I)-D2*T(L,I)
800   CONTINUE
900   CONTINUE
2000  CONTINUE
      T(M1,M2)=T(M1,M2)/T(M1,M1)
      T(M1,M1)=1.000
C
2100  J=M+1
2200  I=J-1
2300  T(I,M2)=T(I,M2)-T(J,M2)*T(I,J)
      T(I,J)=0.000
      I=I-1
      IF (I.GE.1) GO TO 2300
      J=J-1
      IF (J.GE.2) GO TO 2200
C
      WRITE(2,2600)
2600  FORMAT(/' COEFFICIENTS: R0, R1, R2, R3, ...'/)
C
      IF (M9.EQ.0) GOTO 2630
      F=SQRT(W(N)/4.000/3.1415900**2)
      WRITE(2,2625) F,R(N)
2625  FORMAT(//' EXTRA CO-ORDINATE:',E15.7,3X,E15.7/)
C     NUMERATOR CONSTANT FOR PROGRAM ZVR FORTRAN
2630  AYX=1.0/T(M1,M2)
C
      DO 2700 I=1,M1
      WRITE(2,2650) T(I,M2)
2650  FORMAT(E17.9)
2700  CONTINUE
C
C     CALCULATION OF T(W**2)-VALUES AS GIVEN BY THE POLINOMIAL
C     DETERMINED
C
      ERR=0.000
      N5=N
      IF (M9.EQ.0) GOTO 2770
      N5=N-1
2770  DO 3000 I=1,N5
C
      P=R(1)
```

```
         DO 2775 J=2,M1
         P=P+T(J,M2)*W(I)**(J-1)
 2775    CONTINUE
         Q(I)=P
         ERR=ERR+(Q(I)-R(I))**2
 3000    CONTINUE
C
         WRITE(2,3050) ERR
 3050    FORMAT(/' ERROR:',D17.7)
         WRITE(2,3100)
 3100    FORMAT(/'      FREQUENCY             T(W**2)              T(W**2)-SPECIFIED'/
        *)
         DO 3500 I=1,N5
         F=SQRT(W(I)/4.000/3.1415900**2)
         WRITE(2,3200) F,Q(I),R(I)
 3200    FORMAT(E14.2,3X,E17.9,3X,E17.9)
 3500    CONTINUE
C
C        CALCULATION OF THE POLES OF THE POLINOMIAL
C        X(W)=R0+0.0+R1(W**2)+0.0+R2(W**4)+0.0+...
C
         TMULT=10.0
         DO 3520 I=1,15
         AXX=T(M1,M2)*TMULT
         IF (AXX.GE.1.0E-9) GOTO 3525
         TMULT=TMULT*10.0
 3520    CONTINUE
C
 3525    NCOEF=M+1
         DO 3550 I=1,NCOEF
         A(2*I-1)=T(M2-I,M2)*TMULT
         IF (I.EQ.NCOEF) GOTO 3550
         A(2*I)=0.0
 3550    CONTINUE
         MM=M*2
         CALL ZPOLR (A,MM,Z,IER)
         WRITE(2,3560)
 3560    FORMAT(/' THE ROOTS ARE:     REAL PART',15X,'IMAGINARY PART'/)
         DO 3570 I=1,MM
         WRITE(2,3565) Z(I)
 3565    FORMAT(1H ,11X,F16.5,6X,F16.5)
 3570    CONTINUE
C
 3600    CONTINUE
C
         STOP
         END
         SUBROUTINE ZPOLR(A,NDEG,Z,IER)
         DIMENSION A(16)
         COMPLEX Z(NDEG)
         COMPLEX RT,FXI,FXIM1,FXIM2,H,XLAM,AA,XNUM,DEN,DFIIM1
         COMPLEX DIM1M2,G,SQR,FRT
C
C        REFERENCE:  ELEMENTARY NUMERICAL ANALYSIS - CONTE, DE BOOR -
C        MCGRAW-HILL
C
         IER=0
         NNN=NDEG+1
         XNDEG=FLOAT(NDEG)
         DO 2 I=1,NDEG
 2       Z(I)=CMPLX(0.0,0.0)
C
         CC=1.0
```

```
         AAA=A(1)/A(NDEG+1)
         BB=AMAX1(AAA,1.0/AAA)
         IF (BB.LE.(10.0**NDEG)) GOTO 4
         CC=EXP(ALOG(A(1)/A(NDEG+1)*10.0**(NDEG))/XNDEG)
         DO 3 I=1,NNN
3        A(I)=A(I)*CC**(I-1)
C
4        CCC=SQRT(ABS(A(1)*A(NDEG+1)))
         DO 5 I=1,NNN
5        A(I)=A(I)*CCC
C
         DMX=1.0E-08
         DMF=1.0E-10
         MAXNIT=150
C
         DO 1000 K=1,NDEG
C        INITIALIZE
         MFLAG=0
         MFLAGD=0
         IF (CABS(Z(K)).NE.0.0) MFLAG=1
         IF (MFLAG.EQ.1) GOTO 8
         DMFT=1.0E-10
8        NIT=0
         NFLAG=0
10       H=CMPLX(0.5,0.0)
         RT=Z(K)+H
         ASSIGN 60 TO NN
         GOTO 210
60       FXIM2=FXI
C
         RT=Z(K)-H
         ASSIGN 120 TO NN
         GOTO 210
120      FXIM1=FXI
C
         FXIM2=FXIM1-FXIM2
         RT=Z(K)
         ASSIGN 260 TO NN
         NFLAG=1
         GOTO 210
180      NFLAG=0
C
         XLAM=CMPLX(-0.5,0.0)
C        COMPUTE NEXT ESTIMATE FOR ROOT
.200     MFLAGD=0
         DFIIM1=FXI-FXIM1
         AA=FXIM2*XLAM
         XNUM=-FXI*CMPLX(2.0,0.0)*(CMPLX(1.0,0.0)+XLAM)
         G=(CMPLX(1.0,0.0)+CMPLX(2.0,0.0)*XLAM)*DFIIM1-XLAM*AA
         SQR=G*G+CMPLX(2.0,0.0)*XNUM*XLAM*(DFIIM1-AA)
         SQR=CSQRT(SQR)
         DEN=G+SQR
         IF (REAL(G)*REAL(SQR)+AIMAG(G)*AIMAG(SQR).LT.0.0)  DEN=G-SQR
         IF (CABS(DEN).EQ.0) DEN=1
         XLAM=XNUM/DEN
         FXIM1=FXI
         FXIM2=DFIIM1
         IF (CABS(H).LE.1.0E-9) NIT=MAXNIT
         IF (CABS(H).LE.1.0E-9) GOTO 205
         H=H*XLAM
         RT=RT+H
205      IF (NIT.LT.MAXNIT) GOTO 210
         IER=IER+1
```

```
        GOTO 800
C
210     NIT=NIT+1
        FXI=CMPLX(A(1),0.0)
        DO 220 I=1,NDEG
220     FXI=FXI*RT+CMPLX(A(1+I),0.0)
        FRT=FXI
        IF (K.LT.2) GOTO 240
        DO 230 I=2,K
        AA=RT-Z(I-1)
        IF (CABS(AA).GT.(DMX*CABS(Z(I-1)))) GOTO 230
        Z(K)=RT+0.001
        GOTO 8
230     FXI=FXI/AA
240     IF (MFLAGD.EQ.1) GOTO 270
        GOTO NN,(60,120,260)
C
C       CHECK FOR CONVERGENCE
260     IF (CABS(H).GT.DMX*CABS(RT)) GOTO 268
        IF (AMAX1(CABS(FXI),CABS(FRT)).LT.DMF) GOTO 800
268     IF (NFLAG.NE.1) GOTO 270
        IF (MFLAG.EQ.0) GOTO 180
        IF (AMAX1(CABS(FRT),CABS(FXI)).LT.DMFT) GOTO 800
        GOTO 180
C
C       CHECK FOR DIVERGENCE
270     IF (CABS(FXI).LE.(1.1*CABS(FXIM1))) GOTO 200
        MFLAGD=1
        H=H/CMPLX(2.0,0.0)
        IF (CABS(H).LE.1.0E-9) GOTO 200
        XLAM=XLAM/CMPLX(2.0,0.0)
        RT=RT-H
        GOTO 205
800     IF (K.EQ.NDEG) GOTO 980
        IF (CABS(Z(K+1)).NE.0.0) GOTO 980
        IF (K.EQ.1) GOTO 810
        IF (CABS(Z(K)).EQ.CABS(Z(K-1))) GOTO 980
810     RA=ABS(AIMAG(RT)/REAL(RT))
        IF (RA.LE.1000.0) GOTO 910
        Z(K+1)=CONJG(RT)
        DMFT=100.0*AMAX1(CABS(FXI),CABS(FRT))
        GOTO 980
910     IF (RA.GE.0.001) GOTO 920
        Z(K+1)=-RT
        DMFT=100.0*AMAX1(CABS(FXI),CABS(FRT))
        GOTO 980
920     Z(K+1)=CONJG(RT)
        IF (NDEG.LE.(K+3)) GOTO 980
        Z(K+2)=-CONJG(RT)
        Z(K+3)=-RT
        DMFT=100.0*AMAX1(CABS(FXI),CABS(FRT))
980     CONTINUE
1000    Z(K)=RT
C
        DO 1200 I=1,NDEG
        Z(I)=Z(I)/CC
1200    CONTINUE
C
        RETURN
        END
```

Appendix B

ZVR FORTRAN

```fortran
C
C
C     THIS PROGRAM DETERMINES THE MINIMUM-IMPEDANCE (OR MINIMUM-
C     ADMITTANCE) FUNCTION ASSOSIATED WITH A GIVEN RESISTANCE (OR
C     CONDUCTANCE) FUNCTION.  THIS FUNCTION MUST BE EVEN (A(W**2)).
C
C     THE INPUT DATA CONSIST OF THE DEGREE OF THE NUMERATOR, ITS
C     COEFFICIENTS, THE NUMBER OF POLES INSIDE THE FIRST QUADRANT OF
C     THE W-PLANE, THESE POLES, THE NUMBER OF POLES ON THE
C     POSITIVE SIDE OF THE IMAGINARY AXIS OF THE W-PLANE, THESE POLES,
C     THE NUMBER OF FREQUENCIES AT WHICH THE VALUE OF THE MINIMUM-
C     IMPEDANCE (-ADMITTANCE) FUNCTION MUST BE CALCULA TED AS A
C     CHECK ON THE RESULTS AND THE FREQUENCIES OF INTEREST.
C
C     THE IMPEDANCE (ADMITTANCE) FUNCTION DETERMINED IS OF THE FORM
C     Q(JW)= (U0+U1*W+U2*W**2+...)/(P0+P1*W+P2*W**2+...)
C     WHERE U0, U1, ..., P0, P1, ... ARE COMPLEX COEFFICIENTS.
C
      REAL*8 R(10)
      COMPLEX*16 W(16),C(16),S(16),P(16),U(16),WX(8)
      COMPLEX*16 CX(6),XX(6),A1,B1,CA,RW
C
C     NUMERATOR FUNCTION
      WRITE(2,100)
100   FORMAT(' DEGREE OF THE NUMERATOR OF R(W**2)?')
      READ(1,*) M9
      WRITE(2,200)
200   FORMAT(' ENTER NUMERATOR COEFFICIENTS IN THE SEQUENCE A0 / A1 / A
     *2 / ...')
      N0=M9+1
C
      DO 220 I=1,10
      R(I)=0.0D0
220   CONTINUE
C
      DO 300 I=1,N0
      READ(1,*) R(I)
300   CONTINUE
C
C     FIRST QUADRANT POLES
      WRITE(2,350)
350   FORMAT(' NUMBER OF POLES INSIDE THE FIRST QUADRANT?')
      READ(1,*) N
      WRITE(2,400)
400   FORMAT(' ENTER POLES.  FORMAT (A1,B1) / (A2,B2) / ...')
      DO 500 I=1,N
      READ(1,*) W(I)
```

```fortran
500     CONTINUE
C
        WRITE(2,510)
510     FORMAT(' NUMBER OF POLES ON THE POSITIVE SIDE OF THE IMAGINARY AXI
       *S '/' OF THE W-PLANE?')
        READ(1,*) N11
        IF (N11.EQ.0) GOTO 545
        WRITE(2,520)
520     FORMAT(' ENTER POLES.  FORMAT (C1,D1) / (C2,D2) / ...')
        DO 540 I=1,N11
        READ(1,*) WX(I)
540     CONTINUE
C
C       GENERATION OF IMAGINARY AXIS POLES
        DO 542 I=1,N11
        WX(N11+I)=DCONJG(WX(I))
542     CONTINUE
C
C       GENERATION OF THE OTHER POLES OF THE RESISTANCE FUNCTION
545     DO 550 I=1,N
        W(N+I)=-DCONJG(W(I))
        W(2*N+I)=DCONJG(W(N+I))
        W(3*N+I)=DCONJG(W(I))
550     CONTINUE
C
        N4=4*N
C       DO 555 I=1,N4
C
        IF (N11.EQ.0) GO TO 565
        N12=2*N11
C
C       RESIDUES OF THE IMAGINARY AXIS POLES
C       RESIDUES OF THE OTHER POLES
565     N3=2*N
        DO 1000 J=1,N3
C
C       DETERMINING THE CONSTANT
        A1=DCMPLX(R(1),0.0D0)
        DO 600 I=3,N0,2
        A1=A1+DCMPLX(R(I),0.0D0)*W(J)**(I-1)
600     CONTINUE
C
C       CALULATION OF THE VALUE OF THE NUMERATOR
        B1=(1.0D0,0.0D0)
        DO 700 I=1,N4
        IF (I.EQ.J) GO TO 700
        B1=B1*(W(J)-W(I))
700     CONTINUE
C
        IF (N11.EQ.0) GO TO 900
        DO 800 I=1,N12
        B1=B1*(W(J)-WX(I))
800     CONTINUE
C
C       RESIDUE
900     C(J)=A1/B1
C
1000    CONTINUE
C
C       RESIDUE
        IF (N11.EQ.0) GO TO 1080
        DO 1050 J=1,N12
C       NUMERATOR
```

```
          A1=DCMPLX(R(1),0.0D0)
          DO 1010 I=3,N0,2
          A1=A1+DCMPLX(R(I),0.0D0)*WX(J)**(I-1)
1010      CONTINUE
C
C         DENOMINATOR
          B1=(1.0D0,0.0D0)
          DO 1020 I=1,N4
          B1=B1*(WX(J)-W(I))
1020      CONTINUE
C
          DO 1030 I=1,N12
          IF (I.EQ.J) GOTO 1030
          B1=(WX(J)-WX(I))*B1
1030      CONTINUE
C
C         RESIDUE
          CX(J)=A1/B1
1050      CONTINUE
C
C         CONSTANT
C
C         W=1
1080      A1=DCMPLX(R(1),0.0D0)
          DO 1100 I=3,N0,2
          A1=A1+DCMPLX(R(I),0.0D0)
1100      CONTINUE
C
C         DENOMINATOR
          B1=(1.0D0,0.0D0)
          DO 1200 I=1,N4
          B1=B1*((1.0D0,0.0D0)-W(I))
1200      CONTINUE
C
          IF (N11.EQ.0) GO TO 1220
          DO 1210 I=1,N12
          B1=B1*((1.0D0,0.0D0)-WX(I))
1210      CONTINUE
C
1220      CA=A1/B1
C
C         CJ/(W-WJ)
C
          A1=(0.0D0,0.0D0)
          DO 1300 I=1,N3
          A1=A1+C(I)/((1.0D0,0.0D0)-W(I))
          A1=A1+DCONJG(C(I))/((1.0D0,0.0D0)-DCONJG(W(I)))
1300      CONTINUE
C
          IF (N11.EQ.0) GO TO 1360
          DO 1350 I=1,N12
          A1=A1+CX(I)/((1.0D0,0.0D0)-WX(I))
1350      CONTINUE
C
1360      RW=CA-A1
C
          IF (N11.EQ.0) GO TO 1480
C
C         IMPEDANCE FUNCTION
C
C         DENOMINATOR (P(I))
1480      K=0
          K1=0
```

```fortran
      N1=2*N+N11
      CALL SMULT(W,WX,S,K,K1,N1,N4,N11)
1500  CONTINUE
C
      N5=2*N+N11+1
      DO 1600 I=1,N5
      P(I)=S(I)
1600  CONTINUE
C
C     NUMERATOR (U(I))
      DO 1650 I=1,N5
      U(I)=(0.0D0,0.0D0)
1650  CONTINUE
C
      DO 2000 K=1,N3
      N1=2*N-1+N11
      K1=0
      N5=N1+1
      CALL SMULT(W,WX,S,K,K1,N1,N4,N11)
1800  CONTINUE
      DO 1900 J=1,N5
      A1=(2.0D0,0.0D0)*C(K)*S(J)
      U(J)=U(J)+A1
1900  CONTINUE
C
2000  CONTINUE
C
      IF(N11.EQ.0) GO TO 2060
      DO 2050 K1=1,N11
      N1=2*N+N11-1
      K=0
      CALL SMULT(W,WX,S,K,K1,N1,N4,N11)
2010  CONTINUE
      DO 2020 J=1,N5
      A1=(2.0D0,0.0D0)*CX(K1)*S(J)
      U(J)=U(J)+A1
2020  CONTINUE
2050  CONTINUE
C
2060  N5=2*N+N11+1
      DO 2100 I=1,N5
      U(I)=U(I)+RW*P(I)
2100  CONTINUE
C
C     RESULTS
      WRITE(2,2200)
2200  FORMAT(/' NUMERATOR COEFFICIENTS  (U0, U1..)',5X,' DENOMINATOR
     *COEF-'/' FICIENTS  (P0, P1..) '/)
      WRITE(2,2230)
2230  FORMAT('  REAL PART          IMAGINARY PART          REAL PART
     * IMAGINARY PART ')
      N6=N1+2
      DO 2300 I=1,N6
      WRITE(2,2250) U(I),P(I)
2250  FORMAT(2(E14.4,3X),6X,2(E14.4,3X))
2300  CONTINUE
C
      CALL SZIN(U,P,N6)
2350  CONTINUE
      STOP
      END
      SUBROUTINE SMULT(WW,WXX,SS,K,K1,N1,N4,N11)
      DIMENSION M(8)
```

```
      COMPLEX*16 WW(16),SS(16),VV(16),A1,WXX(8)
C
C     DUPLICATE POLES
2910  L1=0
      N8=N1-N11
      IF (K1.EQ.0) GOTO 2915
      N8=N1-N11+1
2915  DO 2950 L=1,N8
      IF (L.NE.K) GO TO 2920
      L1=1
2920  VV(L)=-WW(L+L1)
2950  CONTINUE
C
      IF (N11.EQ.0) GO TO 2965
      L1=0
      N13=N11
      IF (K1.EQ.0) GOTO 2952
      N13=N11-1
2952  IF (N13.EQ.0) GOTO 2965
      DO 2960 L=1,N13
      IF (L.NE.K1) GO TO 2955
      L1=1
2955  VV(N8+L)=-WXX(L+L1)
2960  CONTINUE
C
C     SS=0
2965  N7=N4+N11
      DO 2970 L=1,N7
      SS(L)=(0.0D0,0.0D0)
2970  CONTINUE
      SS(N1+1)=(1.0D0,0.0D0)
C
      DO 2980 I=1,8
      M(I)=0
2980  CONTINUE
C
C     GENERATION OF ALLOWABLE COMBINATIONS AND MULTIPLICATION OF TERMS
C
      DO 3800 MO=1,N1
C
C     INISIALISASIE
      K9=1
      DO 3000 I=1,MO
      M(I)=K9
      K9=K9+1
3000  CONTINUE
C
C     LAST ELEMENT AND MULTIPLICATION
      MOO=MO
3050  DO 3100 I=MOO,N1
      M(MO)=I
C
      A1=(1.0D0,0.0D0)
      DO 3080 L=1,MO
      A1=A1*VV(M(L))
3080  CONTINUE
      SS(N1-MO+1)=SS(N1-MO+1)+A1
3100  CONTINUE
C
3120  IF ((MO-1).EQ.0) GO TO 3800
C
      IF (M(MO-1).EQ.(N1-1)) GO TO 3150
      M(MO-1)=M(MO-1)+1
```

```
       M00=M(M0-1)+1
       GO TO 3050
C
 3150  IF ((M0-2).EQ.0) GO TO 3800
C
       IF (M(M0-2).EQ.(N1-2)) GO TO 3200
       M(M0-2)=M(M0-2)+1
       M(M0-1)=M(M0-2)+1
       M00=M(M0-1)+1
       GO TO 3050
C
 3200  IF ((M0-3).EQ.0) GO TO 3800
C
       IF(M(M0-3).EQ.(N1-3)) GO TO 3220
       M(M0-3)=M(M0-3)+1
       M(M0-2)=M(M0-3)+1
       M(M0-1)=M(M0-2)+1
       M00=M(M0-1)+1
       GO TO 3050
C
 3220  IF((M0-4).EQ.0) GO TO 3800
C
       IF (M(M0-4).EQ.(N1-4)) GO TO 3240
       M(M0-4)=M(M0-4)+1
       M(M0-3)=M(M0-4)+1
       M(M0-2)=M(M0-3)+1
       M(M0-1)=M(M0-2)+1
       M00=M(M0-1)+1
       GO TO 3050
C
 3240  IF ((M0-5).EQ.0) GO TO 3800
C
       IF (M(M0-5).EQ.(N1-5)) GO TO 3800
       M(M0-5)=M(M0-5)+1
       M(M0-4)=M(M0-5)+1
       M(M0-3)=M(M0-4)+1
       M(M0-2)=M(M0-3)+1
       M(M0-1)=M(M0-2)+1
       M00=M(M0-1)+1
       GO TO 3050
C
 3800  CONTINUE
C
       RETURN
       END
C
       SUBROUTINE SZIN(UU,PP,NE)
       COMPLEX*16 UU(16),PP(16),TELL,NOEM,ANTW
       REAL*8 F(30)
C
C      CALCULATION OF IMPEDANCE AT SPECIFIED FREQUENCIES.
C
       WRITE(2,3900)
 3900  FORMAT(//' NUMBER OF FREQUENCIES AT WHICH THE IMPEDANCE (ADMITTANC
      *E IS '/' TO BE CALCULATED?   (EVEN NUMBER) ')
       READ(2,*) N01
       IF (N01.EQ.0) GO TO 4500
C
       WRITE(2,4000)
 4000  FORMAT(' ENTER FREQUENCIES: F1,F2 / F3,F4 / ... '/)
       DO 4100 I=1,N01,2
       READ(2,*) F(I),F(I+1)
       F(I)=F(I)*2.0D0*3.14159D0
       F(I+1)=F(I+1)*2.0D0*3.14159D0
 4100  CONTINUE
       WRITE(2,4150)
 4150  FORMAT(/)
C
       DO 4300 I=1,N01
       TELL=UU(1)
       NOEM=PP(1)
       DO 4200 J=2,NE
```

```
      TELL=TELL+UU(J)*DCMPLX((F(I)**(J-1)),0.0D0)
      NOEM=NOEM+PP(J)*DCMPLX((F(I)**(J-1)),0.0D0)
4200  CONTINUE
      ANTW=TELL/NOEM
      F(I)=F(I)/2.0D0/3.14159D0
      WRITE(2,4250) F(I),ANTW
4250  FORMAT(' F  R-IN  X-IN:   ',E16.2,5X,2(E14.4,3X))
4300  CONTINUE
C
4500  CONTINUE
      WRITE(2,4600)
4600  FORMAT(/)
C
      RETURN
      END
```

Appendix C

LSM FORTRAN

```
C
C
C
C     THIS PROGRAM DETERMINES THE OUTPUT IMPEDANCE (OR ADMITTANCE) OF
C     A LOSSLESS IMPEDANCE MATCHING NETWORK THAT WILL MATCH A RESIS-
C     TIVE SOURCE TO A COMPLEX LOAD IN A PRESCRIBED WAY.  WHERE IT IS
C     THEORETICALLY IMPOSSIBLE TO ACHIEVE THE SPECIFICATIONS THE OUT-
C     PUT IMPEDANCE OF THE OPTIMUM NETWORK WILL BE DETERMINED.
C
C     THE OUTPUT RESISTANCE (CONDUCTANCE) AND REACTANCE (SUSCEPTANCE)
C     OF THE MATCHING NETWORK ARE MODELLED AS PIECE-WISE LINEAR FUNC-
C     TIONS IN THE PROGRAM.  THE IMPEDANCE (ADMITTANCE) FUNCTION DETER-
C     MINED IS OF MINIMUM FORM.
C
C     THE DATA TO BE ENTERED INTO THE PROGRAM CONSIST OF THE NUMBER
C     OF FREQUENCIES WHERE THE LOAD IMPEDANCE (ADMITTANCE) IS TO BE
C     SPECIFIED, THE IMPEDANCE (ADMITTANCE) AND REQUIRED TRANSDUCER
C     POWER GAIN AT THE SPECIFIED FREQUENCIES, THE NUMBER OF FRE-
C     QUENCIES WHERE THE SLOPE OF THE RESISTANCE (CONDUCTANCE) FUNC-
C     TION IS TO CHANGE, THE DC VALUE OF THE RESISTANCE (CONDUCTANCE)
C     (0.0 IN THE CASE OF BANDPASS NETWORKS, THE SOURCE RESISTANCE
C     WHEN A LOWPASS NETWORK IS TO BE DESIGNED), THE CHANGE IN RESIS-
C     TANCE AT EACH INCREMENT FREQUENCY AND THE NUMBER OF ITERATIONS
C     TO BE DONE IN THE OPTIMIZATION PROCESS.
C
C     WHEN THE GAIN-BANDWIDTH PRODUCT CORRESPONDING TO THE LOAD IMPE-
C     DANCE AND THE TRANSDUCER POWER GAIN SPECIFIED IS LIMITED, IT
C     WILL BE FOUND THAT THE POSITION OF THE LAST INCREMENT FREQUENCY
C     IS CRITICAL.  THIS PROGRAM HAS THE FACILTITY TO INCREASE THE
C     LAST INCREMENT FREQUENCY BY A SPECIFIED AMOUNT FOR A SPECIFIED
C     NUMBER OF TIMES.  THE BEST SOLUTION WILL THEN BE DETERMINED
C     AUTOMATICALLY.  IF A DETAILED PRINTOUT OF THE RESULTS IS REQUIRED
C     THE NUMBER ONE MUST BE ENTERED INTO THE PROGRAM AT THE APPRO-
C     PRIATE POINT.
C
C     THE OUTPUT DATA OF THE PROGRAM CONSIST OF THE RESISTANCE (CON-
C     DUCTANCE) AND REACTANCE (SUSCEPTANCE) AT THE VARIOUS FREQUENCIES,
C     THE CORRESPONDING TRANSDUCER POWER GAIN COMPARED TO THE INPUT
C     SPECIFICATIONS AND THE LEAST SQUARE ERROR CORRESPONDING TO THE
C     SOLUTION.
C
C
C
C
      DOUBLE PRECISION BY4(2),PR1(10),ALP,ALPS
      DOUBLE PRECISION GW(20),BW(20),GO(20),EW(20)
      DOUBLE PRECISION DM(20,4),ERR2,ERFL
      DOUBLE PRECISION DBR(10,2),RB22(10)
```

```
      DOUBLE PRECISION AW(10),BV(10),EMAT(3,2),CE,CL,CEOP,CED
      DOUBLE PRECISION FW(10),FWM(10,10),EOFW(10)
      DOUBLE PRECISION FWMS(10,10),EOFWS(10),FWMS1(10,10)
      DOUBLE PRECISION W2,RO,DET,DET0,ERR,E2,E3,E4,F00
      DOUBLE PRECISION BX1,FX1,FX2,FX3,BY1,BY2,BY3,CEC
      DOUBLE PRECISION AAA,BY5,GT1,Z3,BWB,FWNI,ERRM,WXX
      DOUBLE PRECISION RX2,RX3,RX4,RX5,BFY,GX1,RX44,ABC
      INTEGER T,Q1,Q,R,S,L5,M4,M3,NB,NW,NC,ND,NE,NTI,NROWS(10,2)
C
      WRITE(2,1010)
1010  FORMAT(/' NUMBER OF FREQUENCIES AT WHICH THE LOAD IMPEDANCE (ADMIT
     *TANCE) IS TO BE SPECIFIED?')
      READ(1,*) NW
C
      WRITE(2,1020)
1020  FORMAT(/' ENTER:  FREQUENCY, R (OR G), X (OR B), GT-REQUIRED / ...
     */ ...')
      DO 1030 I=1,NW
      READ(1,*) DM(I,1),DM(I,2),DM(I,3),DM(I,4)
      WRITE(2,1025) DM(I,1),DM(I,2),DM(I,3)
1025  FORMAT(' F  R-G  X-B:',3(E16.8,3X))
      DM(I,1)=DM(I,1)*2.0D0*3.141592654D0
1030  CONTINUE
C
1035  WRITE(2,1040)
1040  FORMAT(/' NUMBER OF INCREMENT FREQUENCIES? R (OR G) AT W=0?')
      READ(1,*) NB,RO
      WRITE(2,1045) RO
1045  FORMAT(/' RO:',F8.2/)
C
      WRITE(2,1050)
1050  FORMAT(/' ENTER:  INCREMENT FREQUENCY, INCREMENT / ... / ...')
      DO 1060 I=1,NB
      READ(1,*) DBR(I,1),DBR(I,2)
      WRITE(2,1055) DBR(I,1),DBR(I,2)
1055  FORMAT(' FB IB:',2(E16.8,3X))
      DBR(I,1)=DBR(I,1)*2.0D0*3.141592654D0
1060  CONTINUE
C
1065  WRITE(2,1070)
1070  FORMAT(/' NUMBER OF ITERATIONS TO BE DONE?  DETAILED PRINTOUT? (1-
     *0)')
      READ(1,*) I1,I2
C
      WRITE(2,1080)
1080  FORMAT(/' AMOUNT BY WHICH THE LAST INCREMENT FREQUENCY MUST BE INC
     *REMENTED (ARITHMATICALLY)?   NUMBER OF ITERATIONS?')
      READ(1,*) FWNI,NTI
      WRITE(2,1090) FWNI
1090  FORMAT(/' INCREMENT FOR FINAL FB:',E18.6)
      FWNI=FWNI*2.0D0*3.141592654D0
C
1100  DBR(NB,1)=DBR(NB,1)-FWNI
      ERRM=100.0D0
      DO 8000 MI=1,NTI
      DBR(NB,1)=DBR(NB,1)+FWNI
C
      ND=NB-1
      ERR2=100.0D0
      I7=0
      I4=0
1200  DO 1210 I=1,NB
      DO 1205 J=1,NB
      FWMS(I,J)=0.0D0
```

```
1205    CONTINUE
        EOFWS(I)=0.0D0
1210    CONTINUE
1215    ERR=0.0D0
C
        IF (I2.EQ.1) GOTO 1218
        IF (I7.EQ.7) GOTO 1218
        GOTO 1225
1218    WRITE(2,1220)
1220    FORMAT(/)
1225    DO 4880 M=1,NW
        W2=DM(M,1)
        AW(1)=1.0D0
        IF (W2.GE.DBR(1,1)) GOTO 1250
        AW(1)=W2/DBR(1,1)
        DO 1230 J=2,NB
        AW(J)=0.0D0
1230    CONTINUE
        GO TO 1275
C
1250    DO 1270 I=2,NB
        AW(I)=1.0D0
        IF (W2.GE.DBR(I,1)) GOTO 1270
        AW(I)=(W2-DBR(I-1,1))/(DBR(I,1)-DBR(I-1,1))
C
        IF (I.EQ.NB) GO TO 1270
        NE=NB-I
        DO 1260 J=1,NE
        AW(I+J)=0.0D0
1260    CONTINUE
        GO TO 1275
1270    CONTINUE
C
1275    GX1=0.0D0
        DO 2100 L=1,NB
        GX1=GX1+AW(L)*DBR(L,2)
2100    CONTINUE
        GW(M)=GX1+RO
C
        BY2=W2/DBR(1,1)+1.0D0
        BY1=W2/DBR(1,1)-1.0D0
        BY3=W2/DBR(1,1)
        BY5=DABS(BY1)
        BFY=BY2*DLOG(BY2)+BY1*DLOG(BY5)-2.0D0*BY3*DLOG(BY3)
        BV(1)=BFY/3.141592654D0
        DO 2200 I=2,NB
        DO 2150 J=1,2
        BY2=W2/DBR(I+J-2,1)+1.0D0
        BY1=W2/DBR(I+J-2,1)-1.0D0
        BY3=W2/DBR(I+J-2,1)
        BY5=DABS(BY1)
        BY4(J)=BY2*DLOG(BY2)+BY1*DLOG(BY5)-2.0D0*BY3*DLOG(BY3)
        BY4(J)=BY4(J)*DBR(I+J-2,1)
2150    CONTINUE
        BV(I)=1.0D0/(DBR(I,1)-DBR(I-1,1))*(BY4(2)-BY4(1))/3.141592654D0
2200    CONTINUE
C
2310    BX1=0.0D0
        DO 2400 J=1,NB
        BX1=BX1+BV(J)*DBR(J,2)
2400    CONTINUE
        BW(M)=BX1
C
        GT1=4.0D0*GW(M)*DM(M,2)
```

```
      GO(M)=GT1/((GW(M)+DM(M,2))**2.0D0+(DABS(BW(M)+DM(M,3)))**2.0D0)
C
      EW(M)=GO(M)/DM(M,4)-1.0D0
      ERR=ERR+(DABS(EW(M)))**2.0D0
      IF (I7.EQ.7) GO TO 4854
      FX3=(GW(M)+DM(M,2))**2.0D0+(DABS(BW(M)+DM(M,3)))**2.0D0
      FX1=4.0D0*DM(M,2)/FX3
      FX1=FX1-8.0D0*GW(M)*DM(M,2)*(GW(M)+DM(M,2))/(FX3**2.0D0)
      FX1=FX1/DM(M,4)
      FX2=-8.0D0*GW(M)*DM(M,2)*(BW(M)+DM(M,3))/(FX3**2.0D0)
      FX2=FX2/DM(M,4)
4320  DO 4400 J=1,ND
      FW(J)=FX1*(AW(J)-AW(NB))+FX2*(BV(J)-BV(NB))
4400  CONTINUE
C
C     CREATE MATRIX
      DO 4500 K=1,ND
      DO 4450 L=1,ND
      FWM(K,L)=FW(K)*FW(L)
4450  CONTINUE
4500  CONTINUE
C
4755  DO 4850 J=1,ND
      DO 4800 K=1,ND
      FWMS(J,K)=FWMS(J,K)+FWM(J,K)
4800  CONTINUE
      EOFWS(J)=EOFWS(J)-EW(M)*FW(J)
4850  CONTINUE
C
4854  FOO=DM(M,1)/2.0D0/3.14159654D0
4855  IF (I2.EQ.1) GOTO 4865
4860  IF (NTI.GT.1) GOTO 4880
4865  IF (I7.EQ.0) GOTO 4880
4870  WRITE(2,4875)FOO,GW(M),BW(M),GO(M),DM(M,4)
4875  FORMAT(' F G B GO GI :',5(D10.3,2X))
C
4880  CONTINUE
C
      IF (I7.EQ.0) GOTO 4884
      WRITE(2,4881)
4881  FORMAT(/)
      DO 4883 M=1,NB
      FOO=DBR(M,1)/2.0D0/3.141592654D0
      WRITE(2,4882) FOO,DBR(M,2)
4882  FORMAT(' FB IB:',2(E16.8,3X))
4883  CONTINUE
C
4884  ERFL=ERR/ERR2
      IF (ERFL.LT.0.9999D0) GOTO 4885
      I4=I1+1
C
4885  ERR2=ERR
C
      IF (I2.EQ.1) GOTO 4886
      IF (NTI.GT.1) GOTO 4890
      IF (I7.EQ.0) GOTO 4890
4886  WRITE (2,4887) ERR
4887  FORMAT(/' ERROR:',D16.8)
C
4890  IF (ERR.GT.ERRM) GOTO 4895
      BWB=DBR(NB,1)
      ERRM=ERR
C
4895  IF (I7.EQ.7) GO TO 7400
```

```
C
5001    DO 5030 M=1,ND
        FWMS(M,NB)=EOFWS(M)
5030    CONTINUE
C
        MRS=0
        DO 5175 M=1,ND
C
        IF (FWMS(M,M).GT.0.0D0) GO TO 5085
        IF (FWMS(M,M).LT.0.0D0) GO TO 5085
C       FIND NON-ZERO ROW
        MRS=MRS+1
        NROWS(MRS,1)=M
5055    M4=M+1
        Q1=M+1
        DO 5060 R=Q1,ND
        IF (FWMS(R,M).EQ.0.QD0) GO TO 5057
        M3=1
        GO TO 5065
5057    M3=-1
        M4=M4+1
5060    CONTINUE
5065    IF (M3.EQ.-1) GO TO 7400
C
        NROWS(MRS,2)=M4
C       REVERSE ROWS
        DO 5070 S=M,NB
        Z3=FWMS(M,S)
        FWMS(M,S)=FWMS(M4,S)
        FWMS(M4,S)=Z3
5070    CONTINUE
C
C       NORMALIZE ROW
5085    E2=FWMS(M,M)
        DO 5100 N=M,NB
        FWMS(M,N)=FWMS(M,N)/E2
5100    CONTINUE
C
        IF (M.EQ.ND) GO TO 5175
C
C       SUBTRACT
        NE=M+1
        DO 5150 N=NE,ND
        E2=FWMS(N,M)
        DO 5125 I=M,NB
        FWMS(N,I)=FWMS(N,I)-E2*FWMS(M,I)
5125    CONTINUE
5150    CONTINUE
C
5175    CONTINUE
C
C       COMPLETE SUBTRACTION
        N=ND
5180    L=N-1
5190    FWMS(L,NB)=FWMS(L,NB)-FWMS(L,N)*FWMS(N,NB)
        FWMS(L,N)=0.0D0
        L=L-1
5200    IF (L.GE.1) GO TO 5190
        N=N-1
5300    IF (N.GE.2) GO TO 5180
C
5340    DO 5350 M=1,ND
        PR1(M)=FWMS(M,NB)
5350    CONTINUE
```

```
C
6235    RX2=-RO
        RX3=1.0D0
6236    RX4=RO
        ALP=0.0D0
        DO 6240 L=1,ND
        RX4=RX4+(DBR(L,2)+RX3*PR1(L))
        RX44=DABS(PR1(L)/DBR(L,2))*RX3
        IF (RX44.LT.ALP) GOTO 6238
        ALP=RX44
6238    IF (PR1(L).GT.0.0D0) GO TO 6240
        IF (RX4.GT.0.0D0) GO TO 6240
6239    RX3=RX3*.3D0
        GO TO 6236
6240    CONTINUE
C
        ALP=0.01D0/ALP
        ALPS=ALP
        DO 6300 M=1,ND
        RB22(M)=DBR(M,2)
        PR1(M)=PR1(M)*RX3
6300    CONTINUE
C
C       DETERMINE OPTIMUM VALUE OF INCREMENT FACTOR CEOP
6350    EMAT(1,2)=ERR2
        EMAT(2,2)=ERR2
        EMAT(3,2)=ERR2
        EMAT(1,1)=0.0D0
        EMAT(2,1)=0.0D0
        EMAT(3,1)=0.0D0
        DO 6500 I9=1,25
        CL=1.5D0
        CE=0.0D0
        DO 6400 J=1,I9
        CE=CE+ALP*(CL**(J-1))
6400    CONTINUE
C
        RX2=-RO
        DO 6450 J=1,ND
        DBR(J,2)=RB22(J)+CE*PR1(J)
        RX2=RX2-DBR(J,2)
6450    CONTINUE
        DBR(NB,2)=RX2
        CALL SCER(DM,DBR,RO,NB,NW,ERR)
C       STORE CE, ERR FOR LAST ITERATION
        EMAT(1,1)=EMAT(2,1)
        EMAT(1,2)=EMAT(2,2)
        EMAT(2,1)=EMAT(3,1)
        EMAT(2,2)=EMAT(3,2)
        EMAT(3,1)=CE
        EMAT(3,2)=ERR
        IF (ERR.GT.EMAT(2,2)) GOTO 6550
6500    CONTINUE
C
6550    IF (I9.GE.3) GOTO 6555
        ALP=ALP/10.0D0
        AAA=ALP/ALPS
        IF (AAA.GE.0.01D0) GOTO 6350
        GOTO 7360
C       QUADRATIC ESTIMATION
6555    DO 6590 K=1,10
        CEOP=(EMAT(2,1)**2-EMAT(3,1)**2)*EMAT(1,2)
        CED=(EMAT(2,1)-EMAT(3,1))*EMAT(1,2)
        CEOP=CEOP+(EMAT(3,1)**2-EMAT(1,1)**2)*EMAT(2,2)
```

```fortran
        CED=CED+(EMAT(3,1)-EMAT(1,1))*EMAT(2,2)
        CEOP=CEOP+(EMAT(1,1)**2-EMAT(2,1)**2)*EMAT(3,2)
        CED=CED+(EMAT(1,1)-EMAT(2,1))*EMAT(3,2)
        IF (CED.NE.0.0D0) GOTO 6556
        CED=1.0D-9
6556    CEOP=0.5D0*CEOP/CED
        RX2=-RO
        DO 6558 I=1,ND
        DBR(I,2)=RB22(I)+CEOP*PR1(I)
        RX2=RX2-DBR(I,2)
6558    CONTINUE
        DBR(NB,2)=RX2
        CALL SCER(DM,DBR,RO,NB,NW,ERR)
        CEC=ERR/EMAT(2,2)
        IF (CEC.GT.1.0D0) GOTO 6560
        IF (CEC.GE.0.98D0) GOTO 7360
6560    IF (CEOP.GT.EMAT(2,1)) GOTO 6570
        IF (ERR.GT.EMAT(2,2)) GOTO 6565
        EMAT(3,1)=EMAT(2,1)
        EMAT(3,2)=EMAT(2,2)
        EMAT(2,1)=CEOP
        EMAT(2,2)=ERR
        GOTO 6590
6565    EMAT(1,1)=CEOP
        EMAT(1,2)=ERR
        GOTO 6590
6570    IF (ERR.LT.EMAT(2,2)) GOTO 6580
        EMAT(3,1)=CEOP
        EMAT(3,2)=ERR
        GOTO 6590
6580    EMAT(1,1)=EMAT(2,1)
        EMAT(1,2)=EMAT(2,2)
        EMAT(2,1)=CEOP
        EMAT(2,2)=ERR
6590    CONTINUE
C
7360    I4=I4+1
        IF (I4.LE.I1) GO TO 1200
C
7370    I7=7
        GO TO 1215
7400    CONTINUE
8000    CONTINUE
        IF (NTI.EQ.1) GOTO 8500
        NTI=1
        DBR(NB,1)=BWB
        BWB=BWB/2.0D0/3.141592654D0
        WRITE(2,8200) BWB
8200    FORMAT(/' BEST INCREMENT FREQUENCY:',D17.7/)
        GOTO 1100
8500    CONTINUE
        STOP
        END
        SUBROUTINE SCER(DM,DBR,RO,NB,NW,ERR)
        DOUBLE PRECISION DM(20,4),DBR(10,2),GW(20),BW(20),GO(20)
        DOUBLE PRECISION EW(20),AW(10),BV(10),W2,RO,GX1,BY1,BY2,BY3,BY5
        DOUBLE PRECISION BFY,BX1,GT1,ERR,BY4(2)
C
        ERR=0.0D0
        DO 2500 M=1,NW
        W2=DM(M,1)
        AW(1)=1.0D0
        IF (W2.GE.DBR(1,1)) GOTO 1250
        AW(1)=W2/DBR(1,1)
```

```
         DO 1230 J=2,NB
         AW(J)=0.0D0
1230     CONTINUE
         GO TO 1275
C
1250     DO 1270 I=2,NB
         AW(I)=1.0D0
         IF (W2.GE.DBR(I,1)) GOTO 1270
         AW(I)=(W2-DBR(I-1,1))/(DBR(I,1)-DBR(I-1,1))
C
         IF (I.EQ.NB) GO TO 1270
         NE=NB-I
         DO 1260 J=1,NE
         AW(I+J)=0.0D0
1260     CONTINUE
         GO TO 1275
1270     CONTINUE
C
1275     GX1=0.0D0
         DO 2100 L=1,NB
         GX1=GX1+AW(L)*DBR(L,2)
2100     CONTINUE
         GW(M)=GX1+RO
C
         BY2=W2/DBR(1,1)+1.0D0
         BY1=W2/DBR(1,1)-1.0D0
         BY3=W2/DBR(1,1)
         BY5=DABS(BY1)
         BFY=BY2*DLOG(BY2)+BY1*DLOG(BY5)-2.0D0*BY3*DLOG(BY3)
         BV(1)=BFY/3.141592654D0
         DO 2200 I=2,NB
         DO 2150 J=1,2
         BY2=W2/DBR(I+J-2,1)+1.0D0
         BY1=W2/DBR(I+J-2,1)-1.0D0
         BY3=W2/DBR(I+J-2,1)
         BY5=DABS(BY1)
         BY4(J)=BY2*DLOG(BY2)+BY1*DLOG(BY5)-2.0D0*BY3*DLOG(BY3)
         BY4(J)=BY4(J)*DBR(I+J-2,1)
2150     CONTINUE
         BV(I)=1.0D0/(DBR(I,1)-DBR(I-1,1))*(BY4(2)-BY4(1))/3.141592654D0
2200     CONTINUE
C
2310     BX1=0.0D0
         DO 2400 J=1,NB
         BX1=BX1+BV(J)*DBR(J,2)
2400     CONTINUE
         BW(M)=BX1
C
         GT1=4.0D0*GW(M)*DM(M,2)
         GO(M)=GT1/((GW(M)+DM(M,2))**2.0D0+(DABS(BW(M)+DM(M,3)))**2.0D0)
C
         EW(M)=GO(M)/DM(M,4)-1.0D0
         ERR=ERR+(DABS(EW(M)))**2.0D0
2500     CONTINUE
         RETURN
         END
```

Appendix D

RCDM FORTRAN

```
C
C
C
C     THIS PROGRAM DETERMINES A FUNCTION (Z(S)) FOR THE INPUT IMPEDANCE
C     OF A LOSSLESS IMPEDANCE MATCHING NETWORK THAT WILL MATCH A COMPLEX
C     SOURCE TO A COMPLEX LOAD IN A PRESCRIBED WAY.  WHERE IT IS
C     THEORETICALLY IMPOSSIBLE TO ACHIEVE THE SPECIFICATIONS THE IN-
C     PUT IMPEDANCE OF THE OPTIMUM NETWORK WILL BE DETERMINED.
C
C     THE RESULTS CONSIST OF THE REQUIRED IMPEDANCE FUNCTION AND A  COM-
C     PARISON BETWEEN THE TRANSDUCER POWER GAIN OBTAINED AND THAT SPECI-
C     FIED.  THE IMPEDANCE FUNCTION DETERMINED IS THAT OF THE INPUT
C     IMPEDANCE OF THE NETWORK TERMINATED IN A 50 OHM LOAD.
C
C     THE DATA TO BE ENTERED INTO THE PROGRAM CONSIST OF THE NUMBER
C     OF FREQUENCIES WHERE THE SOURCE AND LOAD IMPEDANCE ARE TO BE
C     SPECIFIED, THESE IMPEDANCES AND THE REQUIRED TRANSDUCER
C     POWER GAIN AT THE SPECIFIED FREQUENCIES, THE DEGREE OF THE RE-
C     FLECTION COEFFICIENT S11(S), THE NUMBER OF ZEROS IT HAS AT
C     THE ORIGIN, THE NUMBER OF ITERATIONS TO BE DONE AND A NUMBER
C     SPECIFYING THE DETAIL OF THE PRINTOUT.
C
C     IT SHOULD BE NOTED THAT A  SUBROUTINE YIELDING MORE ACCURATE
C     APPROXIMATIONS TO THE ROOTS OF A POLYNOMIAL IS SOMETIMES RE-
C     QUIRED.  THE SUBROUTINE TO BE REPLACED IS ZPOLR(GG,MM,POLES,IER),
C     WHERE GG(20) IS THE MATRIX CONTAINING THE COEFFICIENTS OF THE
C     POLYNOMIAL OF WHICH THE ROOTS ARE TO BE DETERMINED WITH
C     GG(1)=XN, GG(2)=XN-1,  ... , GG(MM+1)=XO, MM IS THE DEGREE OF
C     THE POLYNOMIAL, POLES THE COMPLEX MATRIX CONTAINING THE POLES AND
C     IER AN ERROR PARAMETER.
C
      DIMENSION PGTT(10,10),TWI(20),CCCC(20,6)
      DOUBLE PRECISION TW(20)
      DIMENSION RSM(20),XSM(20),RLM(20),XLM(20),WW(20)
      DIMENSION PH(10),PG(10),PHD(10),GG(20),PGT(10)
      DIMENSION ZNUM(10),ZDEN(10),NROWS(10,2),PHT(10)
      DOUBLE PRECISION FW(10),FWM(10,10),EOFW(10),EW(20)
      DOUBLE PRECISION FWMS(10,10),EOFWS(10),PR1(10)
      DOUBLE PRECISION EMAT(3,2),ALP,ALPS,CL,CE,F,ERR2
      COMPLEX SL(20),SS(20),POLES(20),POLHP(10),PGO(20),PGN(20)
      COMPLEX ACC1,ACC2,CC1,CPH,CPHC,CPG,SSS,SLL,CPGC
      INTEGER T,Q1,Q,R,S,L5,M4,M3,NB,NW,NC,ND,NE,NTI,NNNN(3)
      DOUBLE PRECISION AD1,ACD1,ACD2,ER,TWDT,TWDI,PHDD,ERR
      DOUBLE PRECISION E2,Z3,ABC,AMN,ERFL
      DOUBLE PRECISION CEOP,CEC,CED,XNW
C
```

```
      WRITE(2,1010)
      WRITE(3,1010)
1010  FORMAT(/' NUMBER OF FREQUENCIES AT WHICH THE SOURCE AND LOAD IMPED
     *ANCE ARE TO'/' BE SPECIFIED?')
      READ(1,*) NW
      XNW=DFLOAT(NW)
      WRITE(2,1015) NW
      WRITE(3,1015) NW
1015  FORMAT(I3)
C
      WRITE(2,1020)
      WRITE(3,1020)
1020  FORMAT(/' ENTER:  F, RS, XS, RL, XL, GT-REQUIRED  ')
      DO 1030 I=1,NW
      READ(1,*) WW(I),RSM(I),XSM(I),RLM(I),XLM(I),TWI(I)
      WRITE(2,1025) WW(I),RSM(I),XSM(I),RLM(I),XLM(I),TWI(I)
      WRITE(3,1025) WW(I),RSM(I),XSM(I),RLM(I),XLM(I),TWI(I)
1025  FORMAT(E10.4,4(F12.6,1X),F7.4)
      WW(I)=WW(I)*2.00*3.141592654
      TWI(I)=TWI(I)
1030  CONTINUE
C
1035  WRITE(2,1040)
      WRITE(3,1040)
1040  FORMAT(/' DEGREE OF S11(S) NUMERATOR (PH(S))?  NUMBER OF ZEROS AT
     *THE ORIGIN?')
      READ(1,*) NDH,KK
      WRITE(2,1042) NDH,KK
      WRITE(3,1042) NDH,KK
1042  FORMAT(I3,I3)
C
      ERR2=100.0D0
      CK=1.0
      IF (KK.EQ.0) GOTO 1048
      DO 1045 I=1,KK
      CK=-CK
1045  CONTINUE
C
1048  NDH1=NDH+1
      WRITE(2,1050)
      WRITE(3,1050)
1050  FORMAT(/' ENTER NUMERATOR COEFFICIENTS: H0 / H1 / H2 / ...HN')
      DO 1060 I=1,NDH1
      READ(1,*) PH(I)
      WRITE(2,1055) PH(I)
      WRITE(3,1055) PH(I)
1055  FORMAT(D22.15)
1060  CONTINUE
C
      KKN=0
      KK0=0
      IF (KK.NE.0) GOTO 1062
      PH(1)=0.0
      KK0=1
1062  IF (KK.NE.NDH) GOTO 1065
      PH(NDH1)=0.0
      KKN=1
C
1065  WRITE(2,1070)
      WRITE(3,1070)
1070  FORMAT(/' NUMBER OF ITERATIONS TO BE DONE?  DETAILED PRINTOUT? (1-
     *)')
      READ(1,*) I1,I2
```

```
       WRITE(2,1072) I1,I2
       WRITE(3,1072) I1,I2
 1072  FORMAT(I3,I3)
       NSCIT=0
C
C      CALCULATE REFLECTION COEFFICIENTS FOR SOURCE AND LOAD IMPEDANCES
       DO 1080 I=1,NW
       SS(I)=CMPLX(RSM(I)-50.0,XSM(I))/CMPLX(RSM(I)+50.0,XSM(I))
       SL(I)=CMPLX(RLM(I)-50.0,XLM(I))/CMPLX(RLM(I)+50.0,XLM(I))
 1080  CONTINUE
C
       M3=1
       I7=0
       I4=0
       ERR1=0.0D0
       NDH2=NDH+2
       DO 1100 I=1,NW
       CCCC(I,1)=WW(I)
       CCCC(I,2)=REAL(SS(I))
       CCCC(I,3)=AIMAG(SS(I))
       CCCC(I,4)=REAL(SL(I))
       CCCC(I,5)=AIMAG(SL(I))
       CCCC(I,6)=TWI(I)
 1100  CONTINUE
       NNNN(1)=NDH
       NNNN(2)=NW
       NNNN(3)=KK
C
 1200  DO 1210 I=1,NDH2
       DO 1205 J=1,NDH2
       FWMS(I,J)=0.0D0
 1205  CONTINUE
       EOFWS(I)=0.0D0
 1210  CONTINUE
C
C      GENERATE POLINOMIAL G(S) WHERE S11(S)=H(S)/G(S)
 1215  ERR=0.0D0
C      G(S)*G(-S)
 1220  CALL SPG(PH,PG,NDH,KK,CK)
C      CALCULATE EW(M)
       ERR=0.0D0
       DO 1240 M=1,NW
       W2=WW(M)
       SSS=SS(M)
       SLL=SL(M)
C
C      CALCULATE TRANSDUCER POWER GAIN CORRESPONDING TO SOLUTION
       AC1=1.0-(REAL(SSS))**2-(AIMAG(SSS))**2
       AC2=1.0-(REAL(SLL))**2-(AIMAG(SLL))**2
       AC3=W2**KK
       AC1=AC1*AC2*AC3**2
C      CALCULATE PH(JW), PG(JW)
       CPH=(0.0,0.0)
       CPG=(0.0,0.0)
       DO 1230 I=1,NDH1
       CPH=CPH+CMPLX(PH(I),0.0)*CMPLX(0.0,W2)**(I-1)
       CPG=CPG+CMPLX(PG(I),0.0)*CMPLX(0.0,W2)**(I-1)
 1230  CONTINUE
       CPHC=CONJG(CPH)
       CPGC=CONJG(CPG)
       ACC2=CPG-CPGC*CMPLX(CK,0.0)*SSS*SLL
       ACC2=ACC2-CPH*SSS+CMPLX(CK,0.0)*CPHC*SLL
       AD1=AC1
```

```
           ACD1=REAL(ACC2)
           ACD2=AIMAG(ACC2)
           TW(M)=AD1/(ACD1**2+ACD2**2)
           TWDI=TWI(M)
           EW(M)=TW(M)/TWDI-1.0D0
           ERR=ERR+(EW(M))**2
1240       CONTINUE
           ERFL=ERR/ERR2
           IF (ERR.LE.(0.9D0*XNW)) GOTO 1245
           NSCIT=NSCIT+1
           XX=WW(NW)**KK/CABS(CPG)
           DO 1242 I=1,NDH1
1242       PH(I)=PH(I)*XX
           IF (NSCIT.LE.5) GOTO 1215
C
1245       IF (ERFL.LT.0.9999D0) GOTO 1250
           I4=I1+1
1250       ERR2=ERR
C
C          PRINT RESULTS
1600       IF (I2.EQ.1) GOTO 1866
           IF (I7.EQ.0) GOTO 1885
C          CALCULATE THE IMPEDANCE FUNCTION
1866       DO 1867 I=1,NDH1
           ZNUM(I)=(PH(I)+PG(I))*50.0
           ZDEN(I)=PG(I)-PH(I)
1867       CONTINUE
C
           WRITE(2,1868)
           WRITE(3,1868)
1868       FORMAT(/'COEFFICIENTS OF THE NUMERATOR AND DENOMINATOR OF THE IMPE
          *DANCE FUNCTION OF THE'/' DESIGNED NETWORK TERMINATED IN 50 OHMS')
           WRITE(2,1869)
           WRITE(3,1869)
1869       FORMAT(/' SEQUENCE OF RESULTS: N0, D0 / N1, D1 / ...'/)
           DO 1871 I=1,NDH1
           WRITE(2,1870) ZNUM(I),ZDEN(I)
           WRITE(3,1870) ZNUM(I),ZDEN(I)
1870       FORMAT('       ',5X,2(D22.15,3X))
1871       CONTINUE
           WRITE(2,1872)
           WRITE(3,1872)
1872       FORMAT(//)
           DO 1880 M=1,NW
           F00=WW(M)/2.0/3.141592654
1873       WRITE(2,1875)F00,TW(M),TWI(M)
           WRITE(3,1875)F00,TW(M),TWI(M)
1875       FORMAT(' F GT GT-OPT:',E15.5,2(F12.6,5X))
C
1880       CONTINUE
C
1882       IF (I2.EQ.1) GOTO 1885
           IF (I7.EQ.0) GOTO 1895
1885       WRITE (2,1887) ERR
           WRITE (3,1887) ERR
1887       FORMAT(/' ERROR:',E16.8)
1895       CONTINUE
           IF (I7.EQ.7) GOTO 7400
C
C          CREATE MATRIX FOR DETERMINING NEW SET OF COEFFICIENTS
C
           NDD1=NDH1-KKN
           ND1=1+KK0
```

```
        NDD2=NDH2-KKN
        DO 2500 I=ND1,NDD1
        DO 2000 J=1,NDH1
        PHT(J)=PH(J)
 2000   CONTINUE
C
C       CHANGE VALUE OF PH(I) WITH 2%
        PHT(I)=PHT(I)*1.01
        CALL SPG(PHT,PGT,NDH,KK,CK)
        DO 2200 K=1,NDH1
        PGTT(K,I)=PGT(K)
 2200   CONTINUE
 2500   CONTINUE
C
        FW(1)=0.0D0
        FW(NDH1)=0.0D0
C       GENERATE F(W)
        DO 3880 M=1,NW
        W2=WW(M)
        SSS=SS(M)
        SLL=SL(M)
        DO 2800 J=ND1,NDD1
        DO 2600 I=1,NDH1
        PHT(I)=PH(I)
        PGT(I)=PGTT(I,J)
 2600   CONTINUE
        PHT(J)=PHT(J)*1.01
C
C       CALCULATE TRANSDUCER POWER GAIN CORRESPONDING TO SOLUTION
C
        AC1=1-(REAL(SSS))**2-(AIMAG(SSS))**2
        AC2=1-(REAL(SLL))**2-(AIMAG(SLL))**2
        AC3=W2**KK
        AC1=AC1*AC2*AC3**2
C       CALCULATE PH(JW), PG(JW)
        CPH=(0.0,0.0)
        CPG=(0.0,0.0)
        DO 2650 I=1,NDH1
        CPH=CPH+CMPLX(PHT(I),0.0)*CMPLX(0.0,W2)**(I-1)
        CPG=CPG+CMPLX(PGT(I),0.0)*CMPLX(0.0,W2)**(I-1)
 2650   CONTINUE
        CPHC=CONJG(CPH)
        CPGC=CONJG(CPG)
        ACC2=CPG-CPGC*CMPLX(CK,0.0)*SSS*SLL
        ACC2=ACC2-CPH*SSS+CMPLX(CK,0.0)*CPHC*SLL
        AD1=AC1
        ACD1=REAL(ACC2)
        ACD2=AIMAG(ACC2)
        TWT=AD1/(ACD1**2+ACD2**2)
        TWDI=TWI(M)
        ER=TWT/TWDI-1.0D0
        PHDD=PH(J)
        FW(J)=(ER-EW(M))/(0.01D0*PHDD)
 2800   CONTINUE
C
C       CREATE MATRIX FWM
        DO 3500 K=1,NDH1
        DO 3450 L=1,NDH1
        FWM(K,L)=FW(K)*FW(L)
 3450   CONTINUE
 3500   CONTINUE
C
 3755   DO 3850 J=1,NDH1
```

```
      DO 3800 K=1,NDH1
      FWMS(J,K)=FWMS(J,K)+FWM(J,K)
 3800 CONTINUE
      EOFWS(J)=EOFWS(J)-EW(M)*FW(J)
 3850 CONTINUE
 3880 CONTINUE
C
C     DETERMINE NEW SOLUTION
 4001 DO 4030 M=1,NDH1
      FWMS(M,NDD2)=EOFWS(M)
 4030 CONTINUE
C
C
 4052 MRS=0
 4053 DO 4175 M=ND1,NDD1
      IF (FWMS(M,M).GT.0.00D0) GO TO 4085
      IF (FWMS(M,M).LT.0.00D0) GO TO 4085
      MRS=MRS+1
      NROWS(MRS,1)=M
C     FIND NON-ZERO ROW
 4055 M4=M+1
      Q1=M+1
      DO 4060 R=Q1,NDD1
      IF (FWMS(R,M).EQ.0.00D0) GO TO 4057
      M3=1
      GO TO 4065
 4057 M3=-1
      M4=M4+1
 4060 CONTINUE
C
 4065 IF (M3.EQ.-1) GOTO 7400
      NROWS(MRS,2)=R
C     REVERSE ROWS
      DO 4070 S=M,NDD2
      Z3=FWMS(M,S)
      FWMS(M,S)=FWMS(R,S)
      FWMS(R,S)=Z3
 4070 CONTINUE
C
C     NORMALIZE ROW
 4085 E2=FWMS(M,M)
      DO 4100 N=M,NDD2
      FWMS(M,N)=FWMS(M,N)/E2
 4100 CONTINUE
C
      IF (M.EQ.NDD1) GO TO 4175
C
C     SUBTRACT
      NE=M+1
      DO 4150 N=NE,NDD1
      E2=FWMS(N,M)
      DO 4125 I=M,NDD2
      FWMS(N,I)=FWMS(N,I)-E2*FWMS(M,I)
 4125 CONTINUE
 4150 CONTINUE
C
 4175 CONTINUE
C     COMPLETE SUBTRACTION
      N=NDD1
 4180 L=N-1
 4190 FWMS(L,NDD2)=FWMS(L,NDD2)-FWMS(L,N)*FWMS(N,NDD2)
      FWMS(L,N)=0.00D0
      L=L-1
```

```
4200    IF (L.GE.1) GO TO 4190
        N=N-1
4300    IF (N.GE.2) GO TO 4180
4340    ALP=0.0D0
        DO 4350 M=ND1,NDD1
        PR1(M)=FWMS(M,NDD2)
        PHT(M)=PH(M)
        PHDD=PH(M)
        F=DABS(PR1(M)/PHDD)
        IF (F.LT.ALP) GOTO 4350
        ALP=F
4350    CONTINUE
        ALP=0.01D0/ALP
        ALPS=ALP
C
C       DETERMINE OPTIMUM VALUE OF INCREMENT FACTOR CEOP
6350    EMAT(1,2)=ERR2
        EMAT(2,2)=ERR2
        EMAT(3,2)=ERR2
        EMAT(1,1)=0.0D0
        EMAT(2,1)=0.0D0
        EMAT(3,1)=0.0D0
        DO 6500 I9=1,25
        CL=1.5D0
        CE=0.0D0
        DO 6400 J=1,I9
        CE=CE+ALP*(CL**(J-1))
6400    CONTINUE
C
        DO 6450 J=ND1,NDD1
        AA=CE*PR1(J)
        PH(J)=PHT(J)+AA
6450    CONTINUE
        CALL SCER(PH,NNNN,CK,CCCC,ERR)
C       STORE CE, ERR FOR LAST ITERATION
        EMAT(1,1)=EMAT(2,1)
        EMAT(1,2)=EMAT(2,2)
        EMAT(2,1)=EMAT(3,1)
        EMAT(2,2)=EMAT(3,2)
        EMAT(3,1)=CE
        EMAT(3,2)=ERR
        IF (ERR.GT.EMAT(2,2)) GOTO 6550
6500    CONTINUE
C
6550    IF (I9.GE.3) GOTO 6555
        ALP=ALP/10.0D0
        AAA=ALP/ALPS
        IF (AAA.GE.0.001D0) GOTO 6350
        GOTO 7360
C       QUADRATIC ESTIMATION
6555    DO 6590 K=1,25
        CEOP=(EMAT(2,1)**2-EMAT(3,1)**2)*EMAT(1,2)
        CED=(EMAT(2,1)-EMAT(3,1))*EMAT(1,2)
        CEOP=CEOP+(EMAT(3,1)**2-EMAT(1,1)**2)*EMAT(2,2)
        CED=CED+(EMAT(3,1)-EMAT(1,1))*EMAT(2,2)
        CEOP=CEOP+(EMAT(1,1)**2-EMAT(2,1)**2)*EMAT(3,2)
        CED=CED+(EMAT(1,1)-EMAT(2,1))*EMAT(3,2)
        IF (CED.NE.0.0D0) GOTO 6556
        CED=1.0D-9
6556    CEOP=0.5D0*CEOP/CED
C
        DO 6558 I=ND1,NDD1
        AA=CEOP*PR1(I)
```

```
      PH(I)=PHT(I)+AA
6558  CONTINUE
      CALL SCER(PH,NNNN,CK,CCCC,ERR)
      CEC=ERR/EMAT(2,2)
      IF (CEC.GT.1.0D0) GOTO 6560
      IF ((ERR/XNW).LE.0.5) GOTO 6559
      IF (CEC.GE.0.999D0) GOTO 7360
      GOTO 6560 .
6559  IF (CEC.GE.0.99D0) GOTO 7360
6560  IF (CEOP.GT.EMAT(2,1)) GOTO 6570
      IF (ERR.GT.EMAT(2,2)) GOTO 6565
      EMAT(3,1)=EMAT(2,1)
      EMAT(3,2)=EMAT(2,2)
      EMAT(2,1)=CEOP
      EMAT(2,2)=ERR
      GOTO 6590
6565  EMAT(1,1)=CEOP
      EMAT(1,2)=ERR
      GOTO 6590
6570  IF (ERR.LT.EMAT(2,2)) GOTO 6580
      EMAT(3,1)=CEOP
      EMAT(3,2)=ERR
      GOTO 6590
6580  EMAT(1,1)=EMAT(2,1)
      EMAT(1,2)=EMAT(2,2)
      EMAT(2,1)=CEOP
      EMAT(2,2)=ERR
6590  CONTINUE
C
7360  I4=I4+1
      IF (I4.LE.I1) GO TO 1200
      I7=7
      GO TO 1200
7400  CONTINUE
C
      WRITE(2,7405)
      WRITE(3,7405)
7405  FORMAT(/)
      DO 7420 I=1,NDH1
      WRITE(2,7410) PH(I),PG(I)
      WRITE(3,7410) PH(I),PG(I)
7410  FORMAT(' PH:',D22.15,' PG:',D22.15)
7420  CONTINUE
      WRITE(2,7425)
      WRITE(3,7425)
7425  FORMAT(//)
      IF (M3.EQ.1) GOTO 7500
      WRITE(2,7450) M
      WRITE(3,7450) M
7450  FORMAT(/' RANK OF ERROR MATRIX LOWER THAN EXPECTED  ZERO ROW:',I1)
7500  CONTINUE
      WRITE(2,7505)
7505  FORMAT(//)
      STOP
      END
      SUBROUTINE SPG(PH,PG,NDH,KK,CK)
      DIMENSION PH(10),PG(10),GG(20)
      DOUBLE PRECISION GGD(20),PHD(10),DK,GGC,TMULT,AXX
      COMPLEX POLES(20),POLHP(10),PGO(10),PGN(10)
C     GENERATE POLINOMIAL G(S) WHERE S11(S)=H(S)/G(S)
C     G(S)*G(-S)
C
      NDH1=NDH+1
```

```fortran
      DO 7000 I=1,20
      GGD(I)=0.0D0
      IF (I.GT.NDH1) GOTO 7000
      PHD(I)=PH(I)
7000  CONTINUE
C
C
      MN=2*NDH+1
      DO 7500 I=1,NDH1
      DO 7200 J=1,NDH1
      DK=-1.0D0
      DO 7100 K=1,J
      DK=-DK
7100  CONTINUE
      GGD(2*NDH+3-I-J)=GGD(2*NDH+3-I-J)+PHD(I)*PHD(J)*DK
7200  CONTINUE
7500  CONTINUE
      GGD(2*NDH+1-2*KK)=GGD(2*NDH+1-2*KK)+CK
C
C     STORE S**2N COEFFICIENT
      GGC=GGD(1)
C
C     SCALE POLINOMIAL GG FOR DETERMINING IT POLES
      TMULT=1.0D0
      DO 7520 I=1,30
      AXX=GGD(1)*TMULT
      IF (AXX.GE.1.0D-9) GOTO 7525
      TMULT=TMULT*10.0D0
7520  CONTINUE
C
7525  IF (TMULT.NE.1.0D0) GOTO 7538
      DO 7535 I=1,30
      AXX=GGD(2*NDH+1)*TMULT
      IF (AXX.LE.1.0D9) GOTO 7535
      TMULT=TMULT/10.0D0
7535  CONTINUE
C
7538  MN=2*NDH+1
      DO 7550 I=1,MN
      GGD(I)=GGD(I)*TMULT
      GG(I)=GGD(I)
7550  CONTINUE
C
C     DETERMINE ROOTS OF POLINOMIAL GG
      MM=NDH*2
      CALL ZPOLR(GG,MM,POLES,IER)
      CONTINUE
C
C     FIND LEFT HAND PLANE POLES
      MM=2*NDH
      II=0
      DO 7600 I=1,MM
      AXY=REAL(POLES(I))
      IF (AXY.GE.0.0) GOTO 7600
      II=II+1
      POLHP(II)=POLES(I)
C
7600  CONTINUE
C
C     CALCULATE POLINOMIAL G(S)
C
      DO 7650 I=1,10
      PGO(I)=(0.0,0.0)
```

```
            PGN(I)=(0.0,0.0)
7650    CONTINUE
C

        J=1
        PGO(1)=(1.0,0.0)
        DO 7800 I=1,NDH
C

        I1=I+1
        DO 7700 L=1,I1
        PGN(L)=-PGO(L)*POLHP(I)
7700    CONTINUE
C

        I2=I+2
        DO 7750 L=2,I1
        PGN(L)=PGN(L)+PGO(L-1)
7750    CONTINUE
C

        DO 7775 L=1,I1
        PGO(L)=PGN(L)
7775    CONTINUE
C
7800    CONTINUE
C
C
7810    NDH1=NDH+1
        DO 7820 I=1,NDH1
        GGCC=GGC
        PG(I)=REAL(PGN(I))*SQRT(ABS(GGCC))
C
7820    CONTINUE
        RETURN
        END
        SUBROUTINE ZPOLR(A,NDEG,Z,IER)
        DIMENSION A(20)
        COMPLEX Z(NDEG)
        COMPLEX RT,FXI,FXIM1,FXIM2,H,XLAM,AA,XNUM,DEN,DFIIM1
        COMPLEX DIM1M2,G,SQR,FRT
C
C       REFERENCE:  ELEMENTARY NUMERICAL ANALYSIS - CONTE, DE BOOR
C       MCGRAW-HILL
C
        IER=0
        NNN=NDEG+1
        XNDEG=FLOAT(NDEG)
        DO 2 I=1,NDEG
2       Z(I)=CMPLX(0.0,0.0)
C
        CC=1.0
        AAA=A(1)/A(NDEG+1)
        BB=AMAX1(AAA,1.0/AAA)
        IF (BB.LE.(10.0**NDEG)) GOTO 4
        CC=EXP(ALOG(A(1)/A(NDEG+1)*10.0**(NDEG))/XNDEG)
        DO 3 I=1,NNN
3       A(I)=A(I)*CC**(I-1)
C
4       CCC=SQRT(ABS(A(1)*A(NDEG+1)))
        DO 5 I=1,NNN
5       A(I)=A(I)*CCC
C
        DMX=1.0E-9
        DMF=1.0E-10
        MAXNIT=200
C
```

```
        DO 1000 K=1,NDEG
C       INITIALIZE
        MFLAG=0
        MFLAGD=0
        IF (CABS(Z(K)).NE.0.0) MFLAG=1
        IF (MFLAG.EQ.1) GOTO 8
        DMFT=1.0E-10
8       NIT=0
        NFLAG=0
10      H=CMPLX(0.5,0.0)
        RT=Z(K)+H
        ASSIGN 60 TO NN
        GOTO 210
60      FXIM2=FXI
C
        RT=Z(K)-H
        ASSIGN 120 TO NN
        GOTO 210
120     FXIM1=FXI
C
        FXIM2=FXIM1-FXIM2
        RT=Z(K)
        ASSIGN 260 TO NN
        NFLAG=1
        GOTO 210
180     NFLAG=0
C
        XLAM=CMPLX(-0.5,0,0)
C       COMPUTE NEXT ESTIMATE FOR ROOT
200     MFLAGD=0
        DFIIM1=FXI-FXIM1
        AA=FXIM2*XLAM
        XNUM=-FXI*CMPLX(2.0,0.0)*(CMPLX(1.0,0.0)+XLAM)
        IF (CABS(XNUM).LE.1.0E-27) XNUM=0.0
        IF (CABS(XLAM).LE.1.0E-27) XLAM=0.0
        G=(CMPLX(1.0,0.0)+CMPLX(2.0,0.0)*XLAM)*DFIIM1-XLAM*AA
        SQR=XLAM*(DFIIM1-AA)
        SQR=G*G+SQR*XNUM*CMPLX(2.0,0.0)
        SQR=CSQRT(SQR)
        DEN=G+SQR
        IF (REAL(G)*REAL(SQR)+AIMAG(G)*AIMAG(SQR).LT.0.0)  DEN=G-SQR
        IF (CABS(DEN).EQ.0) DEN=1
        XLAM=XNUM/DEN
        FXIM1=FXI
        FXIM2=DFIIM1
        IF (CABS(H).LE.1.0E-9) NIT=MAXNIT
        IF (CABS(H).LE.1.0E-9) GOTO 205
        H=H*XLAM
        RT=RT+H
205     IF (NIT.LT.MAXNIT) GOTO 210
        IER=IER+1
        GOTO 800
C
210     NIT=NIT+1
        FXI=CMPLX(A(1),0.0)
        DO 220 I=1,NDEG
220     FXI=FXI*RT+CMPLX(A(1+I),0.0)
        FRT=FXI
        IF (K.LT.2) GOTO 240
        DO 230 I=2,K
        AA=RT-Z(I-1)
        IF (CABS(AA).GT.(DMX*CABS(Z(I-1)))) GOTO 230
        Z(K)=RT+0.001
        GOTO 8
```

```
230    FXI=FXI/AA
240    IF (MFLAGD.EQ.1) GOTO 270
       GOTO NN,(60,120,260)
C
C      CHECK FOR CONVERGENCE
260    IF (CABS(H).GT.DMX*CABS(RT)) GOTO 268
       IF (AMAX1(CABS(FXI),CABS(FRT)).LT.DMF) GOTO 800
268    IF (NFLAG.NE.1) GOTO 270
       IF (MFLAG.EQ.0) GOTO 180
       IF (AMAX1(CABS(FRT),CABS(FXI)).LT.DMFT) GOTO 800
       GOTO 180
C
C      CHECK FOR DIVERGENCE
270    IF (CABS(FXI).LE.(1.1*CABS(FXIM1))) GOTO 200
       MFLAGD=1
       H=H/CMPLX(2.0,0.0)
       IF (CABS(H).LE.1.0E-9) GOTO 200
       XLAM=XLAM/CMPLX(2.0,0.0)
       RT=RT-H
       GOTO 205
800    IF (K.EQ.NDEG) GOTO 980
       IF (CABS(Z(K+1)).NE.0.0) GOTO 980
       IF (K.EQ.1) GOTO 810
       IF (CABS(Z(K)).EQ.CABS(Z(K-1))) GOTO 980
810    RA=ABS(AIMAG(RT)/REAL(RT))
       IF (RA.LE.1000.0) GOTO 910
       Z(K+1)=CONJG(RT)
       DMFT=100.0*AMAX1(CABS(FXI),CABS(FRT))
       GOTO 980
910    IF (RA.GE.0.001) GOTO 920
       Z(K+1)=-RT
       DMFT=100.0*AMAX1(CABS(FXI),CABS(FRT))
       GOTO 980
920    Z(K+1)=CONJG(RT)
       IF (NDEG.LE.(K+3)) GOTO 980
       Z(K+2)=-CONJG(RT)
       Z(K+3)=-RT
       DMFT=100.0*AMAX1(CABS(FXI),CABS(FRT))
980    CONTINUE
1000   Z(K)=RT
C
       DO 1200 I=1,NDEG
       Z(I)=Z(I)/CC
1200   CONTINUE
C
       RETURN
       END
       SUBROUTINE SCER(PH,NNNN,CK,CCCC,ERR)
       DIMENSION PH(10),PG(10),TWI(20),CCCC(20,6),WW(20)
       DOUBLE PRECISION TW(20)
       DOUBLE PRECISION AD1,TWDI,EW(20),ERR
       COMPLEX SS(20),SL(20),ACC1,ACC2,CC1,CPH,CPHC,CPGC,CPG,SSS,SLL
       INTEGER NNNN(3)
C
       NDH=NNNN(1)
       NW=NNNN(2)
       KK=NNNN(3)
       DO 1000 I=1,NW
       WW(I)=CCCC(I,1)
       SS(I)=CMPLX(CCCC(I,2),CCCC(I,3))
       SL(I)=CMPLX(CCCC(I,4),CCCC(I,5))
       TWI(I)=CCCC(I,6)
1000   CONTINUE
C
```

```fortran
C
C     GENERATE POLINOMIAL G(S) WHERE S11(S)=H(S)/G(S)
C
      NDH1=NDH+1
      NDH2=NDH+2
1215  ERR=0.0D0
C     G(S)*G(-S)
1220  CALL SPG(PH,PG,NDH,KK,CK)
C     CALCULATE EW(M)
      ERR=0.0D0
      DO 1240 M=1,NW
      W2=WW(M)
      SSS=SS(M)
      SLL=SL(M)
C
C     CALCULATE TRANSDUCER POWER GAIN CORRESPONDING TO SOLUTION
      AC1=1.0-(REAL(SSS))**2-(AIMAG(SSS))**2
      AC2=1.0-(REAL(SLL))**2-(AIMAG(SLL))**2
      AC3=W2**KK
      AC1=AC1*AC2*AC3**2
C     CALCULATE PH(JW), PG(JW)
      CPH=(0.0,0.0)
      CPG=(0.0,0.0)
      DO 1230 I=1,NDH1
      CPH=CPH+CMPLX(PH(I),0.0)*CMPLX(0.0,W2)**(I-1)
      CPG=CPG+CMPLX(PG(I),0.0)*CMPLX(0.0,W2)**(I-1)
1230  CONTINUE
      CPHC=CONJG(CPH)
      CPGC=CONJG(CPG)
      ACC2=CPG-CPGC*CMPLX(CK,0.0)*SSS*SLL
      ACC2=ACC2-CPH*SSS+CMPLX(CK,0.0)*CPHC*SLL
      AD1=AC1
      ACD1=REAL(ACC2)
      ACD2=AIMAG(ACC2)
      TW(M)=AD1/(ACD1**2+ACD2**2)
      TWDI=TWI(M)
      EW(M)=TW(M)/TWDI-1.0D0
      ERR=ERR+(EW(M))**2
1240  CONTINUE
      RETURN
      END
```

Appendix E

S-PARAMETER EXPRESSIONS RELEVANT TO THE DESIGN OF RF AND MICROWAVE AMPLIFIERS

Load Stability Circle

Center:

$$C_L = \frac{s_{11} \Delta^* - s_{22}^*}{|\Delta|^2 - |s_{22}|^2} \tag{E.1}$$

Radius:

$$R_L = \left| \frac{s_{12} s_{21}}{|\Delta|^2 - |s_{22}|^2} \right| \tag{E.2}$$

Source Stability Circle

Center:

$$C_s = \frac{s_{22} \Delta^* - s_{11}^*}{|\Delta|^2 - |s_{11}|^2} \tag{E.3}$$

Radius:

$$R_s = \left| \frac{s_{12} s_{21}}{|\Delta|^2 - |s_{11}|^2} \right| \tag{E.4}$$

where

$$\Delta = s_{11} s_{22} - s_{12} s_{21} \tag{E.5}$$

Sterne Stability Factor

$$K = \frac{1 - |s_{11}|^2 - |s_{22}|^2 + |s_{11} s_{22} - s_{12} s_{21}|^2}{2 |s_{12} s_{21}|} \tag{E.6}$$

Maximum Operating and Available Power Gain of an Inherently Stable Transistor and the Corresponding Terminations

$$G_{max} = \left| \frac{s_{21}}{s_{12}} \right| \; [K - \sqrt{K^2 - 1}) \tag{E.7}$$

$$S_{s-opt} = C_1^* [B_1 - (B_1^2 - 4|C_1|^2)^{1/2}]/(2|C_1|^2) \qquad (E.8)$$

$$S_{L-opt} = C_2^* [B_2 - (B_2^2 - 4|C_2|^2)^{1/2}]/(2|C_2|^2) \qquad (E.9)$$

where

$$B_1 = 1 + |s_{11}|^2 - |s_{22}|^2 - |\Delta|^2 \qquad (E.10)$$

$$B_2 = 1 + |s_{22}|^2 - |s_{11}|^2 - |\Delta|^2 \qquad (E.11)$$

$$C_1 = s_{11} - s_{22}^* \Delta \qquad (E.12)$$

$$C_2 = s_{22} - s_{11}^* \Delta \qquad (E.13)$$

Circle of Constant Mismatch for a Voltage Source (Source Reflection Parameter S_s) Terminated in a Passive Load (Reflection Parameter S_L)

Transducer Power Gain

$$G_T = \frac{[1 - |S_L|^2][1 - |S_s|^2]}{|1 - S_s S_L|^2} \qquad (E.14)$$

CIRCLE

Center:

$$C_M = \frac{S_s^* G_T}{1 - |S_s|^2 [1 - G_T]} \qquad (E.15)$$

Radius:

$$R_M = \frac{[1 - |S_s|^2]\sqrt{1 - G_T}}{1 - |S_s|^2 (1 - G_T)} \qquad (E.16)$$

Constant Operating Power Gain Circle

Center:

$$C_w = \frac{g_w (s_{22}^* - \Delta^* s_{11})}{1 + g_w (|s_{22}|^2 - |\Delta|^2)} \qquad (E.17)$$

Radius:

$$R_w = \frac{(1 - 2K|s_{12} s_{21}|g_w + |s_{12} s_{21}|^2 g_w^2)^{1/2}}{1 + g_w (|s_{22}|^2 - |\Delta|^2)} \qquad (E.18)$$

where

$$g_w = G_w / |s_{21}|^2 \qquad (E.19)$$

and

$$G_w = \frac{|s_{21}|^2 (1 - |S_L|^2)}{1 - |s_{11}|^2 + |S_L|^2 (|s_{22}|^2 - |\Delta|^2) - 2 \, Re \, (C_2 S_L)} \qquad (E.20)$$

Constant Available Power Gain Circle

Center:

$$C_{AV} = \frac{g_{AV}(s_{11}^* - \Delta^* s_{22})}{1 + g_{AV}(|s_{11}|^2 - |\Delta|^2)} \tag{E.21}$$

Radius:

$$R_{AV} = \frac{(1 - 2K|s_{12}s_{21}|\,g_{AV} + |s_{12}s_{21}|^2\,g_{AV}^2)^{1/2}}{|1 + g_{AV}(|s_{11}|^2 - |\Delta|^2)|} \tag{E.22}$$

where

$$g_{AV} = G_{AV}/|s_{21}|^2 \tag{E.23}$$

and

$$G_{AV} = \frac{|s_{21}|^2(1 - |S_s|^2)}{1 - |s_{22}|^2 + |S_s|^2(|s_{11}|^2 - |\Delta|^2) - 2\,Re\,(C_1 S_s)} \tag{E.24}$$

Constant Gain Circles for a Transistor with $s_{12} = 0$

Transducer Power Gain

$$G_{T\text{-}u} = G_1\,|s_{21}|^2\,G_2 \tag{E.25}$$

where

$$G_1 = \frac{1 - |S_s|^2}{|1 - S_s s_{11}|^2} \tag{E.26}$$

$$G_2 = \frac{1 - |S_L|^2}{|1 - S_L s_{22}|^2} \tag{E.27}$$

CONSTANT G_1 CIRCLE

Center:

$$C_{G1} = \frac{g_1 s_{11}^*}{1 - |s_{11}|^2(1 - g_1)} \tag{E.28}$$

Radius:

$$R_{G1} = \frac{\sqrt{1 - g_1}\,(1 - |s_{11}|^2)}{1 - |s_{11}|^2(1 - g_1)} \tag{E.29}$$

where

$$g_1 = G_1 / G_{1\text{-}max} = G_1(1 - |s_{11}|^2) \tag{E.30}$$

CONSTANT G_2 CIRCLE

Center:

$$C_{G2} = \frac{g_2 \, s_{22}^*}{1 - |s_{22}|^2 \, (1 - g_2)} \qquad \text{(E.31)}$$

Radius:

$$R_{G2} = \frac{\sqrt{1 - g_2}\,(1 - |s_{22}|^2)}{1 - |s_{22}|^2 \, (1 - g_2)} \qquad \text{(E.32)}$$

where

$$g_2 = G_2 / G_{2-max} = G_2 \, (1 - |s_{22}|^2) \qquad \text{(E.33)}$$

CONSTANT NOISE FIGURE CIRCLE

Center:

$$C_F = \frac{S_{s-opt-F}}{1 + N} \qquad \text{(E.34)}$$

Radius:

$$R_F = \frac{\sqrt{N^2 + N\,(1 - S_{s-opt-F}{}^2)}}{1 + N} \qquad \text{(E.35)}$$

where

$$N = |1 + S_{s-opt-F}|^2 \, (F - F_{min}) / (4 \, r_n) \qquad \text{(E.36)}$$

$$r_n = R_n / Z_0 \qquad \text{(E.37)}$$

$$F = F_{min} + \frac{4 \, R_n}{Z_0} \, \frac{|S_s - S_{s-opt-F}|^2}{|1 + S_{s-opt-F}|^2 \, (1 - |S_s|^2)} \qquad \text{(E.38)}$$

The Equivalent Source Reflection Parameter and Transducer Power Gain for a Given Constant Operating Power Gain Circle

$$S_{s-OUT} = S_{L-opt}{}^* \qquad \text{(E.39)}$$

[Refer to (E.9)]

$$G_T = \frac{A_w}{2} \, [\sqrt{1 + 4/A_w{}^2} - 1] \qquad \text{(E.40)}$$

where

$$A_w = \frac{|C_w|^2}{R_w{}^2 \, |S_{s-OUT}|^2} \, [1 - |S_{s-OUT}|^2]^2 \qquad \text{(E.41)}$$

The Equivalent Load Reflection Parameter and Transducer Power Gain for a Given Constant Available Power Gain Circle

$$S_{L-IN} = S_{s-opt}* \tag{E.42}$$

[Refer to (E.8)]

$$G_T = \frac{A_{AV}}{2} [\sqrt{1 + 4/A_{AV}^2} - 1] \tag{E.43}$$

where

$$A_{AV} = \frac{|C_{AV}|^2}{R_{AV}^2 |S_{L-IN}|^2} [1 - |S_{L-IN}|^2]^2 \tag{E.44}$$

The Equivalent Load Reflection Parameter and Transducer Power Gain for a Given Constant Noise Figure Circle

$$S_{L-IN} = S_{s-opt-F}* \tag{E.45}$$

[Refer to (E.38)]

$$G_T = \frac{A_F}{2} [\sqrt{1 + 4/A_F^2} - 1] \tag{E.46}$$

where

$$A_F = \frac{|C_F|^2}{R_F^2 |S_{L-IN}|^2} [1 - |S_{L-IN}|^2]^2 \tag{E.47}$$

Appendix F

SYZ BASIC

```
1 REM                         APPENDIX F
2 REM
3 REM                         SYZ BASIC
4 REM
5 DIM S(8,4,2),D(4,2),E(4,2),F(4,2),P(2),Y(2)
10 REM
11 REM: Y TO Z CONVERSION:    GOSUB 780
12 REM: Z TO Y CONVERSION:    GOSUB 780
13 REM: S TO Y CONVERSION:    GOSUB 2040      (50 OHM NORMALIZING IMPEDANCES)
14 REM: Y TO S CONVERSION:    GOSUB 1830      (50 OHM NORMALIZING IMPEDANCES)
15 REM: Y TO T CONVERSION:    GOSUB 1400
16 REM: T TO Y CONVERSION:    GOSUB 1720
19 REM
20 REM: THE PARAMETERS ARE STORED AND MANIPULATED IN MAT D(4,2)
21 REM: D(1,1) + JD(1,2) =    X11 + JY11
22 REM: D(2,1) + JD(2,2) =    X12 + JY12
23 REM: D(3,1) + JD(3,2) =    X21 + JY21
24 REM: D(4,1) + JD(4,2) =    X22 + JY22
25 REM
26 REM:   FOR MATRIX MULTIPLICATION USE GOSUB 1100
27 REM:   MAT D = MAT E   *    MAT F
30 REM
31 REM:   FOR MULTIPLYING TWO COMPLEX NUMBERS USE GOSUB 350
32 REM:   C1 + JC2 = (A1+JA2) (B1+JB2)
35 REM
36 REM:   FOR DIVIDING TWO COMPLEX NUMBERS USE GOSUB 380
37 REM:   C1 + JC2 = (A1+JA2)/(B1+JB2)
40 REM
41 REM:   DETERMINE THE INVERSE OF A COMPLEX NUMBER BY USING GOSUB 690
42 REM:   C1 + JC2 = (1.0+J0.0)/(A1+JA2)
50 REM
51 REM:   START INSTRUCTIONS AT STATEMENT 5000
52 REM
53 REM:   USE MAT S TO READ AND STORE DATA
60 REM
61 MATD=ZER(4,2)
62 MATE=ZER(4,2)
63 MATF=ZER(4,2)
64 MATP=ZER(2)
65 MATY=ZERO(2)
66 mats=zero(8,4,2)
100 REM
110 GOTO 5000
340 REM
345 REM:   C=A*B
350 C1=A1*B1-A2*B2
360 C2=A1*B2+A2*B1
370 RETURN
```

```
380 REM
390 REM:   C=A/B
400 C1=A1*B1+A2*B2
410 C2=-A1*B2+A2*B1
420 C1=C1/(B1*B1+B2*B2)
430 C2=C2/(B1*B1+B2*B2)
440 RETURN
450 REM
460 REM:   INTERCHANGE F(2,-) AND  F(3,-)
480 A1=F(2,1)
485 A2=F(2,2)
490 F(2,1)=F(3,1)
495 F(2,2)=F(3,2)
500 F(3,1)=A1
505 F(3,2)=A2
520 RETURN
540 STOP
550 REM
555 REM:   INTERCHANGE D(1,-)  AND  D(4,-)
560 A1=D(1,1)
565 A2=D(1,2)
570 D(1,1)=D(4,1)
575 D(1,2)=D(4,2)
580 D(4,1)=A1
585 D(4,2)=A2
600 RETURN
630 REM
635 REM:   INTERCHANGE D(2,-) AND D(3,-)
640 A1=D(2,1)
645 A2=D(2,2)
650 D(2,1)=D(3,1)
655 D(2,2)=D(3,2)
660 D(3,1)=A1
665 D(3,2)=A2
680 RETURN
690 REM
700 REM:   C=1/A
710 C1=A1
720 C2=-A2
730 C1=C1/(A1*A1+A2*A2)
740 C2=C2/(A1*A1+A2*A2)
750 RETURN
760 REM
780 REM:   INVERSE OF D(4,2)
800 A1=D(1,1)
805 A2=D(1,2)
810 B1=D(4,1)
815 B2=D(4,2)
830 GOSUB350
840 M=C1
850 N=C2
870 A1=D(2,1)
875 A2=D(2,2)
880 B1=D(3,1)
885 B2=D(3,2)
900 GOSUB350
910 M=M-C1
920 N=N-C2
930 FOR I=1 TO 2
940 D(1+I,1)=-D(1+I,1)
950 D(1+I,2)=-D(1+I,2)
960 NEXTI
970 GOSUB 560
980 FOR I=1 TO 4
```

```
990 A1=D(I,1)
1000 A2=D(I,2)
1010 B1=M
1020 B2=N
1030 GOSUB 400
1040 D(I,1)=C1
1050 D(I,2)=C2
1060 NEXTI
1070 RETURN
1080 REM
1090 STOP
1100 REM:     D(4,2)=E(4,2)*F(4,2)
1110 GOSUB 480
1120 FOR I=1 TO 3 STEP 2
1130 A1=E(I,1)
1140 A2=E(I,2)
1150 B1=F(1,1)
1160 B2=F(1,2)
1170 GOSUB350
1180 P(1)=C1
1190 P(2)=C2
1200 B1=F(3,1)
1210 B2=F(3,2)
1220 GOSUB350
1230 Y(1)=C1
1240 Y(2)=C2
1250 A1=E(I+1,1)
1260 A2=E(I+1,2)
1270 B1=F(2,1)
1280 B2=F(2,2)
1290 GOSUB 350
1300 D(I,1)=P(1)+C1
1310 D(I,2)=P(2)+C2
1320 B1=F(4,1)
1322 B2=F(4,2)
1323 P(1)=D(I,1)
1330 P(2)=D(I,2)
1340 GOSUB 350
1350 D(1+I,1)=Y(1)+C1
1360 D(1+I,2)=Y(2)+C2
1370 Y(1)=D(1+I,1)
1371 Y(2)=D(I+1,2)
1372 NEXT I
1380 RETURN
1390 REM
1400 REM:       Y TO T CONVERSION      (D(4,2))
1410 GOSUB 560
1420 GOSUB 640
1440 A1=D(2,1)
1445 A2=D(2,2)
1450 B1=D(3,1)
1455 B2=D(3,2)
1470 GOSUB 350
1480 D(3,1)=-C1
1490 D(3,2)=-C2
1510 A1=D(1,1)
1515 A2=D(1,2)
1520 B1=D(4,1)
1525 B2=D(4,2)
1540 GOSUB 350
1550 D(3,1)=D(3,1)+C1
1560 D(3,2)=D(3,2)+C2
1570 P(1)=-D(2,1)
1580 P(2)=-D(2,2)
```

```
1590 D(2,1)=1
1600 D(2,2)=0
1610 FOR I=1 TO 4
1630 A1=D(I,1)
1635 A2=D(I,2)
1640 B1=P(1)
1645 B2=P(2)
1660 GOSUB 400
1670 D(I,1)=C1
1680 D(I,2)=C2
1690 NEXT I
1700 RETURN
1710 REM
1720 REM:      T TO Y CONVERSION       (D(4,2))
1730 FOR I=1 TO 2
1740 D(1,I)=-D(1,I)
1750 D(4,I)=-D(4,I)
1760 NEXT I
1770 GOSUB 640
1780 GOSUB 1400
1790 GOSUB 640
1810 RETURN
1830 REM:      Y TO S CONVERSION       (D(4,2))
1840 FOR I=1 TO 4
1850 FOR K=1 TO 2
1860 E(I,K)=D(I,K)
1870 F(I,K)=-D(I,K)
1880 NEXTK
1890 NEXTI
1900 FOR I=1 TO 4 STEP 3
1910 D(I,1)=E(I,1)+.02
1920 F(I,1)=F(I,1)+.02
1921 E(I,1)=D(I,1)
1930 NEXTI
1940 GOSUB780
1990 GOSUB 2620
2000 GOSUB1100
2020 RETURN
2030 REM
2040 REM:     S TO Y CONVERSION       (D(4,2))
2050 FOR I=1 TO 4
2060 FOR K=1 TO 2
2070 E(I,K)=-D(I,K)
2080 NEXT K
2090 NEXT I
2100 FOR I=1 TO 4 STEP 3
2110 D(I,1)=D(I,1)+1.0
2120 E(I,1)=E(I,1)+1.0
2130 NEXT I
2140 GOSUB 780
2150 FOR I=1 TO 4
2160 FOR K=1 TO 2
2170 F(I,K)=D(I,K)*.02
2180 NEXT K
2190 NEXT I
2200 GOSUB 1100
2220 RETURN
2247 FOR I1=1 TO N5
2248 FOR I=1 TO 8
2249 Q(I1,I)=O(I1,I)
2250 NEXT I
2251 NEXT I1
2252 RETURN
2430 FORI=1TO4
```

```
2440 D(I,1)=Q(N1,2*I-1)
2450 D(I,2)=Q(N1,2*I)
2460 NEXTI
2470 RETURN
2480 REM
2500 FORI=1TO4
2510 F(I,1)=Q(N1,2*I-1)
2520 F(I,2)=Q(N1,2*I)
2530 NEXTI
2540 RETURN
2550 REM
2560 FORI=1TO4
2570 E(I,1)=Q(N1,2*I-1)
2580 E(I,2)=Q(N1,2*I)
2590 NEXTI
2600 RETURN
2610 REM
2615 REM:       MAT E  =  MAT D
2620 FORI=1TO4
2630 E(I,1)=D(I,1)
2640 E(I,2)=D(I,2)
2650 NEXTI
2660 RETURN
2670 REM
2675 REM:       MAT F  =  MAT D
2680 FORI=1TO4
2690 F(I,1)=D(I,1)
2700 F(I,2)=D(I,2)
2710 NEXTI
2720 RETURN
5000 REM
5005 REM
5010 REM
5015 REM
5020 REM
5025 REM
5030 REM
5035 REM
5040 REM
5045 REM
8000 END
```

INDEX

Admittance plane 279,285,287,296
Algorithm 205,230-231
All-pass function 183
Alumina 63
Alumina substrate 268
Amplifier stability 275-284
Attenuation constant 45,63-64
Augmented two-port 11,13,18,19
Auto-transformer 125
Available power 23
Available power gain (G_A) 3,22,288-289, 297-298,359-360
Al-product 58-59,147,149
Balanced 1:4 transmission-line transformer 129,131-132,142,152-153, 158-160
Balanced 1:9 transmission-line transformer 129,131,159
Balanced amplifier 314-316,318-319
Balanced current 126,135,136,138
Balun core 56-57,125,149,159,95,100,120
Bandpass network 170,172,174,176-179, 183,189,196-197,203,221,234
Bandwidth 70-73,89-90,92-94,95, 100, 112,117,120,127,143,154-159,180-189, *passim*
Boltzman's constant 294ff
Bonding wire inductor 242-243,246, 248-249,271
Boundary conditions 137-138
Bounds 241-246,248-267
Building block 135,138-139,143
Butterworth network 174-175
Capacitor 37-40,42,55,65,73-74,79, 81,85,89,91-93,107-108,112,120,122, 161,170-171,173-174,183-189,226, 223,241,251-256,262
Cascaded networks 9,36,70,91-92,192
Center 360-361
Ceramic capacitors 38,40
Chamfer 268
Characteristic impedance 125,132,143, 149-152,159,191-192,195,241-245, 250,252,254-257,259,260,263-265,267, *passim*

Chebyshev response 171,178,180-181, 183,185-188,236
Chip capacitors 38,39,65,251,272
Circle 284-289,292ff,360-363
Circular loop 246,271-272
Circular spiral inductor 246-247
Coaxial cable 59-61,67,125,143,159
Combiner 134-135,155
Commensurate distributed network 189-193
Compensation 269
Compensation (compensating) capacitors 149-151,156-157,159-160
Compensation (compensating) inductors 149-151
Compensation element 77,78,111,149-150, *passim*
Compensation frequency 150,153-155
Compensation technique 241,263, 265-271
Complex frequency plane 26-29
Conductor length 248-250
Configurations
    of transmission-line transformers 127-135,158
Constant gain circle 217-218,223, 225,227
Constant mismatch 284-285,292,360
Constraints 17,19-21,29-30,171, 180,213,225,230
Continued fractionation 175,212
Conversion of S-parameters 24
Copper losses 41,63,97-98,102,120
Correction factor 246
Coupled coil 95-123
Coupling factor (k) 95,98,100-102, 104-106,109-110,115-116,118-120,137
Critical value of coupling factor (k_c) 110,117
Current S-parameters 12
Current gain (A_I) 3
Current series feedback 281
Cut-off frequency 130,143,150-151, 153-156,158-159,180,236
Darlington synthesis 170,172-176,192
Demagnetizing force 96

Diagram of *S*-parameters relationships 12
Dielectirc material 39,60,64
Dielectric constant 39,40,241,249-250, 261,263-264,267
Dielectric losses 64,65
Discontinuity 241,263-270
Dispersion 63
Dissipation factor 39,40,64,65
Distributed equivalent 252-257,262
Distributed response 191
Divider (*see* Splitter)
Double matching 197,205-215,231
Dual 86,115,154
Dualism 74-75
Dynamic range 275,293, *passim*
Eddy-current losses 41,46-47,97
Effective capacitance 37,61,65
Effective dielectric constant 61-63
Effective inductance 41
Equivalent circuit 95,97-101,106,110-111,115,118,120,144-147,157, 162-163,165,169,171,190,235
Faraday's law 95
Feedback 281
Fingers 251
Flux density 47-49,57-59,97,147-148,156
4:9 transmission-line transformer 129
4:25 transmission-line transformer 129-130
Four-element matching networks 89-90,94
Fringing capacitance 263-264,267
Gain circle 217-218,221-224
Gain-bandwidth constraints 170-172, 180-182,186,196,198,203,204,211, 235,297
Gain-bandwidth product 203
Gap-capacitor 251
Gradient vector 202,228
Grid-search 197,215-216,231
Gunn diode 310,323
Highpass network 173-174,176,183, 208-209,212,221,231

Hybrid transformer 134,154
Hysteresis losses 41,47,97,102
Impatt diode 310
Impedance ratio 97
Incident component 9-10,14,16-17, 21,24-27
Incident current 10-11,14-17,25,34
Incident voltage 10-11,14-17,34
Increment frequency 165-166, 199-200,203-204,211
Increment vector 200,202-204
Indefinite admittance matrix 5-6,24
Indefinite scattering matrix 24-26, 29-30,36
Inductor 37,40-59,66-67,73-74,79,81, 85,89,91-93,148-149,161,170-171, 173-175,179,189,226,233,241-242, 245-250,257,271
Initial values 202-203,207,213-214, 233-234
Inner diameter 247
Input admittance (Y_{IN}) 3,4,92,105,118, 137-139,142-144,146,150,159,162, 165,166,169,177-178,192-195,204, 217-218,221,227,233,235,238,244, 276,287, *passim*
Input reflection parameter 197,207,217
Insertion loss 245,271
Insertion losses 70,90-92
Instantaneous power 96
Interdigital capacitor 251-252,272
Intermodulation distortion 300
Intermodulation products 298-300
Interstage matching network 308
Isolation 134-135
Iterative techniques 161-162, 170-171,195-231
Kronecker delta 30
Kuroda's identities 172,190,193-195,237
LC-transformer 170,172,174-175, 177-180
LSM FORTRAN 203-205,208-210, 337-344

Leakage inductance 95,97-99,103,125

Least-square optimization 198-204, 207,216,228

Line length 126,147-149,151,153, 155-156,158-159,189

Line length 243-245,252-257,259, 263-265,272

Linearity 293,298-300

Line-segment algorithm 197-205

Line-segment technique 198-205

Linville stability factor 278-280,300,302

Locus 215-218,277,285,287

Loss tangent 39,60,64

Lossless networks 9,29-31,36,217, 218,231, *passim*

Lowpass network 174,176,183,189,191, 199,204,207,221,235-237

L-section 69-70,76-81,90,92-93,170, 172,177-179,195,233-234

L-section configurations 76-81

Lumped 242-252,257,271-273

Magnetic core 126-127,136,147-149,158

Magnetic coupling 125-126

Magnetic material 37,41,47-50,56-59,66, 97-98,103,126,135-137,141,156

Magnetizing inductance (L_{11}) 95,97-99, 101-103,120,127,135,138-139,142, 144-148,156-158

Maximum gain 181,182,186-187,230, 237

Maximum relative deviation (MRD) 205, 216,228-230

Mica capacitors 38,65

Microstrip 60-64

Microstrip bend 265-266,269

Microstrip discontinuity 263-271

Minimum insertion loss 205,228, 233,235,236

Minimum-admittance function 162-163, 165,169,198,203,211

Minimum-impedance function 162-165,168,198,201,203,209,211-212

Minimum-reactance functions 26

Multistage amplifier 297-298,300-310

Mutual inductance (M) 98,120

Noise 322ff

Noise figure 296-298,304-306

Noise performance 293-298

Non-unilateral 291-298

Normalized S-parameters 12,34

Normalized components 10,14-21,27,34

Normalizing impedance matrix 10

Normalizing impedances 9-11,13,16-19, 21,23-24,26-28,31,34,92,217

Norton's identifies 172,190,193-195

Open-ended lines 189,190-191,195

Open-ended stub 253,256-257,261, 263-266,270,272-273

Operating power gain 285-288,292, 359-360

Optimum characteristic impedance 125, 143-144,151-152,154,158-159

1:16 transmission-line transformer 130

1:4 push-pull transformer 133

1:4 transmission-line 125,127-128, 133,138-139,143,157

1:9 push-pull transformer 133

1:9 transmission-line transformer 128-131,133

Oscillation 276

Outer diameter 247

Output admittance (Y_0; Y_{OUT}) 3,24,276, *passim*

PI-sections 69-70,81-89,92-94,177-179, 233,242,257-263,272

PLNM FORTRAN 165-167,208,211

Parallel double-tuned transformer 109-115,121

Parallel networks 4

Parallel resonance 70-76,79,93, 102,110,115,188

Parasitic absorption 162,171-173, 186-189

Parasitic capacitance 40-43,51,66,95, 97,100-101,108-118,125,241,271

Parasitic effects 241,251,263-266, 268,272-273

Parasitic inductance 37-40,65

Permeability 97,100

Phase velocity 62-63
Phase-shift 241, *passim*
Planar form 190
Plate capacitance 251
Poles 26,27,70,88,100-101,163,165-169,
    173-175,184,211,237
Positive slope 162
Power amplifier 300,316-318
Primary current 96
Primary voltage 96
Propagation constant 45
Proximity effect 41,46-47,51,65
Quality factor (Q) 38-43,51-58,65-66,
    70,72-74,79,81,88,90-94,103,106-108,
    115-117,120-121,156,170,179,186-
    188,216, 221,223-225,230,244-247,
    249,250,271-272
    Q of a circuit 71-76,79,81,90,93-94,
    107,110,120
    Q of transformation 79,81-85,88-90,
    93,112,170,197,215-216,218-224,
    228-231
Quarter wavelength transformer 318
RCDM FORTRAN 207,208,214
RF choke 246
Radius 360-361
Reactance 248,252,255-257,259,260
Reactive load 171,182-189,198 203,
    205-215,245
Reactive source 171,182-189,198,
    205-215
Real-frequency 195-197
Reflected component 9-10,14-17,21,27
Reflected current 10,14-17,25,34
Reflected voltage 10,14-17,34
Reflection amplifier 311-314
Reflection coefficient 173-174,176,
    180,193-194,197,213
Reflection coefficient technique
    205-208,215
Reflection parameter 244,287,360
Relative bandwidth 129
Relative efficiency 146-147
Resistance function 163-167,197,199,211
Resistance transformation 161,170

Resistive loading 281
Resistor 242
Resonance 188, *passim*
Resonant frequency 37-38,41-43,52-56,
    65-66,71,73,79,93-94
Resonating section 218,221,226,227
Rexolite substrate 268
Ribbon inductor 245-246
Richard's transformation 172,189-193,
    237
Right-angled bend 268
Ripple 144-145,154-157,178-181,183-
    186,236-237, *passim*
Round-wire inductor 246,271
Saturation 122
Scaling factor 195,207-209,212-213,229
Scattering (S-) parameters 1,9-35,
    118-120,206-207,275,283,307,315,
    320,322,359-363
    Physical interpretation 9,17-19,34
Secondary current 96
Secondary voltage 96
Self-capacitance 42,51,53-55,107
Semi-infinite functions 199-200
Semi-lowpass response 170
Semi-rigid coaxial cable 60-61,135
Series double-tuned transformer
    115-117,121
Series inductor 242,248,257
Series networks 7
Series resonance 72-77,79,188
Series transmission line 241-245,258,
    261,271-272
Short-circuited lines 189-191,194-195
Shunt capacitor 252-257,272
Shunt inductor 248,252
Sign 173-175,183-184,186,224,228
Signal-to-noise ratio 295
Silicon 64
Single-matching 209-213
Single-stage 287
Skin-effect 41,44-46,51,60,65,242
Smith chart 19,284-285,296
Solenoidal coil 41,46-47,51-56,66,
    91,135-136,156,158

Solenoidal inductor 245,248-250,272
Source resistance 170,172-175,198-205
Spacing 247,251
Splitter 134-135,154-155,158-159
Square spiral inductor 242-243,
    245-246,248-249
Stability 275
Stability circle 278
Stability factor 280-284
Stabilizing resistor 280-284
Stacked cores (*see* toroidal cores) 100
Step discontinuity 264-265,272
Steps in width 264-265
Sterne stability factor 280,359
Strip inductor 245-246,248-249,271
SYZ FORTRAN 364-368
Tangent function 252-253
Taper 267,270,318
Tapped-coil 103-109,120-121
Tap-point 104-107,109
Teflon 63
Temperature coefficient 40
Termination 276,359-360
Thin-film 241,242,251
T-junction 263,269-270
Topology 171,174-175,183,186,216,
    230-232,235
Toroid 56-57,59,67,95,120
Toroidal core 56-59,66,100,120,125,
    147-149,158-159
Transducer power gain (G_T) 19,34,112,
    114,116-117,138-141,161,170-174,
    180,182-184,187-188,192,195,198,
    202-207,209,213,217,221-222,230,233,
    236,275,304-306, *passim*
Transformation 243,260,262-263
Transformation diagram 80,85,88
Transformation distance 177,179
Transformation element 77-81,85,111
Transformation formulas 188
Transformation ratios 127-129,158
Transformer 161,172,175,197
Transforming section 70,79,81,90,
    93-94,113
Transistor 5,23,33,35,69

Transmission lines 37,59-65,67
Transmission matrix 9,192,242,257
Transmission parameters 1,8-9,17-19,
    31-32,35-36
Transmission-line equivalent 257-263
T-section 69-70,81-82,86-89,92-94,
    100,122,177-179,233,242,257-263,272
Tunability 275,289-290,304-305
Turn ratio 97-98,120
Twisted pairs 60,65
Unbalanced current 126-127,134-137,
    139,158
Unbalanced-to-balanced
    1:1 transmission-line transformer
        131-132,141-142,144-146,152-153,158
    1:4 transmission-line transformer
        132-133,139,143-145,150,153,157-159
    4:1 transmission-line transformer 132
Unilateral 275,290-291
Unilateral transducer power gain 22
Unit element 192-193,195
Unitary 29-30
VSWR 275,302,304-305,308-309
Voltage S-parameters 12
Voltage gain (A_V) 3,18,23
Voltage-shunt feedback 281
Weighting factor (r_k) 199-200,251
Winding dots 97
Wire gauges 43-44
Y-parameters 1-5,8-9,24,33,35,275,278
Z-parameters 1,6-9,24,98,118,120,122
ZVR FORTRAN 165,167-168,208,211,
    330-336
Zero of transmission 170,172,174,
    186,206-207
Zeros 27,70,101,164-166,173,175,
    182,184,203,237